Current Topics in Behavioral Neurosciences

About this series

Current Topics in Behavioral Neurosciences provides critical and comprehensive discussions of the most significant areas of behavioral neuroscience research, written by leading international authorities. Each volume offers an informative and contemporary account of its subject, making it an unrivalled reference source. Titles in this series are available in both print and electronic formats.

With the development of new methodologies for brain imaging, genetic and genomic analyses, molecular engineering of mutant animals, novel routes for drug delivery, and sophisticated cross-species behavioral assessments, it is now possible to study behavior relevant to psychiatric and neurological diseases and disorders on the physiological level. The *Behavioral Neurosciences* series focuses on "translational medicine" and cutting-edge technologies. Preclinical and clinical trials for the development of new diagnostics and therapeutics as well as prevention efforts are covered whenever possible.

Roger A.H. Adan • Walter H. Kaye
Editors

Behavioral Neurobiology of Eating Disorders

Springer

Editors
Prof. Dr. Roger A.H. Adan
Rudolf Magnus Institute of Neuroscience
Department of Neuroscience and Pharmacology
Universiteitsweg 100
3508 AB Utrecht
Netherlands
r.a.h.adan@umcutrecht.nl

Prof. Dr. Walter H. Kaye
University of California, San Diego
Department of Psychiatry
La Jolla Drive 8950
La Jolla, CA 92037
Suite C207
USA
whkaye@gmail.com

ISSN 1866-3370 e-ISSN 1866-3389
ISBN 978-3-642-15130-9 e-ISBN 978-3-642-15131-6
DOI 10.1007/978-3-642-15131-6
Springer Heidelberg Dordrecht London New York

Cover illustration: Artistic representation of oscillatory synchrony and timing of neurons in networks by Gyorgy Buzsaki

Cover design: WMXDesign GmbH, Heidelberg, Germany

Printed on acid-free paper

Springer is part of Springer Science+Business Media (www.springer.com)

Preface

This book describes the current state of the art of the neurobiology of eating disorders and provides the latest insights and ideas in the eating disorders field. Important breakthroughs in the last decade have improved our understanding of the etiology of eating disorders. Due to neuropsychological studies, we now understand better which alterations underlie disturbed cognitive processing in anorexia nervosa and now that these are being combined with imaging and genetics we start to understand the underlying neural circuitry and genetic pathways. Interestingly when combining the novel insights from the different chapters in the book, it emerges, for instance, that anorexia nervosa is accompanied by a deficit in reward processing and impaired behavioral flexibility which is driven by a high degree of cognitive control. Clearly, alterations in the dopamine system and an overactive dorsal neurocircuit (including for instance the dorsolateral prefrontal cortex) mediating cognitive processing are implicated, which may also underlie poor decision making. A new concept arises that anorexia nervosa is a neurodevelopmental striatocortical disorder. Because there are animal models that mimic traits of pathological eating behavior, it is possible that such models, when combined with advances in genetic research, will further contribute to unraveling the molecular pathways underlying eating disorders. That is, studies using animals and the application of genetic variation to explain different responses in human imaging studies may give us a better understanding of what goes wrong where in the brain of eating disorder patients. Finally, the treatments currently available for eating disorders in general, and anorexia nervosa in particular, are inadequate. Perhaps the most important aspect of new insights into how symptoms are coded in the brain is that this provides new targets for developing more effective therapies. Because of the critical need to advance treatment, we have included several chapters from pioneers in this field. In summary, based on the knowledge described in this book, we hope that people in the eating disorder field get inspired to explore new horizons for therapeutic interventions that are based on the new insights.

Utrecht, The Netherlands — Roger A.H. Adan
La Jolla, CA, USA — Walter H. Kaye

Contents

Contributors

Roger A.H. Adan
Department of Neuroscience and Pharmacology, Rudolf Magnus Institute of Neuroscience, University Medical Centre Utrecht, Str. 5.203, P.O.B. 85060 3508 AB Utrecht, The Netherlands and Altrecht Eating Disorders Rintveld, Altrecht Mental Health Institute, Zeist, The Netherlands, r.a.h.adan@umcutrecht.nl

Ursula F. Bailer
Division of Biological Psychiatry, Department of Psychiatry and Psychotherapy, Medical University of Vienna, Vienna, Austria and Eating Disorder Treatment and Research Program, Department of Psychiatry, University of California, San Diego, La Jolla, CA, USA

Kenneth R. Bruce
Eating Disorders Program, Douglas University Institute, 6875 LaSalle Blvd, Montreal (Verdun), QC, Canada H4H 1R3 and Department of Psychiatry, McGill University, Montreal, QC, Canada and Department of Psychology, McGill University, Montreal, QC, Canada

Cynthia M. Bulik
Department of Psychiatry, University of North Carolina at Chapel Hill, 101 Manning Drive, CB #7160, Chapel Hill, NC 27599-7160, USA, cbulik@med.unc.edu

S. Cardona Cano
Department of Neuroscience and Pharmacology, Rudolf Magnus Institute of Neuroscience, University Medical Centre Utrecht, Str. 5.203, P.O.B. 85060, 3508 AB Utrecht, The Netherlands

David A. Collier
MRC Social Genetic and Developmental Psychiatry Centre, Institute of Psychiatry, King's College London, P82, Room 1.11, De Crespigny Park, Denmark Hill, London SE5 8AF, UK, david.collier@kcl.ac.uk

Kristen M. Culbert
Department of Psychology, Michigan State University, 43 Psychology Building, East Lansing, MI 48824-1116, USA, culbertk@msu.edu

Alain Dagher
McGill University, Montreal, QC, Canada

U.N. Danner
Altrecht Eating Disorders Rintveld, Altrecht Mental Health Institute, Zeist, The Netherlands and Department of Clinical and Health Psychology, Utrecht University, Utrecht, The Netherlands

Annemarie van Elburg
Rintveld Center for Eating Disorders, Altrecht Mental Health Institute, Oude Arnhemseweg 260, Zeist, Utrecht 3705BK and Department of Child and Adolescent Psychiatry, University Medical Center, Utrecht, The Netherlands, a.van.elburg@altrecht.nl

Guido K.W. Frank
Developmental Brain Research Program, Department of Psychiatry, University of Colorado Denver, The Children's Hospital, Gary Pavilion A036/B-130, 13123 East 16th Avenue, Aurora, CO 80045, USA and Department of Neuroscience, University of Colorado Denver, The Children's Hospital, Gary Pavilion A036/B-130, 13123 East 16th Avenue, Aurora, CO 80045, USA, Guido.Frank@ucdenver.edu

Hans-Christoph Friederich
Department of General Internal Medicine and Psychosomatics, Medical Hospital, University of Heidelberg, Im Neuenheimer Felds 410, 69120, Heidelberg, Germany, hans-christoph.friederich@med.uni-heidelberg.de

Julie L. Fudge
Department of Psychiatry, University of Rochester Medical Center, Rochester, NY, USA and Department of Neurobiology and Anatomy, University of Rochester Medical Center, Rochester, NY, USA

Patricia Groleau
Eating Disorders Program, Douglas University Institute, 6875 LaSalle Blvd, Montreal (Verdun), QC, Canada H4H 1R3 and Department of Psychiatry, McGill

University, Montreal, QC, Canada and Department of Psychology, McGill University, Montreal, QC, Canada

Remco Havermans
Faculty of Psychology and Neuroscience, Department of Clinical Psychological Science, Maastricht University, P.O. Box 616, 6200 MD Maastricht, The Netherlands

Sietske G. Helder
MRC Social Genetic and Developmental Psychiatry Centre, Institute of Psychiatry, King's College London, P82, Room 1.11, De Crespigny Park, Denmark Hill, London SE5 8AF, UK, sietske.helder@kcl.ac.uk

David Herzog
Massachusetts General Hospital and Harvard Medical School, Harvard University, Cambridge, MA, USA

Wolfgang Herzog
Department of General Internal Medicine and Psychosomatics, Medical Hospital, University of Heidelberg, Im Neuenheimer Felds 410, 69120, Heidelberg, Germany, wolfgang.herzog@med.uni-heidelberg.de

J.J.G. Hillebrand
Department of Neuroscience and Pharmacology, Rudolf Magnus Institute of Neuroscience, University Medical Centre Utrecht, Str. 5.203, P.O.B. 85060 3508 AB Utrecht, The Netherlands and Altrecht Eating Disorders Rintveld, Altrecht Mental Health Institute, Zeist, The Netherlands

Anita Jansen
Faculty of Psychology and Neuroscience, Department of Clinical Psychological Science, Maastricht University, P.O. Box 616, 6200 MD Maastricht, The Netherlands

Martien J.H. Kas
Department of Neuroscience and Pharmacology, Rudolf Magnus Institute of Neuroscience, University Medical Centre Utrecht, Str. 5.203, P.O.B. 85060, 3508 AB Utrecht, The Netherlands, m.j.h.kas@umcutrecht.nl

Walter H. Kaye
Eating Disorder Treatment and Research Program, Department of Psychiatry, University of California, San Diego, La Jolla, CA, USA, whkaye@gmail.com

Kelly L. Klump
Department of Psychology, Michigan State University, 107B Psychology Building, East Lansing, MI 48824-1116, USA, klump@msu.edu

Lisa Rachelle Riso Lilenfeld
Clinical Psychology Program, Argosy University, 1550 Wilson Boulevard, Suite 600, Arlington, VA 22209, Washington, DC, USA, LLilenfeld@argosy.edu

Tammy Lin
Duke University, Durham, NC, USA

James Lock
Department of Psychiatry and Behavioral Sciences, Stanford University School of Medicine, Stanford, CA, USA

Suzanne E. Mazzeo
Department of Psychology, Virginia Commonwealth University, PO Box 842018, Richmond, VA 23284-2018, USA, semazzeo@vcu.edu

Rhonda Merwin
Duke University Medical Center, Durham, NC, USA

Palmiero Monteleone
Department of Psychiatry, University of Naples SUN, Largo Madonna delle Grazie, 80138 Naples, Italy, monteri@tin.it

Ashley Moskovich
Duke University, Durham, NC, USA

Anna Oldershaw
Section of Eating Disorders, Institute of Psychiatry, King's College London, De Crespigny Park, London SE5 8AF, UK

Martin Paulus
Laboratory of Biological Dynamics and Theoretical Medicine, Department of Psychiatry, University of California, San Diego, La Jolla, CA, USA and Psychiatry Service, San Diego Veterans Affairs Health Care System, San Diego, CA, USA

Sarah E. Racine
Department of Psychology, Michigan State University, 43 Psychology Building, East Lansing, MI 48824-1116, USA, racinesa@msu.edu

Anne Roefs
Faculty of Psychology and Neuroscience, Department of Clinical Psychological Science, Maastricht University, P.O. Box 616, 6200 MD Maastricht, The Netherlands

Ulrike Schmidt
Section of Eating Disorders, Institute of Psychiatry, King's College London, De Crespigny Park, London SE5 8AF, UK, u.schmidt@iop.kcl.ac.uk

Nicolette Siep
Faculty of Psychology and Neuroscience, Department of Clinical Psychological Science, Maastricht University, P.O. Box 616, 6200 MD Maastricht, The Netherlands, Nicolette.Siep@maastrichtuniversity.nl

Howard Steiger
Eating Disorders Program, Douglas University Institute, 6875 LaSalle Blvd, Montreal (Verdun), QC, Canada H4H 1R3, stehow@douglas.mcgill.ca and Department of Psychiatry, McGill University, Montreal, QC, Canada and Department of Psychology, McGill University, Montreal, QC, Canada

Eric Stice
Oregon Research Institute, Eugene, OR, USA, estice@ori.org

Kate Tchanturia
Department of Psychological Medicine, King's College London, London, UK, Kate.Tchanturia@iop.kcl.ac.uk

Laura M. Thornton
Department of Psychiatry, University of North Carolina at Chapel Hill, 101 Manning Drive, CB #7160, Chapel Hill, NC 27599-7160, USA, laurathornton@verizon.net

L.A.W. Verhagen
Department of Neuroscience and Pharmacology, Rudolf Magnus Institute of Neuroscience, University Medical Centre Utrecht, Str. 5.203, P.O.B. 85060, 3508 AB Utrecht, The Netherlands

Angela Wagner
Department of Psychiatry, University of California, San Diego, La Jolla, CA, USA

Sonja Yokum
Oregon Research Institute, Eugene, OR, USA

David Zald
Vanderbilt University, Nashville, TN, USA

Nancy L. Zucker
Duke University Medical Center, Durham, NC, USA and Duke University, Durham, NC, USA, Zucke001@mc.duke.edu

Part I
Cognition and Emotions in Eating Disorders

Personality and Temperament

Lisa Rachelle Riso Lilenfeld

Contents

Abstract The assessment of personality and temperament in the context of eating disorders (EDs) poses unique challenges because of the physiological symptoms and sequelae of these illnesses. Four models of the relationship between personality and EDs are presented, along with a discussion of the different methodological designs which can evaluate these models. Current data support the likelihood that neuroticism and perfectionism are risk factors for EDs. Perfectionism and the related obsessive–compulsive personality disorder may also share a common cause with anorexia nervosa. High harm avoidance and low self-directedness also characterize all EDs, though more data are needed to confirm their role as risk factors; importantly however, this combination of traits may diminish one's ability to cope with stressful life events. At the other end of the spectrum, considering impulsivity multidimensionally may be important to understanding the role of this personality trait in EDs, though existing data do not yet allow for conclusions regarding its role as a risk factor versus a consequence of the ED. All of the

L.R.R. Lilenfeld
Clinical Psychology Program, Argosy University, 1550 Wilson Boulevard, Suite 600, Arlington VA 22209, Washington DC, USA
e-mail: LLilenfeld@argosy.edu

R.A.H. Adan and W.H. Kaye (eds.), *Behavioral Neurobiology of Eating Disorders*, Current Topics in Behavioral Neurosciences 6, DOI 10.1007/7854_2010_86

identified traits that may be risk factors are also exacerbated as a consequence of having, or having had, an ED. Finally, the role of personality disorders in influencing the course and outcome of EDs is still unclear. A fruitful avenue for future research in this area is to utilize personality and temperament to classify individuals in a way that allows for the best chance of identifying genetic loci that confer increased risk for EDs.

Keywords Classification · Eating disorder (ED) · Family study · Genetic · Harm avoidance · Impulsivity · Neuroticism · Obsessive–compulsive · Perfectionism · Personality · Risk factor · Self-directedness · Temperament

1 Unique Challenges in the Assessment of Personality and Temperament in the Field of Eating Disorders

Any clinician specializing in the field of eating disorders (EDs) knows that certain personality traits and temperamental styles are very common in this population, such as a restrained, cautious, perfectionstic style or in other cases, an impulsive, labile, dramatic style, for example. However, careful empirical investigation of the exact nature of the relationship between temperament, personality, and EDs is actually quite complex due to the unique challenges accompanying these illnesses. Temperament refers to primarily genetically determined ways of responding to the environment, which, in combination with experience, forms the basis for the later development of personality traits (Kim 2009). Personality encompasses those dispositional qualities that guide behavior and adaptation to the environment, approximately half of the total variance of which is explained by additive genetic factors (Kim 2009). While there are serious challenges in the assessment of personality more generally, such as the nature of the informant (e.g., the individual versus a family member), what assessment measures are used, hotly debated personality disorder diagnostic criteria, as well as controversy regarding the utility of dimensional versus categorical approaches (e.g., Skodol and Bender 2009), there are some additional rather daunting challenges when attempting to accurately assess and understand the role of personality among those who have EDs (Vitousek and Stumpf 2005). While it is helpful for a clinician to understand what temperamental style and personality traits are present at the current moment in a given individual, from a scientific standpoint the primary interest is in understanding the nature of the relationship between personality/temperament and the ED in order to elucidate our understanding of the etiology of these illnesses.

One of these challenges is the difficulty in disentangling personality traits that are present premorbidly (i.e., a true "trait") versus those that are a consequence of having had (i.e., a "scar") or currently having (i.e., a "state effect") an ED, which is not unique to the ED field (e.g., Kendler et al. 1993). However, it is particularly problematic for this field due to what we know about the profound impact of semistarvation (Keys et al. 1950) and chaotic eating patterns (Vitousek and Stumpf 2005) upon personality. Another challenge is that of potentially biased reporting of apparent premorbid traits due to recall biases that are typical of all humans (e.g., Drews and Greenland 1990). Again, however, recall may be particularly flawed in this population given the impact of cognitive impairment and dysregulation that result from EDs. It can be nearly impossible to expect even the most reliable "informant" to be able to accurately recall personality traits prior to the onset of illness (Zimmerman 1994). A third challenge is the nature of the sampling when studying personality traits among those with EDs; although, again not specific to the field of EDs, we have evidence that pathological personality traits are likely to be exaggerated in treatment-seeking samples compared with population-based samples (Perkins et al. 2005). This is critical to keep in mind when interpreting the results of studies where recruitment occurred in clinical settings, which are the majority in our field thus far.

2 Models and Methodology to Examine the Relationship Between Personality and Eating Disorders

A number of models of the relationship between personality and EDs have been empirically tested, although often the model under examination has not been made explicit. There are four theoretical models that have been examined in other fields of psychopathology (e.g., Clark et al. 1994; Klein and Riso 1993) for which we have enough data to evaluate them in the ED field. They are described in more detail elsewhere (Lilenfeld et al. 2006) and will be only briefly summarized here. The "predispositional" model describes a relationship in which the personality trait or disorder of interest precedes and, in fact, increases the risk of developing an ED. The personality pathology and eating pathology are assumed to be independent, distinct constructs in this model. By contrast, the "common cause" model assumes again that the personality construct and ED are independent conditions, yet in this case are caused by some common underlying factor(s). A third model is the "complication model," which posits that the personality trait or disorder is a result of either having had (in which case we would refer to the personality trait as a "scar") or currently having an ED (in which case we would refer to the personality trait as a "state effect" and would be assumed to disappear when the ED remitted). Technically, this model states that the personality trait is a product of the ED, though another possibility is that a premorbid trait can get exacerbated or otherwise altered by currently having or previously having had an ED. A final model that has been examined in the ED field, although again often not explicitly, is the "pathoplasty model." This is not a causal model but rather suggests that the ED and personality trait(s), both once established, modify the course and presentation of the other. As with the prior model, some combination with the predispositional model is possible, where even if a personality trait was indeed a predisposing (i.e., risk) factor for an ED, it can additionally have pathoplastic effects over time where it impacts the manifestation of the ED.

Numerous methodological designs have been utilized to evaluate these models of the relationship between personality or temperament and EDs. Again, they are described in more detail elsewhere (Lilenfeld et al. 2006) but will be summarized here. A prospective design is the ideal test of the predispositional model: that is, one must conduct an assessment (in this case, of personality) prior to the development of the condition of interest (i.e., before the ED develops) and then again after the individual changes on the outcome of interest (i.e., after the ED has developed). A true "risk factor" must precede the development of the condition, which is only discernable through a prospective design or the identification of genetic markers (Jacobi et al. 2004). Because of the relatively low base rates of these illnesses, community-based prospective studies are extremely challenging to perform. Thus, retrospective designs have often been used instead (Anderluh et al. 2009; Fairburn et al. 1997, 1998, 1999), which, however, are vulnerable to the biases inherent in retrospective reporting.

Another less than ideal alternative methodological approach that has been used to test the predispositional model is a recovered study design where individuals who have recovered from an ED (though definitions of recovery have not been universally agreed upon, which poses an additional challenge) are studied as a proxy for a premorbid state since they are assumedly no longer experiencing the physiological effects of the ED. The problem is that, while one can be reasonably certain that there are no confounding "state effects" as would be found when studying an individual who is ill, one cannot be certain that there are no "scar" effects remaining that are being mistakenly interpreted as premorbid risk factors. Comparing ill individuals to those who have recovered to those who are never ill is, however, a useful way to distinguish "scar" effects from "state" effects (e.g., Kaye et al. 1998).

Family studies are a useful methodological design that can avoid many of these problems. For instance, a combination of a family and recovered study design, in which patterns of personality disorders in family members of currently ill, recovered, and never ill probands are examined, can discriminate well between common cause and complication models (Lilenfeld et al. 2000).

In addition to family studies, genetic studies are a powerful methodological approach that can be used to evaluate both the common cause and the predispositional models. Multivariate twin studies examine associations between personality traits/disorders and EDs among monozygotic compared to dizygotic twin pairs to gather information regarding the extent to which there may be an underlying shared causal factor(s), as well as to what extent the nature of the common cause is determined by shared genes and shared environment. The role of genes and environment in the determination of personality, as well as illnesses such as EDs, is certainly a complicated one, involving a complex relationship between genes and environment (Bulik 2005). Additional genetic methodological approaches that have been used to examine the predispositional model include linkage studies (e.g., Bacanu et al. 2005) and association studies (e.g., Rybakowski et al. 2006).

Rapid developments in the field of behavioral genetics will now allow for the utilization of more powerful genome-wide association studies (e.g., The Wellcome Trust Case Control Consortium 2007) to identify potential risk alleles, which will likely improve our understanding of the role of temperament and personality in the development of EDs.

3 Current Knowledge Regarding the Relationship of Personality and Temperament and Eating Disorders

Cross-sectional studies of personality and temperament among individuals currently ill with EDs abound, with much having been written about numerous such traits, most commonly perfectionism, neuroticism, harm avoidance, impulsivity, as well as obsessive–compulsive (OCD) and borderline personality disorder (BPD) (e.g., Cassin and von Ranson 2005). However, cross-sectional research does little to illuminate the potentially complex relationship between the ED and various temperament or personality traits and disorders.

Despite the challenges inherent in conducting prospective research with this population, several such studies have been done, giving us insight as to personality and temperament traits that are likely to be predisposing risk factors for EDs. While certainly not specific to EDs, neuroticism or negative emotionality has emerged as a likely risk factor (Bulik et al. 2006; Cervera et al. 2003; Ghaderi and Scott 2000; Killen et al. 1996; Stice 2002; Lilenfeld et al. 2006). The most impressive such prospective study of anorexia nervosa thus far was of a very large population-based sample of Swedish twins in which premorbidly assessed neuroticism was found to be strongly predictive of the later development of anorexia nervosa (Bulik et al. 2006).

In addition to neuroticism, another moderately heritable (Tozzi et al. 2004; Wade and Bulik 2007) personality trait, namely perfectionism, has also been found to predict EDs when assessed prospectively (Tyrka et al. 2002) and retrospectively (Fairburn et al. 1997, 1998), suggesting that it may be another risk factor. Elevated perfectionism also has been found to persist after recovery (e.g., Stein et al. 2002). It has further been suggested that it may be the interaction between perfectionism and other risk factors that makes one especially vulnerable to developing an ED. Specifically, it has been shown that an interaction among perfectionism, perceived weight status, and self-esteem may prospectively predict bulimic symptom development in young women (Bardone-Cone et al. 2006; Vohs et al. 1999), although the nature of the perfectionism (e.g., "adaptive," such as having high personal standards vs. "maladaptive," such as being overly self-critical) appears important in this relationship (Pearson and Gleaves 2006; Mazzeo et al. 2006; Bardone-Cone et al. 2009), as others have not replicated these findings (Steele et al. 2007; Shaw et al. 2004). In addition to functioning as a possible risk factor, perfectionism is likely to also contribute to the maintenance of the disorder once it is established (e.g., Santonastaso et al. 1999). Also, similar to neuroticism, while perfectionism is not a risk factor that is specific to EDs (Bardone-Cone et al. 2007), certain aspects of perfectionism appear to be especially relevant to the development of EDs, namely self-oriented perfectionism (Tissot and Crowther 2008; Castro-Fornieles et al. 2007) and concern over mistakes (e.g., Bulik et al. 2003). Powerful twin study methodology has implicated a common cause model for anorexia nervosa and a temperamental style involving perfectionism (Wade et al. 2008). Finally, perfectionism has also been examined in the context of a pathoplastic model, namely to evaluate to what extent this particular personality trait may predict the course and outcome of the ED. Indeed, lower levels of perfectionism appear to be associated with better response to treatment and better outcome at followup (e.g., Sutandar-Pinnock et al. 2003; Anderluh et al. 2009). Thus, perfectionism is undoubtedly of great importance in elucidating the causes of EDs, as there is evidence for the role of this personality trait as a risk factor for EDs, sharing a common cause with anorexia nervosa, and influencing the course and outcome of EDs.

The most common measure used to assess temperament in the field of EDs is the Temperament and Character Inventory (TCI; Cloninger et al. 1993). Mostly cross-sectional studies have been conducted and have concluded that, while some diagnostic subtype differences may be present, the most robust findings are that a harm-avoidant temperament (i.e., anxious and fearful) and low levels of the character trait of self-directedness (i.e., immature, experiences difficulty pursuing goals, and believes that one's behavior is determined by external influences outside of one's control) describe individuals across ED diagnoses (Fassino et al. 2004). This finding has also been replicated in a sample of women recovered from EDs compared to never-ill women (Klump et al. 2004). The combination of high harm avoidance and low self-directedness is thought to diminish the ability to cope with stressful life events (Fassino et al. 2004). As previously discussed, findings from a recovered study design leave open the possibility that these traits are premorbid factors that increase risk for an ED, yet such a conclusion cannot be definitive without a prospective design. Interestingly, harm avoidance has been proposed as a potential mediator, which may explain the pathway from obstetric complications to the development of an ED (Favaro et al. 2008).

A related construct and accompanying diagnosis that has been examined both as a potential risk factor as well as potentially sharing a common cause with anorexia nervosa in particular is obsessive–compulsive personality disorder (OCPD). While the elevated rates of OCPD in individuals with EDs, especially anorexia nervosa, is not under dispute (e.g., Sansone et al. 2005), the question of interest is what underlies this comorbidity. Although personality disorders are assumed to be long-standing, generally unremitting conditions, this has not garnered much empirical support when longitudinal investigations have been conducted, as personality disorder rates have been found to drastically decrease when individuals emerge from an acutely ill state (Vrabel et al. 2010). While this is supportive of a "state effect" complication model, it is also likely that OCPD is not simply a complication of the ED, as rates remain elevated beyond what is found in the general population even after ED recovery (Matsunaga et al. 2000). Using a quasi-prospective design, OCPD traits have been found to occur at elevated rates premorbidly among those who later developed anorexia nervosa compared to those who did not (Rastam 1992), suggesting that obsessive–compulsive personality traits may be a predisposing factor for the development of anorexia nervosa. Findings from less robust designs (e.g., retrospective studies) support this conclusion as well (Anderluh et al. 2003). One small longitudinal study found that OCPD traits were most strongly predictive of an increase in ED symptoms over several years in college among African American (compared to Caucasian and Asian American) young women specifically (Lilenfeld et al. 2008), which highlights the potential interaction between this personality trait and other individual or environmental factors, which merits further study.

Another methodological design that has been utilized to understand the nature of the relationship between OCPD and anorexia nervosa is a family study design, in which lifetime prevalence rates of OCPD have been examined among the relatives of individuals with anorexia nervosa, some of whom have OCPD themselves and some of whom do not. Two separate family studies both found the same pattern of findings, namely that OCPD is comparably elevated among the relatives of individuals with restricting-type anorexia nervosa who themselves do not have comorbid OCPD as it is among the relatives of probands with comorbid restricting-type anorexia nervosa and OCPD (Strober et al. 2007; Lilenfeld et al. 1998). This pattern of findings is supportive of a common cause model, suggesting that OCPD and the restricting subtype of anorexia nervosa share a common underlying vulnerability. They each are likely to have distinct causes as well, but these results suggest that there is at least some shared cause that gives rise to both conditions.

In addition to being examined in a common cause model, OCPD has been examined in the context of a pathoplastic model, which means to evaluate to what extent this comorbid condition may predict the course and outcome of the ED. It is still not clear to what extent OCPD, in addition to all personality disorders examined with the exception of histrionic personality disorder, may be associated with poorer outcome and response to treatment (Crane et al. 2007; Steinhausen 2002). A prospective study of severe anorexia nervosa in adolescents found that obsessive–compulsive personality traits were indeed predictive of worse outcome over a 10–15-year period (Strober et al. 1997). Similarly, an 18-year quasi-prospective study found that premorbid OCPD was predictive of poor outcome (Wentz et al. 2009). However, another 5-year prospective study of women with bulimia nervosa or ED not otherwise specified showed that the natural course of these disorders was not influenced by the presence or severity of co-occurring personality disorder pathology, including OCPD (Grilo et al. 2007).

On what might be considered the other end of the spectrum are the personality traits of impulsivity and the diagnosis of BPD. Impulsive personality traits and BPD specifically have long been associated with EDs, particularly those that involve purging (e.g., Claes et al. 2006a, b). However, the association between EDs and BPD has been criticized because many of these studies utilize potentially unrepresentative samples (e.g., patients solely recruited from clinical settings), frequently lack control groups, use questionable assessment techniques, and employ problematic diagnostic criteria that overlap among symptoms of BPD, ED, and depression (Pope and Hudson 1989). More recent reviews have concluded that BPD is the most common comorbid personality disorder among those who have bulimia nervosa, as well as among those with binge eating/purging type anorexia nervosa (Sansone et al. 2005). The more interesting question, however, is what the explanation is for this association. Unfortunately, there have been no rigorous investigations of the potential role of BPD specifically as a true risk factor or sharing a common cause with EDs.

Instead, the pathoplastic model has probably been the most commonly examined with regard to the relationship between BPD and EDs. While it has long been assumed that BPD predicted worse outcome among individuals with EDs (e.g., Rosenvinge and Mouland 1990), the validity of this conclusion has come into question. It appears that the relationship between BPD and outcome may not be at all specific to EDs but may instead be related to general psychiatric symptomatology (Bruce and Steiger 2005; Grilo et al. 2007).

In contrast to BPD, the relationship between EDs and the specific personality trait of impulsivity has been examined much more extensively. As with perfectionism, impulsivity can be conceptualized as a multidimensional trait; however, this has rarely been done in the field of EDs and may in fact explain some of the contradictory findings. On the basis of the limited existing longitudinal research, it cannot be concluded that impulsivity is a true risk factor for the development of EDs, even when bulimic symptom increase specifically is being predicted (Wonderlich et al. 2004; Stice and Agras 1998). Behavioral constructs and diagnoses thought to be associated with impulsivity (e.g., substance abuse; Bulik et al. 2004; Lilenfeld et al. 1997) have some predictive validity (Stice 2002), but not impulsivity itself, when examined as a trait in the traditional sense.

To address this limitation of typically treating impulsivity unidimensionally in the field of EDs, a meta-analysis examined the relationship between bulimic symptoms and various specific aspects of impulsivity (Fischer et al. 2008). These included sensation seeking, lack of planning, lack of persistence, and the tendency to act impulsively when distressed ("negative urgency"; Whiteside and Lynam 2001). The largest effect size in predicting bulimic symptoms was found for negative urgency. Thus, this may be one of the more useful aspects of impulsivity to target for future research. Because all of the data used for this meta-analysis were cross-sectional, it does not help to illuminate to what extent impulsivity plays a role in any one or more of the explanatory models of the relationship between impulsivity and EDs.

4 Using Personality and Temperament for Classification and the Search for Genetic Susceptibility Loci

In addition to furthering our understanding of the relationship between personality/temperament and EDs, a promising advancement in our field has been to utilize personality and temperament constructs to "sort" individuals with EDs in a way that may be most clinically meaningful and also allow for the best chance of identifying genetic loci that confer increased risk for the disorder (e.g., Bulik et al. 2007; Lilenfeld et al. 2001). The current DSM (Diagnostic and Statistical Manual for Mental Disorders) classification system is designed to allow for clinical utility (e.g., efficient communication among clinicians, the basis for insurance reimbursement decisions) as well as research on the etiology and treatment of meaningful psychopathological entities. Sometimes, these two goals are at odds with each other. While many areas of psychopathology in the DSM struggle in this regard, there is probably no area in a more unsatisfactory state than the field of EDs as a result of frequent diagnostic crossover (e.g., Eddy et al. 2002) and the preponderance of ED "not otherwise specified" diagnoses in clinical settings (e.g., Fairburn and Bohn 2005). Rather than depending upon the DSM to guide the advancement of knowledge in our field, there has instead been a movement to utilize personality and temperament variables in order to classify individuals with EDs in a way that is most meaningful and potentially useful to understanding etiology. An added specific benefit would be to utilize this new classification scheme to guide the search for genetic susceptibility loci, as there are not likely to be genes that confer risk for a given ED diagnosis, but more likely those that confer risk for specific traits, which, in combination with other risk factors, give rise to an ED.

Several "clusters" based upon personality and temperament variables have emerged. One is an impulsive, emotionally dysregulated cluster. A second is a compulsive, emotionally constricted cluster. A third is a more normal or "high functioning" cluster (Steiger et al. 2009; Wonderlich et al. 2005, 2007a, b; Thompson-Brenner et al. 2008a, b; Claes et al. 2006b). These personality/temperament clusters are relatively independent from diagnostic groupings and, while this is an empirical question to be tested, they are more likely to demonstrate stability over time than what is observed in our current diagnostic classification system. They are also likely to be more useful in the search for etiological factors, as well as other meaningful clinical issues such as prognosis, comorbid psychopathology, family history, treatment history, daily experiences in the natural environment, and overall functioning (Holliday et al. 2006; Westen and Harnden-Fischer 2001; Thompson-Brenner and Westen 2005; Thompson-Brenner et al. 2008a, b; Wonderlich et al. 2005, 2007a, b). Most of these studies have utilized cluster analytic statistical techniques (e.g., Wonderlich et al. 2005). Latent class analysis is a related statistical approach that allows one to identify "classes" of related cases based upon multivariate categorical personality/temperament data. Taxometric analytic techniques are similar, yet are designed to "carve nature at its joints" and discern whether these classes are truly qualitatively distinct from each other. Attempts have been made to use such statistical techniques to derive empirically based ED phenotypes based upon symptom and other clinical data (e.g., Keel et al. 2004), but we are just beginning to apply such techniques to generate empirically based phenotypes based upon personality and temperament variables (Wilksch and Wade 2009; Steiger et al. 2009; Jacobs et al. 2009; Bulik et al. 2005).

5 Conclusions and Future Directions

Early work in the area of personality and temperament began with observations of higher than average levels of certain traits among individuals ill with ED. Such cross-sectional observations then importantly moved from solely examining individuals in clinical settings to those in the community. The field then progressed toward examining such traits in those who recovered from EDs in an effort to identify a presumed "premorbid" glimpse of what the individual may have looked like outside of an acutely ill state. Retrospective studies, and better yet a number of true prospective studies, have been conducted in which traits identified before the onset of illness were identified. More recently, the family and genetic study methodologies have been utilized to examine the relationship between personality traits and EDs. All of these different study designs (i.e., cross-sectional, recovered, retrospective, prospective, family, genetic studies) have been utilized to evaluate various models of the relationship between EDs and personality traits/disorders. These models include the predispositional model (i.e., the personality trait causes the ED), the common cause model (i.e., the personality trait and the ED share a common cause), the complication model (i.e., the ED causes the personality trait), and the pathoplasty model (i.e., the personality trait and the ED each modify the course and presentation of each other).

The most robust findings from the existing literature suggest that likely predisposing (i.e., risk) factors for the development of EDs are neuroticism and perfectionism. The nature of the perfectionism (e.g., "adaptive" versus "maladaptive") is likely to be relevant in understanding the elevated risk. High harm avoidance and low self-directedness are also potential risk factors for EDs, though more prospective or genetic data are needed to confirm this. The diagnosis of OCPD is known to occur at high rates in individuals with EDs, especially restricting-type anorexia nervosa, as well as their first-degree relatives, regardless of whether the individuals themselves have OCPD. These data are most suggestive of a common cause underlying restricting-type anorexia nervosa and OCPD. Nearly all of these identified traits appear to be accentuated by being in an ill state. The data are inconsistent with regard to the extent to which certain personality traits (e.g., impulsivity) or disorders (e.g., BPD) impact the course and response to treatment. As with perfectionism, the extent to which future studies are able to utilize a multidimensional approach to the study of the personality trait of impulsivity may help to clarify the situation.

One of the most promising areas for further advancement in this field is to utilize the personality- and temperament-based "classes" identified through advanced statistical analytic techniques to guide future work in the search for etiological factors, particularly in the expanding field of behavioral and molecular genetics in the search for susceptibility genes for EDs. Utilization of personality- and temperament-based classes to search for etiological factors is much more likely to be fruitful than the majority of current classification schemes based upon diagnostic symptoms.

References

Anderluh MB, Tchanturia K, Rabe-Hesketh S, Treasure J (2003) Childhood obsessive compulsive personality traits in adult women with eating disorders: defining a broader eating disorder phenotype. Am J Psychiatry 160:242–247

Anderluh MB, Tchanturia K, Rabe-Hesketh S, Collier D, Treasure J (2009) Lifetime course of eating disorders: design and validity testing of a new strategy to define the eating disorders phenotype. Psychol Med 39:105–114

Bacanu S, Bulik CM, Klump KL et al (2005) Linkage analysis of anorexia and bulimia nervosa cohorts using selected behavioral phenotypes as quantitative traits or covariates. Am J Med Genet B Neuropsychiatr Genet 139B:61–68

Bardone-Cone AM, Abramson LY, Vohs KD, Heatherton TF, Joiner TE Jr (2006) Predicting bulimic symptoms: an interactive model of self-efficacy, perfectionism, and perceived weight status. Behav Res Ther 44:27–42

Bardone-Cone AM, Wonderlich SA, Frost RO et al (2007) Perfectionism and eating disorders: current status and future directions. Clin Psychol Rev 27:384–405

Bardone-Cone AM, Weishuhn AS, Boyd CA (2009) Perfectionism and bulimic symptoms in African American college women: Dimensions of perfectionism and their interactions with perceived weight status. J Couns Psychol 56:266–275

Bruce KR, Steiger H (2005) Treatment implications of axis-II comorbidity in eating disorders. Eat Disord 13:93–108

Bulik CM (2005) Exploring the gene-environment nexus in eating disorders. J Psychiatry Neurosci 30:335–339

Bulik CM, Tozzi F, Anderson C, Mazzeo SE, Aggen S, Sullivan PF (2003) The relation between eating disorders and components of perfectionism. Am J Psychiatry 160:366–368

Bulik CM, Klump KL, Thornton L et al (2004) Alcohol use disorder comorbidity in eating disorders: a multicenter study. J Clin Psychiatry 65:1000–1006

Bulik CM, Bacanu S, Klump KL et al (2005) Selection of eating-disorder phenotypes for linkage analysis. Am J Med Genet B Neuropsychiatr Genet 139B:81–87

Bulik CM, Sullivan PF, Tozzi F, Furberg H, Lichtenstein P, Pedersen NL (2006) Prevalence, heritability, and prospective risk factors for anorexia nervosa. Arch Gen Psychiatry 63: 305–312

Bulik CM, Hebebrand J, Keski-Rahkonen A et al (2007) Genetic epidemiology, endophenotypes, and eating disorder classification. Int J Eat Disord 40:S52–S60

Cassin SE, von Ranson KM (2005) Personality and eating disorders: a decade in review. Clin Psychol Rev 25:895–916

Castro-Fornieles J, Gual P, Lahortiga F et al (2007) Self-oriented perfectionism in eating disorders. Int J Eat Disord 40:562–568

Cervera S, Lahortiga F, Martínez-González MA, Gual P, Irala-Estévez J, Alonso Y (2003) Neuroticism and low self-esteem as risk factors for incident eating disorders in a prospective cohort study. Int J Eat Disord 33:271–280

Claes L, Nederkoorn C, Vandereycken W, Guerrieri R, Vertommen H (2006a) Impulsiveness and lack of inhibitory control in eating disorders. Eat Behav 7:196–203

Claes L, Vandereycken W, Luyten P, Soenens B, Pieters G, Vertommen H (2006b) Personality prototypes in eating disorders based on the big five model. J Pers Disord 20:401–416

Clark LA, Watson D, Mineka S (1994) Temperament, personality, and the mood and anxiety disorders. J Abnorm Psychol 103:103–116

Cloninger CR, Svrakic DM, Przybeck TR (1993) A psychobiological model of temperament and character. Arch Gen Psychiatry 50:975–990

Crane AM, Roberts ME, Treasure J (2007) Are obsessive-compulsive personality traits associated with a poor outcome in anorexia nervosa? A systematic review of randomized controlled trials and naturalistic outcome studies. Int J Eat Disord 40:581–588

Drews CD, Greenland S (1990) The impact of differential recall on the results of case-control studies. Int J Epidemiol 19:1107–1112

Eddy KT, Keel PK, Dorer DJ, Delinsky SS, Franko DL, Herzog DB (2002) Longitudinal comparison of anorexia nervosa subtypes. Int J Eat Disord 31:191–201

Fairburn CG, Welch SL, Doll HA, Davies BA (1997) Risk factors for bulimia nervosa: A community-based case-control study. Arch Gen Psychiatry 54:509–517

Fairburn CG, Doll HA, Welch SL, Hay PJ, Davies BA, O'Connor ME (1998) Risk factors for binge eating disorder: a community-based, case-control study. Arch Gen Psychiatry 55:425–432

Fairburn CG, Cooper Z, Doll HA, Welch SL (1999) Risk factors for anorexia nervosa: three integrated case-control comparisons. Arch Gen Psychiatry 56:468–476

Fairburn CG, Bohn K (2005) Eating disorder NOS (EDNOS): an example of the troublesome "not otherwise specified" (NOS) category in DSM-IV. Behav Res Ther 43:691–701

Fassino S, Amianto F, Gramaglia C, Facchini F, Abbate-Daga G (2004) Temperament and character in eating disorders: ten years of studies. Eat Weight Disord 9:81–90

Favaro A, Tenconi E, Santonastaso P (2008) The relationship between obstetric complications and temperament in eating disorders: a mediation hypothesis. Psychosom Med 70:372–377

Fischer S, Smith GT, Cyders MA (2008) Another look at impulsivity: a meta-analytic review comparing specific dispositions to rash action in their relationship to bulimic symptoms. Clin Psychol Rev 28:1413–1425

Ghaderi A, Scott B (2000) The big five and eating disorders: a prospective study in the general population. Eur J Pers 14:311–323

Grilo CM, Pagano ME, Skodol AE et al (2007) Natural course of bulimia nervosa and of eating disorder not otherwise specified: 5-year prospective study of remissions, relapses, and the effects of personality disorder psychopathology. J Clin Psychiatry 68:738–746

Holliday J, Uher R, Landau S, Collier D, Treasure J (2006) Personality pathology among individuals with a lifetime history of anorexia nervosa. J Pers Disord 20:417–430

Jacobi C, Hayward C, de Zwaan M, Kraemer HC, Agras WS (2004) Coming to terms with risk factors for eating disorders: application of risk terminology and suggestions for a general taxonomy. Psychol Bull 130:19–65

Jacobs MJ, Roesch S, Wonderlich SA et al (2009) Anorexia nervosa trios: behavioral profiles of individuals with anorexia nervosa and their parents. Psychol Med 39:451–461

Kaye WH, Greeno CG, Moss H et al (1998) Alterations in serotonin activity and psychiatric symptoms after recovery from bulimia nervosa. Arch Gen Psychiatry 55:927–935

Keel PK, Fichter M, Quadflieg N et al (2004) Application of a latent class analysis to empirically define eating disorder phenotypes. Arch Gen Psychiatry 61:192–200

Kendler KS, Neale MC, Kessler RC, Heath AC, Eaves LJ (1993) A longitudinal twin study of personality and major depression in women. Arch Gen Psychiatry 50:853–862

Keys A, Brozek J, Henschel A, Mickelsen O, Taylor HL (1950) The biology of human starvation. University of Minnesota Press, Minnesota

Killen JD, Taylor CB, Hayward C et al (1996) Weight concerns influence the development of eating disorders: A 4-year prospective study. J Consult Clin Psychol 64:936–940

Klein DN, Riso LP (1993) Psychiatric disorders: problems of boundaries and comorbidity. In: Costello CG (ed) Basic issues in psychopathology. Guildford, New York

Klump KL, Strober M, Bulik CM et al (2004) Personality characteristics of women before and after recovery from an eating disorder. Psychol Med 34:1407–1418

Kim Y-K (ed) (2009) Handbook of behavioral genetics. Springer, New York

Lilenfeld LR, Kaye WH, Greeno CG et al (1997) Psychiatric disorders in women with bulimia nervosa and their first-degree relatives: effects of comorbid substance dependence. Int J Eat Disord 22:253–264

Lilenfeld LR, Kaye WH, Greeno CG et al (1998) A controlled family study of anorexia nervosa and bulimia nervosa: psychiatric disorders in first-degree relatives and effects of proband comorbidity. Arch Gen Psychiatry 55:603–610

Lilenfeld LR, Stein D, Bulik CM et al (2000) Personality traits among current eating disordered, recovered, and never ill first-degree female relatives of bulimic and control women. Psychol Med 30:1399–1410

Lilenfeld LR, Devlin B, Bulik CM et al (2001) Deriving behavioural phenotypes in an international, multi-centre study of eating disorders. Psychol Med 31:635–645

Lilenfeld LR, Wonderlich S, Riso LP, Crosby R, Mitchell J (2006) Eating disorders and personality: a methodological and empirical review. Clin Psychol Rev 26:299–320

Lilenfeld LR, Jacobs CH, Woods AM, Picot AK (2008) A prospective study of obsessive-compulsive and borderline personality traits, race and disordered eating. Eur Eat Disord Rev 16:124–132

Matsunaga H, Kaye WH, McConaha C, Plotnicov K, Pollice C, Rao R (2000) Personality disorders among subjects recovered from eating disorders. Int J Eat Disord 27:353–357

Mazzeo SE, Landt MC, Jones I et al (2006) Associations among postpartum depression, eating disorders, and perfectionism in a population-based sample of adult women. Int J Eat Disord 39:202–211

Pearson CA, Gleaves DH (2006) The multiple dimensions of perfectionism and their relation with eating disorder features. Pers Individ Dif 41:225–235

Perkins PS, Klump KL, Lacono WG, McGue M (2005) Personality traits in women with anorexia nervosa: Evidence for a treatment-seeking bias? Int J Eat Disord 37:32–37

Pope HG Jr, Hudson JI (1989) Are eating disorders associated with borderline personality disorder? A critical review. Int J Eat Disord 8:1–9

Rastam M (1992) Anorexia nervosa in 51 Swedish adolescents: premorbid problems and comorbidity. Academy of Child and Adolescent Psychiatry 31:819–829

Rosenvinge JH, Mouland SO (1990) Outcome and prognosis of anorexia nervosa: a retrospective study of 41 subjects. Br J Psychiatry 156:92–97

Rybakowski F, Slopien A, Dmitrzak-Weglarz M, Czerski P, Rajewski A, Hauser J (2006) The 5-HT2A-1438 A/G and 5-HTTLPR polymorphisms and personality dimensions in adolescent anorexia nervosa: association study. Neuropsychobiology 53:33–39

Sansone RA, Levitt JL, Sansone LA (2005) The prevalence of personality disorders among those with eating disorders. Eat Disord 13:7–21

Santonastaso P, Friederici S, Favaro A (1999) Full and partial syndromes in eating disorders: a 1-year prospective study of risk factors among female students. Psychopathology 32:50–56

Shaw HE, Stice E, Springer DW (2004) Perfectionism, body dissatisfaction, and self-esteem in predicting bulimic symptomatology: lack of replication. Int J Eat Disord 36:41–47

Skodol AE, Bender DS (2009) The future of personality disorders in DSM-V? Am J Psychiatry 166:388–391

Steele A, Corsini N, Wade TD (2007) The interaction of perfectionism, perceived weight status, and self-esteem to predict bulimic symptoms: the role of 'benign' perfectionism. Behav Res Ther 45:1647–1655

Steiger H, Richardson J, Schmitz N, Israel M, Bruce KR, Gauvin L (2009) Trait-defined eating-disorder subtypes and history of childhood abuse. Int J Eat Disord. doi:10.1002/eat.20711

Stein D, Kaye WH, Matsunaga H et al (2002) Eating-related concerns, mood, and personality traits in recovered bulimia nervosa subjects: a replication study. Int J Eat Disord 32:225–229

Steinhausen H (2002) The outcome of anorexia nervosa in the 20th century. Am J Psychiatry 159:1284–1293

Stice E (2002) Risk and maintenance factors for eating pathology: a meta-analytic review. Psychol Bull 128:825–848

Stice E, Agras WS (1998) Predicting onset and cessation of bulimic behaviors during adolescence: a longitudinal grouping analysis. Behav Ther 29:257–276

Strober M, Freeman R, Morrell W (1997) The long-term course of severe anorexia nervosa in adolescents: survival analysis of recovery, relapse, and outcome predictors over 10-15 years in a prospective study. Int J Eat Disord 22:339–360

Strober M, Freeman R, Lampert C, Diamond J (2007) The association of anxiety disorders and obsessive compulsive personality disorder with anorexia nervosa: Evidence from a family study with discussion of nosological and neurodevelopmental implications. Int J Eat Disord 40: S46–S51
Sutandar-Pinnock K, Woodside DB, Carter JC, Olmsted MP, Kaplan AS (2003) Perfectionism in anorexia nervosa: a 6-24-month follow-up study. Int J Eat Disord 33:225–229
The Wellcome Trust Case Control Consortium (2007) Genome-wide association study of 14, 000 cases of seven common diseases and 3,000 shared controls. Nature 447:661–678
Thompson-Brenner H, Westen D (2005) Personality subtypes in eating disorders: validation of a classification in a naturalistic sample. Br J Psychiatry 186:516–524
Thompson-Brenner H, Eddy KT, Franko D et al (2008a) A personality classification system for eating disorders: a longitudinal study. Compr Psychiatry 49:551–560
Thompson-Brenner H, Eddy KT, Satir DA, Boisseau CL, Westen D (2008b) Personality subtypes in adolescents with eating disorders: validation of a classification approach. J Child Psychol Psychiatry 49:170–180
Tissot AM, Crowther JH (2008) Self-oriented and socially prescribed perfectionism: risk factors within an integrative model for bulimic symptomatology. J Soc Clin Psychol 27:734–755
Tozzi F, Aggen SH, Neale BM et al (2004) The structure of perfectionism: a twin study. Behav Genet 34:483–494
Tyrka AR, Waldron I, Graber JA, Brooks-Gunn J (2002) Prospective predictors of the onset of anorexic and bulimic syndromes. Int J Eat Disord 32:282–290
Vitousek KM, Stumpf RE (2005) Difficulties in the assessment of personality traits and disorders in eating-disordered individuals. Eat Disord 13:37–60
Vohs KD, Bardone AM, Joiner TE Jr, Abramson LY, Heatherton TF (1999) Perfectionism, perceived weight status, and self-esteem interact to predict bulimic symptom development. J Abnorm Psychol 108:695–700
Vrabel KR, Rø Ø, Martinsen EW, Hoffart A, Rosenvinge JH (2010) Five-year prospective study of personality disorders in adults with longstanding eating disorders. Int J Eat Disord 43:22–28
Wade TD, Bulik CM (2007) Shared genetic and environmental risk factors between undue influence of body shape and weight on self-evaluation and dimensions of perfectionism. Psychol Med 37:635–644
Wade TD, Tiggemann M, Bulik CM, Fairburn CG, Wray NR, Martin NG (2008) Shared temperament risk factors for anorexia nervosa: a twin study. Psychosom Med 70:239–244
Wentz E, Gillberg IC, Anckarsäter H, Gillberg C, Råstam M (2009) Adolescent-onset anorexia nervosa: 18-year outcome. Br J Psychiatry 194:168–174
Westen D, Harnden-Fischer J (2001) Personality profiles in eating disorders: rethinking the distinction between axis I and axis II. Am J Psychiatry 158:547–562
Whiteside SP, Lynam DR (2001) The five factor model and impulsivity: using a structural model of personality to understand impulsivity. Pers Individ Dif 30:669–689
Wilksch SM, Wade TD (2009) An investigation of temperament endophenotype candidates for early emergence of the core cognitive component of eating disorders. Psychol Med 39:811–821
Wonderlich SA, Connolly KM, Stice E (2004) Impulsivity as a risk factor for eating disorder behavior: assessment implications with adolescents. Int J Eat Disord 36:172–182
Wonderlich SA, Crosby RD, Joiner TE Jr et al (2005) Personality subtyping and bulimia nervosa: psychopathological and genetic correlates. Psychol Med 35:649–657
Wonderlich SA, Crosby RD, Engel SE, Mitchell JE, Smyth J, Miltenberger R (2007a) Personality-based clusters in bulimia nervosa: differences in clinical variables and ecological momentary assessment. J Pers Disord 21:340–357
Wonderlich SA, Joiner TE Jr, Keel PK, Williamson DA, Crosby RD (2007b) Eating disorder diagnoses: empirical approaches to classification. Am Psychol 62:167–180
Zimmerman M (1994) Diagnosing personality disorders: a review of issues and research methods. Arch Gen Psychiatry 51:225–245

Cognitions and Emotions in Eating Disorders

Nicolette Siep, Anita Jansen, Remco Havermans, and Anne Roefs

Contents

Abstract The cognitive model of eating disorders (EDs) states that the processing of external and internal stimuli might be *biased* in mental disorders. These biases, or cognitive errors, systematically distort the individual's experiences and, in that way, maintains the eating disorder. This chapter presents an updated literature review of experimental studies investigating these cognitive biases. Results indicate that ED patients show biases in attention, interpretation, and memory when it comes to the processing of food-, weight-, and body shape-related cues. Some recent studies show that they also demonstrate errors in general cognitive abilities such as set shifting, central coherence, and decision making. A future challenge is whether cognitive biases and processes can be manipulated. Few preliminary

---

N. Siep (✉), A. Jansen, R. Havermans, and A. Roefs
Department of Clinical Psychological Science, Faculty of Psychology and Neuroscience, Maastricht University, P.O. Box 616, 6200 MD Maastricht, The Netherlands
e-mail: Nicolette.Siep@maastrichtuniversity.nl

R.A.H. Adan and W.H. Kaye (eds.), *Behavioral Neurobiology of Eating Disorders*,
Current Topics in Behavioral Neurosciences 6, DOI 10.1007/7854_2010_82

studies suggest that an attention retraining and training in the cognitive modulation of food reward processing might be effective strategies to change body satisfaction, food cravings, and eating behavior.

Keywords Attention · Body image · Central coherence · Cognitive bias · Craving · Decision-making · Food reward · Interpretation · Memory · Retraining · Set shifting

1 Introduction

Imagine that you are sleeping in your bedroom. In the middle of the night you suddenly awake of a loud noise downstairs in the living room. You think "o no, there is a burglar in my house." You feel extremely anxious. You do not dare to go downstairs to have a look in the living room; that is why you stay where you are, trembling, and waiting for what is coming. Now imagine the following situation. You are sleeping again. In the middle of the night you suddenly awake of a loud noise downstairs in the living room. You think "that stupid cat! What did she knock over this time?" You feel a bit irritated but soon continue your sweet dreams. These examples make clear that it is not what is really happening that causes one to feel anxious, irritated, or happy. It is how one interprets what is happening that determines one's feelings. In eating disorders (EDs), it is not one's actual appearance or body weight that causes a problem, but one's evaluation of it. This is a key assumption of the cognitive model of EDs.

2 Cognitive Biases

The cognitive model of EDs has its roots in the cognitive model of psychopathology or mental disorders that was formulated by Beck in 1964. Beck theorized that the processing of external events or internal stimuli is *biased* in mental disorders. These biases, or cognitive errors, systematically distort the individual's experiences. The model points out that eating psychopathology arises from maladaptive knowledge structures (e.g., schemas) that are involved in the allocation of attention, in memory, and in the interpretation of incoming information (Hargreaves and Tiggemann 2002; Williamson et al. 2004). Activation of these knowledge structures causes disorder relevant information to be processed in a biased manner, resulting in a range of cognitive biases in attention, judgment, and memory (Williamson et al. 2004).

In line with Beck's (1976) cognitive specificity hypothesis, Vitousek and Hollon (1990) proposed two decades ago that ED patients consider their own weight and shape as the predominant referents for inferring personal value. Nowadays, this is also referred to as "overevaluation" (Fairburn 2008). These cognitions about weight and shape in relationship to the self are organized into structures referred to as *weight-related self-schemata*. The central premise of the cognitive model is that these schemata are the core cognitive component of EDs. The operation of the schemata might cause and/or maintain EDs by producing systematic errors in weight and shape information processing.

The cognitive model specifies three main cognitive errors in EDs: attention bias, interpretation bias, and memory bias, and some recent studies suggest impairments in general cognitive processing. In this chapter, we will first discuss the main cognitive biases and general impairments in cognitive processing that are demonstrated in EDs. We then will focus on the question whether the manipulation of cognitive processing affects ED psychopathology.

2.1 Attention Bias

An attention bias refers to the tendency to selectively attend to disorder relevant stimuli (Mathews and MacLeod 2005). According to the cognitive model of EDs, ED patients are more likely to give priority to cues pertaining to body and food-related information than to neutral cues, in comparison to healthy controls. The cognitive model proposes that, with progression of the ED, a phobic orientation toward the body, high-calorie foods, and weight gain develops (Williamson et al. 1999). The hyperattention to disorder relevant cues is presumed to maintain EDs: it might lead to the avoidance of cues that elicit anxiety and negative affect. This avoidance will immediately decrease anxiety, but it will also prevent its extinction. Therefore, in the long run, anxiety and negative effect will maintain or even increase and the ED will continue. Several paradigms have been used to test the attention bias hypothesis in EDs, including the modified Stroop task, the dot-probe task, the visual search task, and eye tracking.

2.1.1 Stroop Task

The Stroop *task* is the most frequently used paradigm to investigate attention bias in ED patients. Neutral words and ED relevant words are printed in different colors (Faunce and Job 2000; Williams et al. 1996). Participants are required to ignore the meaning of the words and to simply report the color in which each word is printed. Disorder relevant words interfere with color naming more than neutral words do, leading to longer color-naming responses for disorder relevant words than for neutral words. This difference in response time for disorder relevant versus neutral words is the "interference effect". Currently, more than 30 studies have evaluated the interference effect for food-, weight-, and shape-related words in ED patients. Confirming the attention bias hypothesis, it was concluded in a meta-analysis that ED patients show increased interference for food-, weight-, and shape-related words (Johansson et al. 2005). Both patients with bulimia nervosa (BN) and anorexia nervosa (AN) showed the Stroop interference for weight- and shape-related words, but AN patients were more interfered by food stimuli relative to BN patients (but see Dobson and Dozois 2004).

Although a large number of studies used the modified Stroop task to demonstrate attention biases in EDs, a number of concerns regarding the use of this task have been raised. In a review of modified Stroop studies in EDs, Lee and Shafran (2004) conclude that Stroop interference with ED-related stimuli is also found in non-ED groups, which brings into question its clinical relevance. More specifically, these biases have been demonstrated in restrained eaters (Francis et al. 1997), hungry participants (Mogg et al. 1998), food deprived participants (Placanica et al. 2002), and in healthy participants who had just finished an appetizer (Overduin et al. 1995). Another, more general, concern about the Stroop task is that little effort has been made to account for the underlying mechanisms of the interference effect (Williams et al. 1996). It is not clear why ED patients show interference. Food, weight, and shape words might trigger anxiety in ED patients, who usually experience strong concerns pertaining to their body weight, shape, and eating, hence making these words especially meaningful and frightening. The anxiety might induce an automatic tendency to avoid further exposure to these words. Interference then might reflect avoidance and not any attention bias. Indeed, De Ruiter and Brosschot (1994) demonstrated that attempts to cognitively avoid the processing of disorder relevant word stimuli also result in increased interference scores.

Thus, the central premise that attention is biased toward food and body stimuli cannot be tested with the modified Stroop task, because it is not clear what the Stroop task precisely measures. Given this uncertainty about the meaning of increased interference scores, no firm conclusions can be drawn about the existence of an attention bias in ED patients using the modified Stroop paradigm. Therefore, alternatives to the Stroop task were used, such as the dot-probe task.

2.1.2 Dot-Probe Task

The dot-probe task (MacLeod et al. 1986) is based on the assumption that individuals respond faster to a small dot that is presented in an attended area of a visual display than to a dot that is presented in an unattended area of a visual display. Pairs of pictures or words are concurrently and briefly presented to participants. One word/picture of each pair is disorder relevant and the other is neutral. When the word/picture pair disappears, a small dot appears in a position previously occupied by one of the two words or pictures. The participant is instructed to push a button as quickly as possible when the dot appears. The attention bias for disorder relevant stimuli is calculated by taking the difference in reaction times to the dot when it replaces the neutral stimulus minus reaction times to the dot when it replaces the disorder relevant one. Faster reactions to dots that replace the disorder relevant stimuli indicate an attention bias.

The first study using the dot-probe task in EDs (Rieger et al. 1998) tested the attention bias for shape-related words. Results showed that the ED group detected a target dot that replaced a thin shape-related word slower than did healthy individuals. In contrast, the ED group detected a target dot that replaced a large shape-related word faster than did healthy individuals. Rieger and colleagues concluded that ED patients are more likely to attend to information consistent with fatness and to ignore information consistent with thinness, which might maintain distorted cognitions about shape and weight.

Rieger et al. (1998) used words, but Mogg and colleagues argue that using words as stimuli provides a relatively fragile index of attention bias, and they suggest using pictures instead (Mogg et al. 2000). Pictorial dot-probe studies were also done to study ED attention biases (Glauert et al. 2010; Lee and Shafran 2008; Shafran et al. 2007). Using the pictorial dot-probe, Shafran et al. (2007) tested (1) whether attention biases for food-, shape-, and weight-related pictures are stronger in ED participants compared to healthy, anxious, and shape-concerned controls, and (2) whether the strength of attention biases is associated with the severity of ED psychopathology. Stimuli included pictures of food, body shape, body weight, and animals. ED participants were faster to respond to a dot when it replaced negative food and weight pictures, and they responded slower to a dot when it replaced positive food pictures, compared to anxious controls and women with high, moderate, and low levels of shape concerns. In study 2, but not in study 1, ED participants were significantly faster to respond to the dot when it replaced negative and neutral shape pictures compared to the controls. No bias was found for positive shape stimuli. These findings only partially support the previous findings by Rieger et al. (1998). Shafran et al. (2007) suggested that biases for body shapes might be less robust because the pictures were of other persons and hence not personally relevant. In addition, their results showed there was a modest relationship between the degree of psychopathology and the extent of attention biases.

Further research showed that with increasing duration of the interstimulus-interval from 500 to 2,000 ms, the attention bias for food and shape pictures in ED patients disappeared, whereas the bias for weight stimuli remained (Lee and Shafran 2008). These findings suggest a pattern of cognitive avoidance for food and shape pictures that increases over time. This is in line with the cognitive model, stating that the initial attention bias for threatening stimuli eventually serves to enable patients to actively avoid these stimuli. Thus, the attention bias might prevent confrontation with the feared stimulus. If there is no exposure to feared stimuli, fear responses will not be able to extinguish, and in that way the ED will persist. Interestingly, Shafran et al. (2007) showed that attention biases for food, weight, and shape pictures in ED patients decreased after cognitive-behavioral treatment.

2.1.3 Visual Search Task

Another method to study attention bias is the visual search task. Treisman and Gelade (1980) devised the first visual search task, in which participants were instructed to locate a simple target (e.g., circle, square) as quickly as possible among an array of several distractors, all with a similar shape but different from the shape of the target. Differences in search performance on the visual search task are proposed to reflect differences in the focus of attention. Hansen and Hansen (1988) developed the face-in-the-crowd visual search task in which participants search for an odd face stimulus in a matrix of face stimuli. This adaptation had important advantages. First, if the odd-one-out stimulus is a disorder relevant stimulus, which is presented among neutral distractors, speeded detection of disorder relevant information can be measured. Second, by making the odd-one-out stimulus a neutral stimulus, which is presented among disorder relevant distractors, distraction by disorder relevant information can be measured (Rinck et al. 2005). The distraction component may arguably be similar to maintained attention (Mogg et al. 2005), or slowed disengagement (Fox et al. 2001) components of attention bias. This task thus allows for the investigation of specific mechanisms of attention bias: speeded detection and increased distraction. To our knowledge, so far only one study has applied the odd-one-out visual search task in ED patients (Smeets et al. 2008). In this study, ED participants showed evidence of speeded detection of body shape words, but not of increased distraction by shape stimuli compared to healthy controls. The opposite pattern of results was found for food-related words: ED participants showed no evidence of speeded detection of high-calorie food words, but there was an increased distraction compared to neutral and low calorie words. The finding that ED patients were more distracted by high-calorie words than controls was explained by craving (Mogg et al. 2005; Smeets et al. 2008). Indeed, a substantial number of studies have found significant correlations between attention bias for craving-related stimuli and levels of subjective craving (Field et al. 2005, 2007; Franken et al. 2000; Rosse et al. 1993, 1997). In a next study, Smeets et al. (2009) experimentally induced craving and a causal link between induced chocolate craving and a bias in the distraction component of attention was found. More specifically, when brought to an elevated state of chocolate craving, chocoholics showed more distraction by chocolate than in the absence of such increased chocolate craving state. This study shows that the suggested relationship between craving and attention bias only holds true for the specific distraction component and not for speeded detection.

2.1.4 Eye Tracking

Changes in attention usually are directly related to eye movements (Henderson et al. 1989; Rayner 1998). Eye tracking has several advantages over reaction time tasks. It provides a more direct indication of attention bias, and it allows not only to measure the initial detection of a stimulus but also changes in the direction of attention. Also the maintenance of attention on relevant stimuli can be directly assessed by measuring the fixation duration on the stimulus.

Eye tracking studies show that ED patients tend to focus on dissatisfying body parts (Freeman et al. 1991; Jansen et al. 2005). Jansen et al. (2005) showed that ED patients allocate their attention more toward their self-identified unattractive body parts than to their self-identified attractive body parts. When looking at the bodies of other persons, the ED patients paid most attention to the other's attractive body parts than to the other's unattractive body parts (upward comparison). Healthy controls, however, did exactly the opposite; they looked more at their own attractive body parts compared to their own unattractive body parts, and they attended more to the other person's unattractive parts than to the other's attractive body parts (downward comparison).

In an additional study, Jansen et al. (2006) showed that control models had a strong positively biased perception of their own attractiveness, whereas ED patients lack this self-serving bias. The cognitive processing in ED patients might have caused this a lack of a *self-serving body-image bias,* since they focus their attention on body parts that are evaluated as unattractive by themselves whereas healthy controls do the opposite. Mulkens and Jansen (2009) demonstrated that increased attention for appearance leads to increased body dissatisfaction in vulnerable participants (highly body dissatisfied participants), whereas healthy controls show increased body satisfaction after increased attention.

It can be concluded that, in line with the cognitive model of EDs, EDs are characterized by an attention bias for high-calorie foods and bodies. Even *if* Stroop studies measure biases in attention, they do not clearly indicate *how* one's attention exactly is biased; is the attention directed at specifically thin bodies, fat bodies, the own body, attractiveness or unattractiveness, and so on. What cognitive processes are involved? And do high-calorie food biases and body biases reflect identical cognitive processes? Use of the dot-probe task suggests that early attention of ED patients is specifically biased toward negative food and body stimuli. The visual search task showed that body stimuli elicit a bias in speeded detection. This task also showed increased distraction by tasty high-calorie food stimuli. It was further demonstrated that craving leads to an attention bias for high-calorie food stimuli – more specifically increased distraction. Eye tracking studies demonstrate that the way one inspects bodies is causal to body (dis)satisfaction. EDs specifically focus on their own negatively evaluated body parts, thereby inducing greater body dissatisfaction.

2.2 *Interpretation Bias*

The interpretation bias refers to the tendency to interpret ambiguous stimuli in a disorder relevant way; in EDs, ambiguous stimuli are expected to be interpreted as weight- or shape-related, and in a negative way, especially when these stimuli refer to the patients themselves. Ambiguous scenarios such as "Two friends are giggling and whispering behind you. What do you think they are saying?" are presented. It is assessed how the participant interprets each scenario and whether the interpretation relates to body weight or shape, either in an open-ended or a forced-choice format.

Cooper (1997) showed that ED patients responded more often with a weight and shape interpretation compared to the healthy controls, both in open-ended and forced-choice negative outcome scenarios referring to themselves. ED patients responded more with positive weight and shape interpretations when the scenarios related to others. Thus, ED patients judge weight and shape to be a more likely explanation for events with a negative outcome and referring to themselves (Cooper 1997; Morrison et al. 2006). Jansen et al. (2007) showed that this was also true for overweight and obese children. Williamson et al. (2000) found that ED patients are able to change interpretations in shape-related situation after explicit instructions. A relevant question for future research is whether retraining interpretations into more positive ones leads to less ED symptoms. It can be concluded that, in line with the cognitive model, EDs are characterized by a self-blaming style, in which they judge weight and shape to be a most likely explanation for negative ambiguous events related to the self. Interestingly, Williamson et al. (2000) showed that it is possible to experimentally manipulate the interpretation bias in ED patients; they demonstrated that interpretations could be made less biased, that is less shape-related (Williamson et al. 2000).

2.3 *Memory Bias*

A memory bias refers to the tendency to recall disorder-specific information more easily. The cognitive model predicts that food, weight, and shape information will be more readily encoded in memory and more easily accessed in recall by ED patients. In general, two types of memory are distinguished: *explicit* and *implicit*. Explicit memory is characterized by conscious recollection or recognition of a previous event or experience. In contrast, implicit memory is exhibited when prior experience facilitates or primes performance on a task, without conscious recollection of the experience.

2.3.1 Explicit Memory Bias

Research has found convincing support for the existence of an explicit memory bias in ED patients, but data for an implicit memory bias are equivocal (Hermans et al. 1998; Hunt and Cooper 2001; King et al. 1991; Sebastian et al. 1996; Pietrowsky et al. 2002). Studies using paradigms such as the *cued and free recall task* and the *directed forgetting paradigm* indicate that EDs show specific explicit ED-related memory biases: EDs show a memory bias for food-, weight-, and shape-related words and not for general emotional words (Hermans et al. 1998; Hunt and Cooper 2001; King et al. 1991; Sebastian et al. 1996; Suslow et al. 2004). It was also found that the memory bias for high-caloric foods was independent of food deprivation in AN participants but not in BN participants (Hunt and Cooper 2001; Pietrowsky et al. 2002; Suslow et al. 2004).

2.3.2 Implicit Memory Bias

To date, as far as the present authors know, only two studies have tested implicit memory in ED patients (Hermans et al. 1998; Johansson et al. 2008). Hermans et al. (1998) used a *word-stem completion task* but found no differences between AN participants and controls. However, Johansson et al. (2008) argued that the word-stem completion task is not a valid measure of implicit memory, because explicit memory can be used to complete the task. A more valid measure is *Jacoby's white noise task* (Jacoby et al. 1988). In the white noise task, participants listen to sentences and repeat them out loud. Subsequently, these sentences are presented again, but now they are intermixed with new sentences not previously heard along with background noise that varies in intensity. The participant is instructed to judge the intensity of the background noise. Participants rate noise as less loud for sentences previously heard compared with new sentences, suggesting implicit memory (Jacoby et al. 1988). In agreement with the implicit memory bias hypothesis, Johansson et al. (2008) showed that ED participants rated background noise for food and shape sentences being less loud, compared to neutral sentences. In conclusion, EDs show an explicit memory bias for food, weight, and shape information, and there are some indications for an implicit memory bias in EDs also (but see Hermans et al. 1998).

3 General Impairments in Cognitive Processing

In addition to studying the biases proposed by the cognitive model, general processing impairments not specifically related to food, weight, or shape have also been investigated. The general cognitive processes of interest are (1) set shifting, (2) central coherence, and (3) decision making.

3.1 Set Shifting

First, cognitive set shifting is considered, which refers to the ability to move back and forth between multiple tasks, operations, or mental sets and is a major component of executive functioning. Problems in set shifting may manifest either as cognitive inflexibility (e.g., concrete and rigid approaches to problem solving) or response inflexibility (e.g., stereotyped behaviors) and have been associated with EDs (Tchanturia et al. 2004a). A systematic meta-analysis shows that both BN and AN participants have problems performing a wide range of cognitive set shifting tasks, for example, the Wisconsin Card Sorting Task (Roberts et al. 2007). Furthermore, it was shown that weight recovery in AN participants did *not* improve cognitive set shifting, indicating that it is a trait and not a state marker of AN (Tchanturia et al. 2004b). Proposing a genetic basis for impaired cognitive set shifting, Holliday et al. (2005) found more set shifting difficulties in healthy sisters of AN patients than in unrelated healthy women. The implication of these findings might be that a therapy improving cognitive flexibility and performance on set shifting tasks in ED patients is beneficial in treatment (Tchanturia et al. 2007).

3.2 *Central Coherence*

The second cognitive ability hypothesized to be impaired in ED is central coherence. A weak central coherence refers to enhanced detailed processing, accompanied by a limited ability to understand context or to "see the big picture." This causes information to be processed in parts, rather than as a whole, which impairs global thinking. Central coherence has been recognized as playing an important role in autism spectrum disorders, and it has now been suggested to also be related to the development of EDs (Happe and Frith 2006; Lopez et al. 2009). Weak central coherence might explain the preoccupation with details and rules observed in many ED patients.

AN patients were also found to have a weaker theory *of mind* compared to healthy controls (Russell et al. 2009; Harrison et al. 2009). Theory of mind refers to the cognitive ability to understand the internal states of others, and a weak theory of mind is characteristic of autism. However, Oldershaw et al. (2009) demonstrated that recovered AN patients performed significantly better than currently ill AN patients when inferring emotions during a theory of mind task, showing that impaired theory of mind is likely to be caused by self-induced starvation. In a study by Lopez et al. (2009), AN patients and healthy controls completed several tests measuring visuospatial and verbal aspects of central coherence (Rey–Osterrieth Complex Figure Test, Embedded Figures Test, Homograph Reading Test, and Sentence Completion Task). Results showed that the AN group scored significantly better on tests requiring local processing and worse on global processing tasks. There was no association between performance and depression, anxiety or degree of starvation. However, weak central coherence was correlated with the number of obsessive-compulsive traits. This fits nicely with the idea of a weak central coherence explaining the anorectic preoccupation with details and rules.

3.3 *Decision Making*

A third cognitive ability that is studied in ED is decision making. ED patients show impaired decision making (Boeka and Lokken 2006; Brand et al. 2007). BN patients performed significantly worse on the Iowa Gambling Task (IGT) and the Game of Dice Task, compared to a control group (Cavedini et al. 2004). There was a significant negative correlation between performance and bulimic symptoms, independent of depressive symptoms (Boeka and Lokken 2006;

Brand et al. 2007; Cavedini et al. 2004). It was also found that AN patients who had better decision-making abilities at the start of a treatment showed significantly greater improvement in nutritional status (Cavedini et al. 2004). Decision making has been linked to increased impulsivity and overeating (Davis et al. 2004; Nederkoorn et al. 2006).

4 Manipulation of Cognitive Processes

It is concluded that ED patients are characterized by cognitive biases and some errors in general cognitive processing. It is not entirely clear whether these biases and impairments in cognitive processes are causes, consequences, or an epiphenomenon of EDs. What happens when the biases are manipulated, for example, by attention retraining? In this section, we discuss some recent studies on the retraining of attention bias and the cognitive modulation of food reward processes.

4.1 Retraining Attention Bias

It was discussed above that ED patients show an attention bias for their own unattractive body parts. A first question is whether this way of looking is causal to greater body dissatisfaction. To test causality, Smeets et al. (2010) experimentally manipulated the way of looking in healthy participants and measured its effects on body satisfaction. Results showed that the way one looks at one's own body causes changes in body satisfaction: healthy participants who were trained to attend to their self-defined unattractive body parts showed significantly decreased body satisfaction after the training. It was also showed that slightly body dissatisfied female students who were trained in attending one's own most attractive body parts showed a significant increase in body satisfaction after the training (Smeets et al. 2010). Thus, the way one inspects bodies is causal to body (dis)satisfaction. Retraining the attention bias for negatively evaluated body parts into increased attention for positively evaluated body parts, that is the way healthy females look, was found to be beneficial; this way of looking increased body satisfaction (Smeets et al. 2010). It is of great interest to find out whether such an attention retraining is clinically useful.

Another retraining study focused on inhibitory control (Houben and Jansen 2010). It was studied whether strengthening inhibitory control can increase resistance to high-calorie food temptations. Chocolate cravers were trained to inhibit their responses to chocolate stimuli, which led to significantly reduced chocolate consumption. These findings suggest that strengthening inhibitory control might be an effective strategy to help regain control over food intake.

4.2 Modulation of Food Reward Processing

Food reward stimulates eating behavior (Toates 1986). The incentive salience of food is evaluated in the dopaminergic corticomesolimbic circuitry and motivates eating behavior in the absence of energy deficits (Berridge 2004; Bindra 1978; Bolles 1972). Whether one actually eats – or not – depends on the interaction between food reward and cognitive control (Appelhans 2009). Is it possible to manipulate one's cognitive control in such a way that one is better able to cope with an environment rich in palatable, readily available high-calorie foods? It was therefore studied whether healthy lean women are able to modulate food reward processing using different types of cognitive control strategies (Siep et al. 2010b). Participants were instructed (1) to suppress food palatability thoughts (i.e., do not think about how tasty the food is), (2) to apply cognitive reappraisal (e.g., think about the health consequences of eating it, like gaining weight), and (3) to upregulate thoughts of food palatability (e.g., think about how good the food smells and tastes). It was investigated whether the three cognitive control strategies changed self-reported food cravings and associated food reward activity in the brain as assessed by functional magnetic resonance imaging (fMRI). Both the cognitive reappraisal and suppression manipulations decreased self-reported food cravings compared to upregulation. The fMRI data suggest that these strategies rely on different neural substrates. Cognitive reappraisal decreased activity in the fusiform gyrus (FG) compared to upregulation and suppression, but did not differ from passive viewing in mesocorticolimbic regions. The FG is involved in the processing of visual cues and their reward values, and determines future reward actions (Murray and Izquierdo 2007). The brain responses during cognitive reappraisal suggest that this strategy prevents food cues from eliciting further reward processing. This proposition is in line with the definition of cognitive reappraisal (Gross and John 2003) as "thinking about the emotion eliciting cue in a way that changes its emotion impact." Suppression decreased activity in the ventral striatum (VS) and ventral tegmental area (VTA), and increased activity in the lateral orbitofrontal cortex (OFC) and anterior prefrontal cortex compared to upregulation and suppression. Although suppression successfully inhibited activity in the VS and VTA, the increased prefrontal cortex activity suggests that this strategy requires increased mental effort compared to cognitive reappraisal. As expected, upregulation increased activity in the mesocorticolimbic circuitry. Together these findings show that people *can* actively up- and downregulate mesocorticolimbic food reward processing, and that craving covaries with the use of cognitive control strategies.

In line with this, one might speculate that AN is associated with extremely successfully applied cognitive strategies to decrease food reward processing. A cognitive effect of rigid dieting might be the downregulation of food reward activity in the brain. AN might develop when highly controlled, rigid, and obsessively dieting adolescent females overcome the natural rewarding value of food (Pinel et al. 2000). To test the hypothesis of downregulated food reward activity in AN, AN patients were instructed to evaluate the palatability of high and low calorie foods and at the same time corticomesolimbic food reward activity was measured (Siep et al. 2010a). Furthermore, it was studied what would happen in the dorsolateral prefrontal cortex (dlPFC) when AN patients focused their attention on a neutral part of the picture while they were simultaneously presented with food pictures. The dlPFC is an area involved in successful self-control (Hare et al. 2009). Interestingly, it was shown that AN patients evaluating high-calorie foods fail to activate two important regions of the corticomesolimbic food reward circuitry: the anterior insular cortex (AIC) and caudal anterior cingulate cortex (cACC). Both the AIC and cACC are involved in craving (Craig 2002, 2003; Naqvi et al. 2007) and in the motivation of reward-related behavior (Walton et al. 2009). Naqvi et al. (2007) showed that smokers who acquire insula damage are likely to stop smoking quite easily. In line with this, it could be hypothesized that a decreased responsiveness of AIC in AN patients allows them to easily stop eating. The AN patients did show increased activity in the right anterior OFC during the palatability evaluation of low calorie foods, supporting the hypothesis of successfully downregulating the food reward processing. In a meta-analysis of neuroimaging studies, Kringelbach and Rolls (2004) concluded that the anterior OFC is involved in the processing of abstract reinforcers such as money. It was concluded that AN patients might evaluate low calorie foods as rewarding but, in contrast to healthy people, their food evaluations appear not to be intuitive (Kaye et al. 2009) or based on palatability (Roefs et al. 2005). To test the proposition of increased dlPFC activity in AN, participants were instructed to focus their attention on a neutral cue while high-calorie foods were simultaneously presented in an unattended part of the stimulus display. This led to a strongly activated dlPFC in AN, but not in healthy controls (Siep et al. 2010a). These findings support earlier ones that show dlPFC hyperactivity in AN, indicating a relatively quick and automatic activation of increased self-control when confronted with high-calorie foods.

5 Cognitive Therapy

According to the cognitive model of EDs, ED patients use their own body weight and shape as the predominant factors for inferring personal value. Dysfunctional weight and shape beliefs flow from cognitive structures referred to as schemata. Activation of these schemata produces systematic errors in information processing, such as attention biases, interpretation biases, and memory biases for food and body stimuli. In this way, a negative self-perception is maintained. ED patients show biases in the processing of food-, weight-, and shape-related cues; some recent studies showed that they also demonstrate errors in general cognitive abilities such as set shifting, central coherence, and decision making.

Cognitive behavior therapy (CBT) is up to now the most effective treatment for EDs (see e.g., Wilson et al. 2007). During the cognitive intervention of CBT, dysfunctional cognitions are challenged and a change in thinking is strived for. The assumption is that when thinking changes, emotions will be more positive and behavior will be less symptomatic. Future studies might focus on the question whether CBT might profit from incorporating attention retraining to improve body image, attention retraining to strengthen inhibitory control, techniques to change the errors in general cognitive abilities, and training in the manipulation of food reward processing. This of course requires large-scale randomized clinical trials with a so termed additive design. But we would like to contend that the development of progressively efficacious treatments for EDs just as much requires elegant experimental studies such as outlined in the present chapter, experiments that increase our knowledge of the basic mechanisms maintaining eating psychopathology.

References

Appelhans BM (2009) Neurobehavioral inhibition of reward-driven feeding: implications for dieting and obesity. Obesity 17:640–647

Beck AT (1964) Thinking and depression: theory and therapy. Arch Gen Psychiatry 10:561–571

Beck AT (1976) Cognitive therapy and the emotional disorders. International Universities Press, New York

Berridge KC (2004) Motivation concepts in behavioral neuroscience. Physiol Behav 81:179–209

Bindra D (1978) How adaptive behavior is produced: a perceptual-motivation alternative to response reinforcement. Behav Brain Sci 79:41–91

Boeka AG, Lokken KL (2006) The Iowa gambling task as a measure of decision making in women with bulimia nervosa. J Int Neuropsychol Soc 12:741–745

Bolles RC (1972) Reinforcement, expectancy, and learning. Psychol Rev 79:394–409

Brand M, Franke-Sievert C, Jacoby GE, Markowitsch HJ, Tuschen-Caffier B (2007) Neuropsychological correlates of decision making in patients with bulimia nervosa. Neuropsychology 21:742–750

Cavedini P, Bassi T, Ubbiali A, Casolari A, Giordani S, Zorzi C et al (2004) Neuropsychological investigation of decision-making in anorexia nervosa. Psychiatry Res 127:259–266

Cooper M (1997) Bias in interpretation of ambiguous scenarios in EDs. Behav Res Ther 35: 619–626

Craig AD (2002) How do you feel? Interoception: the sense of the physiological condition of the body. Nat Rev Neurosci 3:655–666

Craig AD (2003) Interoception: the sense of the physiological condition of the body. Curr Opin Neurobiol 13:500–505

Davis C, Levitan RD, Muglia P, Bewell C, Kennedy JL (2004) Decision-making deficits and overeating: a risk model for obesity. Obesity Res 12:929–935

De Ruiter C, Brosschot JF (1994) The emotional Stroop interference effect in anxiety: attentional bias or cognitive avoidance? Behav Res Ther 32:315–319

Dobson KS, Dozois DJ (2004) Attentional biases in EDs: a meta-analytic review of Stroop performance. Clin Psychol Rev 23:1001–1022

Fairburn CG (2008) Cognitive behavior therapy and EDs. Guilford, New York

Faunce GJ, Job RF (2000) The Stroop colour-naming task and addictive behaviour: some recommendations. Addiction 95:1438–1442

Field M, Mogg K, Bradley BP (2005) Craving and cognitive biases for alcohol cues in social drinkers. Alcohol Alcohol 40:504–510

Field M, Duka T, Eastwood B, Child R, Santarcangelo M, Gayton M (2007) Experimental manipulation of attentional biases in heavy drinkers: do the effects generalise? Psychopharmacology 192:593–608

Fox E, Russo R, Bowles R, Dutton K (2001) Do threatening stimuli draw or hold visual attention in subclinical anxiety? J Exp Psychol Gen 130:681–700
Francis JA, Stewart SH, Hounsell S (1997) Dietary restraint and the selective processing of forbidden and non-forbidden food words. Cogn Ther Res 21:633–646
Franken IH, Kroon LY, Hendriks VM (2000) Influence of individual differences in craving and obsessive cocaine thoughts on attentional processes in cocaine abuse patients. Addict Behav 25:99–102
Freeman R, Touyz S, Sara G, Rennie C, Gordon E, Beumont P (1991) In the eye of the beholder: processing body shape information in anorexic and bulimic patients. Int J Eat Disord 10:709–714
Glauert R, Rhodes G, Fink B, Grammer K (2010) Body dissatisfaction and attentional bias to thin bodies. Int J Eat Disord 43:42–49
Gross JJ, John OP (2003) Individual differences in two emotion regulation processes: implications for affect, relationships, and well-being. J Pers Soc Psychol 85:348–362
Hansen CH, Hansen RD (1988) Finding the face in the crowd: an anger superiority effect. J Pers Soc Psychol 54:917–924
Happe F, Frith U (2006) The weak coherence account: detail-focused cognitive style in autism spectrum disorders. J Autism Dev Disord 36:5–25
Hare TA, Camerer CF, Rangel A (2009) Self-control in decision-making involves modulation of the vmPFC valuation system. Science 324:646–648
Hargreaves D, Tiggemann M (2002) The effect of television commercials on mood and body dissatisfaction: the role of appaerance-schema activation. J Soc Clin Psychol 21: 287–308
Harrison A, Sullivan S, Tchanturia K, Treasure J (2009) Emotion recognition and regulation in anorexia nervosa. Clin Psychol Psychother 16:348–356
Henderson JM, Pollatsek A, Rayner K (1989) Covert visual attention and extrafoveal information use during object identification. Percept Psychophys 45:196–208
Hermans D, Pieters G, Eelen P (1998) Implicit and explicit memory for shape, body weight, and food-related words in patients with anorexia nervosa and nondieting controls. J Abnorm Psychol 107:193–202
Holliday J, Tchanturia K, Landau S, Collier D, Treasure J (2005) Is impaired set-shifting an endophenotype of anorexia nervosa? Am J Psychiatry 162:2269–2275
Houben K, Jansen A (2010) Training inhibitory control: a recipe for resisting sweet temptations. Manuscript submitted for publication
Hunt J, Cooper M (2001) Selective memory bias in women with bulimia nervosa and women with depression. Behav Cogn Psychother 29:93–102
Jacoby LL, Allan LG, Collins JC, Larwill LK (1988) Memory influences subjective experience: noise judgment. J Exp Psychol Learn Mem Cogn 14:240–247
Jansen A, Nederkoorn C, Mulkens S (2005) Selective visual attention for ugly and beautiful body parts in EDs. Behav Res Ther 43:183–196
Jansen A, Smeets T, Martijn C, Nederkoorn C (2006) I see what you see: the lack of a self-serving body-image bias in EDs. Br J Clin Psychol 45:123–135
Jansen A, Smeets T, Boon B, Nederkoorn C, Roefs A, Mulkens S (2007) Vulnerability to interpretation bias in overweight children. Psychol Health 22:561–574
Johansson L, Ghaderi A, Andersson G (2005) Stroop interference for food- and body-related words: a meta-analysis. Eat Behav 6:271–281
Johansson L, Ghaderi A, Hallgren M, Andersson G (2008) Implicit memory bias for eating- and body appearance-related sentences in EDs: an application of Jacoby's white noise task. Cogn Behav Ther 37:135–145
Kaye WH, Fudge JL, Paulus M (2009) New insights into symptoms and neurocircuit function of anorexia nervosa. Nat Rev Neurosci 10:573–584
King GA, Polivy J, Herman CP (1991) Cognitive aspects of dietary restraint: effects on person memory. Int J EDs 10:313–321

Kringelbach ML, Rolls ET (2004) The functional neuroanatomy of the human orbitofrontal cortex: evidence from neuroimaging and neuropsychology. Prog Neurobiol 72:341–372
Lee M, Shafran R (2004) Information processing biases in EDs. Clin Psychol Rev 24:215–238
Lee M, Shafran R (2008) Processing biases in EDs: the impact of temporal factors. Int J EDs 41:372–375
Lopez C, Tchanturia K, Stahl D, Treasure J (2009) Weak central coherence in EDs: a step towards looking for an endophenotype of EDs. J Clin Exp Neuropsychol 31:117–125
MacLeod C, Mathews A, Tata P (1986) Attentional bias in emotional disorders. J Abnorm Psychol 95:15–20
Mathews A, MacLeod C (2005) Cognitive vulnerability to emotional disorders. Annu Rev Clin Psychol 1:167–195
Mogg K, Bradley BP, Hyare H, Lee S (1998) Selective attention to food-related stimuli in hunger: are attentional biases specific to emotional and psychopathological states, or are they also found in normal drive states? Behav Res Ther 36:227–237
Mogg K, Bradley BP, Dixon C, Fisher S, Twelftree H, McWilliams A (2000) Trait anxiety, defensiveness, and selective processing of treat: an investigation using two measures of attentional bias. Pers Individ Dif 28:1063–1077
Mogg K, Field M, Bradley BP (2005) Attentional and approach biases for smoking cues in smokers: an investigation of competing theoretical views of addiction. Psychopharmacology 180:333–341
Morrison T, Waller G, Lawson R (2006) Attributional style in the EDs. J Nerv Ment Dis 194:303–305
Mulkens S, Jansen A (2009) Mirror gazing increases attractiveness in satisfied, but not in dissatisfied women: a model for body dysmorphic disorder? J Behav Ther Exp Psychiatry 40:211–218
Murray EA, Izquierdo A (2007) Orbitofrontal cortex and amygdala contributions to affect and action in primates. Ann NY Acad Sci 1121:273–296
Naqvi NH, Rudrauf D, Damasio H, Bechara A (2007) Damage to the insula disrupts addiction to cigarette smoking. Science 315:531–534
Nederkoorn C, Havermans H, Roefs A, Smulders FTY, Jansen A (2006) Impulsivity in obese women. Appetite 47:253–256
Oldershaw A, Hambrook D, Tchanturia K, Treasure J, Schmidt U (2009) Emotional theory of mind and emotional awareness in recovered anorexia nervosa patients. Psychosomatic Med 72:73–79
Overduin J, Jansen A, Louwerse E (1995) Stroop interference and food intake. Int J EDs 18:277–285
Pietrowsky R, Krug R, Fehm HL, Born J (2002) Food deprivation fails to affect preoccupation with thoughts of food in anorectic patients. Br J Clin Psychol 41:321–326
Pinel JP, Assanand S, Lehman DR (2000) Hunger, eating, and ill health. Am Psychol 55:1105–1116
Placanica JL, Faunce GJ, Soames Job RF (2002) The effect of fasting on attentional biases for food and body shape/weight words in high and low ED Inventory scorers. Int J EDs 32:79–90
Rayner K (1998) Eye movements in reading and information processing: 20 years of research. Psychol Bull 124:372–422
Rieger E, Schotte DE, Touyz SW, Beumont PJ, Griffiths R, Russell J (1998) Attentional biases in EDs: a visual probe detection procedure. Int J EDs 23:199–205
Rinck M, Reinecke A, Ellwart T, Heuer K, Becker ES (2005) Speeded detection and increased distraction in fear of spiders: evidence from eye movements. J Abnorm Psychol 114:235–248
Roberts ME, Tchanturia K, Stahl D, Southgate L, Treasure J (2007) A systematic review and meta-analysis of set-shifting ability in EDs. Psychol Med 37:1075–1084
Roefs A, Stapert D, Isabella LA, Wolters G, Wojciechowski F, Jansen A (2005) Early associations with food in anorexia nervosa patients and obese people assessed in the affective priming paradigm. Eat Behav 6:151–163
Rosse RB, Miller MW, Hess AL, Alim TN, Deutsch SI (1993) Measures of visual scanning as a predictor of cocaine cravings and urges. Biol Psychiatry 33:554–556

Rosse RB, Johri S, Kendrick K, Hess AL, Alim TN, Miller M et al (1997) Preattentive and attentive eye movements during visual scanning of a cocaine cue: correlation with intensity of cocaine cravings. J Neuropsychiatry Clin Neurosci 9:91–93

Russell TA, Schmidt U, Doherty L, Young V, Tchanturia K (2009) Aspects of social cognition in anorexia nervosa: affective and cognitive theory of mind. Psychiatry Res 168:181–185

Sebastian SB, Williamson DA, Blouin DC (1996) Memory bias for fatness stimuli in the EDs. Cogn Ther Res 20:275–286

Shafran R, Lee M, Cooper Z, Palmer RL, Fairburn CG (2007) Attentional bias in EDs. Int J EDs 40:369–380

Siep N, Roefs A, Roebroeck A, Havermans R, Bonte ML, Jansen A (2010a) Decreased mesocorticolimbic food reward processing and increased dorsolateral cognitive control in Anorexia Nervosa patients. Paper submitted for publication

Siep N, Roefs A, Roebroeck A, Havermans R, Bonte ML, Jansen A (2010b). The regulation of mesocorticolimbic food reward processes through cognitive control. Paper submitted for publication

Smeets E, Roefs A, van Furth E, Jansen A (2008) Attentional bias for body and food in EDs: increased distraction, speeded detection, or both? Behav Res Ther 46:229–238

Smeets E, Roefs A, Jansen A (2009) Experimentally induced chocolate craving leads to an attentional bias in increased distraction but not in speeded detection. Appetite 53:370–375

Smeets E, Jansen A, Lindelauf T, Roefs A (2010) Selective attention for unattractive body parts causes body dissatisfaction. Paper submitted for publication

Suslow T, Ohrmann P, Lalee-Mentzel J, Donges US, Arolt V, Kersting A (2004) Incidental learning of food and emotional words in women with anorexia nervosa. Eat Weight Disord 9:290–295

Tchanturia K, Anderluh MB, Morris RG, Rabe-Hesketh S, Collier DA, Sanchez P et al (2004a) Cognitive flexibility in anorexia nervosa and bulimia nervosa. J Int Neuropsychol Soc 10:513–520

Tchanturia K, Morris RG, Anderluh MB, Collier DA, Nikolaou V, Treasure J (2004b) Set shifting in anorexia nervosa: an examination before and after weight gain, in full recovery and relationship to childhood and adult OCPD traits. J Psychiatry Res 38:545–552

Tchanturia K, Davies H, Campbell IC (2007) Cognitive remediation therapy for patients with anorexia nervosa: preliminary findings. Ann Gen Psychiatry 6:14

Toates F (1986) Motivational systems. Cambridge University Press, Cambridge, MA, USA

Treisman AM, Gelade G (1980) A feature-integration theory of attention. Cogn Psychol 12:97–136

Vitousek K, Hollon S (1990) The investigation of schematic content and processing in EDs. Cogn Ther Res 14:191–214

Walton ME, Groves J, Jennings KA, Croxson PL, Sharp T, Rushworth MF et al (2009) Comparing the role of the anterior cingulate cortex and 6-hydroxydopamine nucleus accumbens lesions on operant effort-based decision making. Eur J Neurosci 29:1678–1691

Williams JM, Mathews A, MacLeod C (1996) The emotional Stroop task and psychopathology. Psychol Bull 120:3–24

Williamson DA, Muller SL, Reas DL, Thaw JM (1999) Cognitive bias in EDs: implications for theory and treatment. Behav Modif 23:556–577

Williamson DA, Perrin L, Blouin DC, Barbin JM (2000) Cognitive bias in EDs: interpretation of ambiguous body-related information. Eat Weight Disord 5:143–151

Williamson DA, White MA, York-Crowe E, Stewart TM (2004) Cognitive-behavioral theories of EDs. Behav Modif 28:711–738

Wilson GT, Grilo CM, Vitousek KM (2007) Psychological treatment of EDs. Am Psychol 62:199–216

Part II
Neural Circuits, Neurotransmitters, and Behavior

Neurocircuity of Eating Disorders

Walter H. Kaye, Angela Wagner, Julie L. Fudge, and Martin Paulus

Contents

W.H. Kaye (✉) and A. Wagner
Department of Psychiatry, University of California, San Diego, CA, USA
e-mail: whkaye@gmail.com

J.L. Fudge
Departments of Psychiatry and Neurobiology and Anatomy, University of Rochester Medical Center, Rochester, NY, USA

M. Paulus
Department of Psychiatry, Laboratory of Biological Dynamics and Theoretical Medicine, University of California, San Diego, CA, USA
and
San Diego Veterans Affairs Health Care System, Psychiatry Service, San Diego, CA, USA

R.A.H. Adan and W.H. Kaye (eds.), *Behavioral Neurobiology of Eating Disorders*,
Current Topics in Behavioral Neurosciences 6, DOI 10.1007/7854_2010_85

Abstract *Objectives*: This chapter reviews brain imaging findings in anorexia and bulimia nervosa which characterize brain circuitry that may contribute to the pathophysiology of eating disorders (EDs).

Summary of recent findings: Recent imaging studies provide evidence of disturbed gustatory processing in EDs which involve the anterior insula as well as striatal regions. These results raise the possibility that individuals with anorexia nervosa have altered appetitive mechanism that may involve sensory, interoceptive, or reward processes. Furthermore, evidence of altered reward mechanisms is supported by studies that suggest that individuals with anorexia nervosa and bulimia nervosa share a trait toward similar anterior ventral striatal pathway dysregulation. This shared trait disturbance of the modulation of reward and emotionality may create a vulnerability for dysregulated appetitive behaviors. However, those with anorexia nervosa may be able to inhibit appetite and have extraordinary self-control because of exaggerated dorsal cognitive circuit function, whereas individuals with bulimia nervosa are vulnerable to overeating when they get hungry, because they have less ability to control their impulses.

Future directions: Current therapeutic interventions have modest success. Better understanding of neurocircuits that may be related to altered appetite, mood, impulse control, and other symptoms underlying the pathophysiology of EDs might improve psychotherapeutic and drug treatment strategies.

Keywords Anorexia nervosa · Appetite regulation · Bulimia nervosa · Brain imaging · Interoceptive awareness · Reward

1 Introduction

The pathophysiology of anorexia nervosa (AN) and bulimia nervosa (BN) is poorly understood. The primary characteristic required for a DSM IV (Diagnostic and Statistical Manual of Mental Disorders) diagnosis of AN and BN is pathological eating: AN must restrict and lose weight, and BN must binge and purge. The complex appetitive symptoms displayed by AN and BN are relatively unique and tend not to be shared with other psychiatric disorders. The stereotypic presentation and relentless expression of these feeding behaviors supports the possibility that they reflect some aberrant function of appetitive pathways. In addition, many individuals with eating disorders (ED) have (1) extremes of behavioral inhibition and dysinhibition; (2) anxiety, depression, and obsessionality; and (3) puzzling symptoms such as body image distortion, perfectionism, and anhedonia. Data support the hypothesis that these behaviors tend to express in concert because they are likely to be encoded in limbic and cognitive circuits known to modulate and integrate neuronal processes related to appetite, emotionality, and cognitive control.

1.1 Confounding Effects of Malnutrition

When malnourished and emaciated, individuals with AN, and to a lesser degree BN, have alterations of brain and peripheral organ function that are arguably more severe than in any other psychiatric disorder; for example, enlarged ventricles and sulci widening (Ellison and Fong 1998), altered brain metabolism in frontal, cingulate, temporal, and parietal regions (Kaye et al. 2006), and widespread neuropeptide, hormonal, and autonomic disturbances (Boyar et al. 1974; Jimerson and Wolfe 2006; Kaye et al. 2009). Determining whether such symptoms are a consequence or a potential cause of pathological feeding behavior or malnutrition is a major methodological problem in the field. It is difficult to study EDs prospectively because of the young age of onset and difficulty in premorbid identification of people who will develop EDs. Neurobiological studies during the acute illness are confounded by the effects of malnutrition. Thus we have used a method of identifying behavioral phenotypes that are independent of the confounding effects of malnutrition by studying women who are recovered AN and BN.

1.2 Vulnerabilities That Create a Risk for Developing AN and BN

Recent studies show that certain childhood temperament and personality traits (Lilenfeld et al. 2006; Stice 2002; Anderluh et al. 2003; Fairburn et al. 1999) such as negative emotionality, harm avoidance, perfectionism, inhibition, drive for thinness, altered interoceptive awareness, and obsessive–compulsive personality create a vulnerability for developing AN and BN. Malnutrition tends to exaggerate these premorbid behavioral traits (Pollice et al. 1997) after the onset of the illness, with the addition of other symptoms that maintain or accelerate the disease process, including exaggerated emotional dysregulation and obsessionality (Godart et al. 2007; Kaye et al. 2004).

1.3 Recovered (REC) AN and BN Subjects

The process of recovery in AN is poorly understood and, in most cases, protracted. Still, approximately 50–70% of affected individuals will eventually have complete or moderate resolution of the illness, often in the early to mid-20s (Wagner et al. 2006a; Steinhausen 2002; Strober et al. 1997). It is important to emphasize that temperament and personality traits such as negative emotionality, harm avoidance and perfectionism, and obsessional behaviors persist after recovery from both AN and BN (Casper 1990; Srinivasagam et al. 1995; Wagner et al. 2006a;

Steinhausen 2002) and are similar to the symptoms described premorbidly in childhood. Compared to the ill state, symptoms in REC AN and BN tend to be mild to moderate, including elevated scores on core ED measures. Interestingly, REC AN and BN tend to be more alike than different on many of these measures, although there are some differences on factors related to impulse control or stimuli seeking, such as novelty seeking (Strober et al. 1997; Wagner et al. 2006a; Lilenfeld et al. 2006).

1.4 Persistent Alterations in ED Found in Brain Imaging Studies After Recovery

Studies from our group found that AN and BN after recovery show normalization of gray and white matter volume (Wagner et al. 2006b) and cerebral blood flow (Frank et al. 2007) and tend to have normal neuropeptide function (Kaye et al. 2009), suggesting that these factors are not the cause of persistent neurobiological disturbances. However, several studies in REC AN showed hypoperfusion of frontal, temporal, parietal, and occipital regions (Rastam et al. 2001; Gordon et al. 1997) as well as of frontal and anterior cingulate cortex (ACC) activation, in response to pictures of food (Uher et al. 2003), suggesting disturbances of limbic and cognitive neural circuits. Many studies suggest that disturbances of limbic and cognitive neural networks occur in a range of psychiatric disorders, such as major depression (Drevets 2001; Tremblay et al. 2005), anxiety disorders (Protopopescu et al. 2005; Stein et al. 2007; Wright et al. 2003), and obsessive–compulsive disorder (OCD) (Insel 1992; Saxena 2003). Specifically, a ventral neurocircuit (Phillips et al. 2003), which includes the amygdala, insula, ventral striatum, and ventral regions of the ACC and the prefrontal cortex (PFC), is necessary for identifying emotional significance of stimuli and for generating affective responses to these stimuli. These regions are also important for automatic regulation and mediation of autonomic responses to emotional stimuli and contexts accompanying the production of affective states. In comparison, a dorsal executive function neurocircuit, which includes the hippocampus, dorsal regions of the caudate, dorsolateral prefrontal cortex (DLPFC), parietal cortex, and other regions, is thought to modulate selective attention, planning, and effortful regulation of affective states. It is possible that the altered emotional regulation or cognition found in all of these syndromes involves aberrant function of these circuits, but perhaps with different patterns on a molecular level (Phillips et al. 2003). In fact, neurobiological disturbances in EDs are different from those found in depression, anxiety, or OCD. For example, decreased 5-HT_{1A} receptor binding has been reported in ill (Drevets et al. 1999; Sargent et al. 2000) and recovered (Bhagwagar et al. 2004) depressed subjects, as well as those with social phobia (Lanzenberger et al. 2007) and panic disorder (Neumeister et al. 2004). However, increased 5-HT_{1A} receptor binding has been found in EDs (Kaye 2008).

1.5 Implications

We hypothesize that behaviors and abnormal physiology that persist after REC are a re-emergence of the vulnerabilities that created a risk for developing an ED. While it is possible that these findings could be "scars" caused by chronic malnutrition, several studies (Bulik et al. 2007) show that these factors are heritable, occur in unaffected family members, and are independent of body weight, which strongly support the argument that they are traits, not scars. Because no agreed-upon definition of recovery from AN or BN presently exists, our research studies employ a definition that emphasizes stable and healthy body weight for more than 1 year, with stable nutrition, relative absence of dietary abnormalities, normal menstruation, and free of medication. Because many individuals with AN and BN cross from one subtype to another over the course of their illness, it is not possible to investigate "pure" subtypes in the ill state. However, we can ascertain whether they had pure or mixed subtypes over the course of their illness once they have recovered. Thus we have studied pure subtypes of AN (REC AN; e.g., restricting- type who never binged or purged) or BN (REC BN; e.g., no history of AN).

2 Appetitive Regulation and AN and BN

Due to the puzzling nature of many ED symptoms, the ED field lags behind other psychiatric disorders in terms of progress in understanding responsible brain circuits and pathophysiology. Although AN and BN are characterized (APA 2000) as EDs, it remains unknown as to whether there is a primary disturbance of appetitive function. The regulation of appetite and feeding are complex phenomena, integrating peripheral signals (gastrointestinal (GI) tract, adipose tissue, hormonal secretion, etc.), hypothalamic factors (neuropeptides), cortical and subcortical processes (reward, emotionality, cognition), and external influences (Rolls 1997; Schwartz et al. 2000; Elman et al. 2006). While it is possible that a disturbance could occur anywhere in this axis in AN and BN, limbic and cortical brain circuits that contribute to appetite are of particular interest because these circuits (1) show persistent altered function after recovery and (2) code for rewarding and emotionality properties of food, homeostatic needs, and cognitive modulation (Elman et al. 2006; Hinton et al. 2004; Kelley 2004).

2.1 Studies of Altered Feeding Behavior in AN and BN

Relatively little data exist on appetite regulation in ED despite the prominent nature of these symptoms. Laboratory studies support clinical observations that individuals with AN dislike high-fat foods (Fernstrom et al. 1994; Drewnowski et al. 1988) and BN tend to binge on sweet and high-fat foods (Kaye et al. 1992; Weltzin et al. 1991). These patterns of responses did not change following weight regain. Other studies (Garfinkel et al. 1978, 1979) reported altered interoceptive disturbances in AN in terms of the absence of satiety aversion to sucrose, and that these disturbances persisted after normalization of weight or failure to rate food as positive when hungry (Santel et al. 2006). In addition, there is evidence (Kaye et al. 2003; Strober 1995; Vitousek and Manke 1994) that there is an anxiety-reducing character to dietary restraint in AN. For BN, negative mood states and hunger may precipitate a binge (Hilbert and Tuschen-Caffier 2007; Smyth et al. 2007; Waters et al. 2001) and overeating may relieve dysphoria and anxiety (Abraham and Beaumont 1982; Kaye et al. 1986; Johnson and Larson 1982). Taken together, these studies support the possibility of an altered response to palatable foods and a dysphoria-reducing aspect to pathological eating.

2.2 Brain Imaging Studies of Feeding Behavior in AN and BN Confirm Alterations in Limbic and Cognitive Circuits

Neuroimaging studies using different techniques in emaciated and malnourished individuals with AN found consistently altered activity in the insula and orbitofrontal cortex (OFC), as well as in mesial temporal, parietal, and the ACC regions as compared to control women (CW) (Ellison et al. 1998; Gordon et al. 2001; Naruo et al. 2000; Santel et al. 2006; Uher et al. 2004). One functional magnetic resonance imaging (fMRI) study (Uher et al. 2003) found that pictures of food stimulated ACC and medial prefrontal cortex (mPFC) activity in both ill and REC AN individuals, but not CW. These findings suggest that hyperactivity of these regions may be a trait marker of AN.

2.3 Neurocircuitry of Appetite Regulation

Sweet taste perception (Fig. 1) is peripherally mediated by tongue receptors (Chandraskekar et al. 2006) through cranial nerves, the nucleus tractus solitarius, and thalamic ventroposterior medial nucleus, to the primary gustatory cortex, which in humans comprise the frontal operculum and the anterior insula (AI) (Ogawa 1994; Scott et al. 1986; Yaxley et al. 1990; Faurion et al. 1999; Schoenfeld et al. 2004). Projections from the primary taste cortex reach the central nucleus of the amygdala and, from there, the lateral hypothalamus and midbrain dopaminergic regions (Simon et al. 2006). The primary taste cortex also projects heavily to the striatum (Chikama et al. 1997; Fudge et al. 2005). The AI is contiguous with the posterior OFC at the operculum. This region is reciprocally connected with the mPFC and

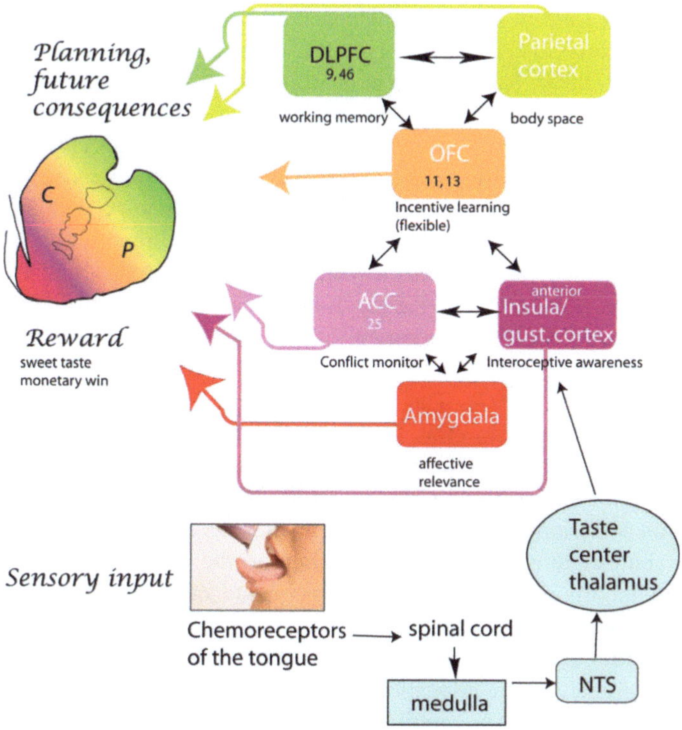

Fig. 1 *Schematic of cortical–striatal pathways with a focus on taste*. Chemoreceptors on the tongue detect a sweet taste. The signal is then transmitted through brainstem and thalamic taste centers to the primary gustatory cortex, which lies adjacent to and is densely interconnected with the anterior insula (AI). The AI is an integral part of the "ventral (limbic) neurocircuit" through its connections with the amygdala, the anterior cingulate cortex, and the orbitofrontal cortex. Afferents from the cortical structures involved in the ventral neurocircuit (AI and interconnected limbic cortices) are directed to the ventral striatum, whereas cortical structures involved in cognitive strategies (the dorsal neurocircuits) send inputs to the dorsolateral striatum. Thus, the sensory aspects of taste are primarily an insula phenomenon, whereas higher cortical areas modulate pleasure, motivation, and cognitive aspects of taste. These aspects are then integrated, resulting in an "eat" or "don't eat" decision. Coding the awareness of pleasant sensation from the taste experience via the AI might be altered in AN patients, tipping the balance of striatal processes away from normal, automatic reward responses mediated by the ventral striatum and toward a more "strategic" approach mediated by the dorsal striatum. The figure links each cortical structure with similarly colored arrows, indicating all cortical structures project to striatum in topographic manner. *ACC* anterior cingulate cortex; *DLPFC* dorsolateral prefrontal cortex; *NTS* nucleus tractus solitarius; *OFC* orbitofrontal cortex

ACC (Carmichael and Price 1996). The ventral striatum receives input from the AI and ACC (Carmichael and Price 1996; Fudge et al. 2005; Haber et al. 1995).

The AI and associated gustatory cortex respond not only to the taste and physical properties of food, but also to its rewarding properties (O'Doherty et al. 2001; Schultz et al. 2000; Small 2002). Some studies argue that the AI provides a representation of food in the mouth which is independent of hunger and, thus, of reward value (Rolls 2005), whereas the OFC computes the hedonic value of food (O'Doherty et al. 2000; Kringelbach et al. 2003; Rolls 2005). Other studies (Small et al. 2001) suggest that the AI and OFC have overlapping representations of sensory and reward/affective processing of taste. The AI is centrally placed to receive information about the salience (both appetitive and aversive) and relative value of the stimulus environment and integrate this information with the effect that these stimuli may have on the body state. The AI has bidirectional connections to the amygdala, nucleus accumbens (Reynolds and Zahm 2005), and OFC (Ongur and Price 2000). The striatum (Kelley 2004) receives inputs from brain regions involved in reward, incentive learning, and emotional regulation, including the ACC, the ventromedial PFC, the OFC, and AI (Fudge et al. 2004, 2005; Haber et al. 2006; Chikama et al. 1997). The OFC is associated with flexible responses to changing stimuli (Izquierdo et al. 2004; Kazama and Bachevalier 2006) such as the incentive value, e.g., whether the animal is hungry (Critchley and Rolls 1996; Hikosaka and Watanabe 2000; Gottfried et al. 2003). Of note, the OFC is highly dependent on 5-HT innervation for flexible reversal learning (Clarke et al. 2007), so that 5-HT abnormalities in ED may contribute to the disturbed inhibitory control (inability to incorporate changing incentive value of stimuli). The information about the interoceptive state processed in the AI is relayed to the ACC, which, as part of the central executive system, can generate an error signal that is critical for conflict monitoring and the allocation of attentional resources (Carter et al. 1999). Thus, interoception involves monitoring the sensations that are important for the integrity of the internal body state and connecting to systems that are important for allocating attention, evaluating context, and planning actions (Paulus and Stein 2006). The role of the AI is thus focused on how the value of stimuli might affect the body state. Thus, these regions play an important role in determining homeostatic appetitive needs when hungry or satiated. In addition, interoceptive sensations are often associated with intense affective and motivational components (Paulus and Stein 2006), and the evaluative component of the signal is highly dependent on the homeostatic state of the individual.

2.4 Gustatory fMRI Studies

Our group (Wagner et al. 2008) administered tastes of 10% sucrose and water in a blind, controlled manner to individuals with REC AN and healthy CW. There were two main findings: (1) Compared to CW, the individuals with REC AN had a significantly reduced blood-oxygen-level dependent (BOLD) response to the blind administration of sucrose or water in the AI (Fig. 1, left insula $p = 0.003$), ACC, and striatal regions; (2) CW, but not individuals with REC AN, showed a positive relationship between self-ratings of pleasantness and the intensity of the signal for sugar in the AI, ventral, and dorsal putamen as well as ACC.

2.5 *Implication*

Appetitive dysregulation in AN and BN is poorly understood. Appetite regulation is a complex process that involves the integration of a wide variety of signals such as energy needs in the body, hedonic attraction to palatable foods, and long-term cognitive concerns about weight. The data reviewed above are the first to localize potential pathology of appetite disturbances in individuals with AN. We hypothesize that REC AN individuals have altered incentive processing in the AI and related regions. AN individuals fail to become appropriately hungry when starved, and thus are able to become emaciated.

3 Does the Anterior Insula Contribute to Altered Interoceptive Awareness in AN?

Do AN individuals have an AI disturbance specifically related to gustatory modulation or a more generalized disturbance related to the integration of interoceptive stimuli? Interoception has long been thought to be critical for self-awareness because it provides the link between cognitive and affective processes and the current body state (Craig 2002; Paulus and Stein 2006). This lack of recognition of the symptoms of malnutrition, diminished insight and motivation to change, and altered central coherence could be related to disturbed AI function.

It is thought that altered interoceptive awareness might be a precipitating and reinforcing factor in AN (Bruch 1962; Fassino et al. 2004; Garner et al. 1983; Lilenfeld et al. 2006). Indeed, many of the symptoms of AN, such as distorted body image, lack of recognition of the symptoms of malnutrition (e.g., a failure to appropriately respond to hunger), and diminished motivation to change, could be related to disturbed interoceptive awareness. In particular, there might be a qualitative change in the way that specific interoceptive information is processed. For example, individuals with AN might experience an aversive visceral sensation when exposed to food or food-related stimuli. This experience might fundamentally alter the reward-related properties of food and result in a bias towards negative emotionality. Moreover, the aversive interoceptive experience associated with food might trigger top-down modulatory processes aimed at anticipating and minimizing the exposure to food stimuli ("harm avoidance"), leading to increased anticipatory processing aimed to reduce the exposure to the aversively valued stimulus. Therefore, individuals with AN might exhibit attenuated responses to the immediate reward-related signal of food (reducing hunger) but show increased responses to the long-term reward signal associated with the goal of weight reduction or other "ideal" cognitive constructs. Finally, the AI has been implicated in risk-prediction errors (Preuschoff and Quartz 2008), suggesting that impairments in insula functioning might lead to anomalous attitudes in a context of uncertainty and thus contribute to harm avoidance. Thus, given the prominent alterations in insula activity in AN patients, one might speculate that these individuals experience an altered sensitivity to or integration of internal body signals. Specifically, the projection of the AI to the anterior cingulate may serve to modulate the degree to which cognitive control is engaged to alter behavior toward poor decision making that does not subserve the homeostatic weight balance but instead results in progressive weight loss.

4 Reward Function in AN and BN

It is also possible that food has little rewarding value to AN and thus may be associated with corresponding responses in the OFC or the striatum. Clinical observations suggest that AN individuals have disturbed reward modulation that affects a wide range of appetitive behaviors – not just food. Individuals with AN have long been noted to be anhedonic and ascetic, and are able to sustain self-denial of food as well as most comforts and pleasures in life (Frank et al. 2005). Reward is one characteristic that differentiates AN and BN, since BN individuals tend to be more impulsive, pleasure and stimuli seeking, and less paralyzed by concerns with future consequences (Cassin and von Ranson 2005). Positive reinforcers or rewards promote selected behaviors, induce subjective feelings of pleasure and other positive emotions, and maintain stimulus–response associations (Thut et al. 1997). Negative reinforcement also plays an essential role by encouraging avoidance or withdrawal behavior, as well as production of negative emotions.

4.1 Altered DA Function in AN and BN

Animal studies indicate that dopamine (DA) in the striatum plays a key role in the optimal response to reward stimuli (Delgado et al. 2000; Montague et al. 2004; Schultz 2004). In fact, genetic, pharmacologic, and physiologic data (Kaye 2008; Bergen et al. 2005; Lawrence 2003; Friederich et al. 2006) show that ill and REC individuals with AN have altered striatal DA function. DA disturbances could contribute to an altered modulation of appetitive behaviors, as well as symptoms of anhedonia, dysphoric mood, and increased motor activity (Halford et al. 2004; Volkow et al. 2002). Because fewer DA studies have been done in BN individuals, it remains uncertain whether they have trait-related DA disturbances (Jimerson et al. 1992; Kaye et al. 1990). In terms of positron emission tomography (PET) studies, our group found that REC AN had increased [^{11}C]raclopride BP_{ND} in the anterior ventral striatum (AVS) (Frank et al. 2005). Because PET measures of [^{11}C]raclopride binding are sensitive to endogenous DA concentrations (Drevets et al. 2001), elevated [^{11}C]raclopride BP_{ND} could indicate either a reduction in intrasynaptic DA concentrations or an elevation of the density and/or affinity of the D2/D3 receptors.

4.2 *BOLD Response to Reward and Punishment Is Altered in AN*

Human neuroimaging studies show that a highly interconnected network of brain areas including OFC, mPFC, amygdala, striatum and DA mid-brain is involved in reward processing of both primary (i.e., pleasurable tastes) (Berns et al. 2001; McClure et al. 2003) and secondary (i.e., money) reinforcers (O'Doherty 2004; Breiter et al. 2001; Delgado et al. 2000; Gehring and Willoughby 2002; Montague et al. 2004). These regions code stimulus–reward value, maintain representations of predicted future reward and future behavioral choice, and may play a role in integrating and evaluating reward prediction to guide decisions. In animals, DA modulates the influence of limbic inputs on striatal activity (Goto and Grace 2005; Montague et al. 2004; Schultz 2004; Yin and Knowlton 2006) and mediates the "binding" of hedonic evaluation of stimuli to objects or acts ("wanting" response) (Berridge and Robinson 1998). It has been postulated that dorsal striatum is engaged by real or perceived stimulus–response outcomes, with DA projections modulating this behavior (Tricomi et al. 2004; O'Doherty et al. 2004).

Because of the DA findings in REC AN individuals (Bergen et al. 2005; Frank et al. 2005; Kaye et al. 1999), our group (Wagner et al. 2007, 2009) performed an event-related fMRI study using a variation of a well-characterized "guessing-game" protocol (Delgado et al. 2000), which is known to activate the AVS with a differential response to positive and negative feedback in healthy volunteers. Importantly, REC AN (Wagner et al. 2007) and REC BN individuals (Wagner et al. 2009) failed to show a differential AVS response to positive and negative monetary feedback when compared to CW, suggesting that both groups have an impaired ability to identify the rewarding/emotional significance of a stimulus. This shared-trait disturbance of the modulation of reward and emotionality may create a vulnerability for dysregulated appetitive behaviors. In contrast, fMRI studies consistently show that ill and REC AN individuals have increased activity in cognitive neural circuits (Zastrow et al. 2009; Wagner et al. 2007), whereas ill and REC BN individuals have diminished or impaired activity in these regions (Marsh et al. 2009; Schienle et al. 2008; Wagner et al. 2009), consistent with enhanced higher order inhibitory function in AN and reduced inhibition in BN. We hypothesize that AN individuals are able to inhibit appetite and have extraordinary self-control, because they have exaggerated dorsal cognitive circuit function, whereas BN individuals are vulnerable to overeating when they get hungry, because they have less ability to control their impulses.

4.3 *Implications*

In summary, AN individuals may have both an impaired ability to identify the emotional significance of a stimulus and an enhanced ability to plan or foresee consequences. Because of AVS pathway dysregulation, REC AN individuals may focus on long-term consequences rather than an immediate response to salient stimuli. In fact, AN individuals tend to have an enhanced ability to pay attention to detail or use a logical/analytic approach, but exhibit worse performance for global strategies in the here and now (Lopez et al. 2008; Strupp et al. 1986). In particular, the most anxious AN individuals may respond in an overly "cognitive" manner to both negative and positive stimuli. Consequently, they may not be able to process information about rewarding outcomes of an action and may have impaired ability to identify emotional significance of the stimuli (Phillips et al. 2003). This may provide an important, new understanding of why it is so difficult to motivate AN individuals to engage in treatment since they may not be able to appreciate rewarding stimuli (Halmi et al. 2005).

5 The Neurocircuitry of AN

Based on the above processes and associated brain areas, our group (Kaye et al. 2009) has begun to assemble a neural systems processing model of AN. Specifically, top-down (cortical) amplification of anticipatory signals related to food such as ghrelin, or stimuli associated with satiety signals (integrated within the insula), could trigger behavioral strategies for avoiding exposure to food. These anticipatory interoceptive stimuli are associated with an aversive body state that resembles some aspects of the physiological state of the body after feeding. This abnormal response to food anticipation might function as a learning signal to further increase avoidance behavior, i.e., to engage in activities aimed at minimizing exposure to food. Specifically, stimuli that predict food intake, such as displays of food or food smells, could generate a "body prediction error," resulting from comparing the current body state with the anticipated body state (e.g., feeling satiated) after feeding. This prediction error would generate a motivational or approach signal in healthy individuals but might lead to an avoidance signal in AN individuals. The dorsal and ventral neurocircuits described earlier might be involved in these processes: The ACC, one of the projection areas of the insular cortex, is important in processing the conflict between available behaviors and outcomes, e.g., "shall I eat this cake and satisfy my hunger now or shall I not eat this cake and stay thin?" (Carter et al. 2000). The OFC, another projection area of the anterior insular cortex (Ongur and Price 2000), can dynamically adjust reward valuation based on the current body state of the individual (Rolls 1996). The DPLFC can switch between competing behavioral programs based on the error signal it receives from the ACC (Kerns et al. 2004).

Although we do not propose that AN is an insula-specific disorder, we speculate that an altered insula response in response to food-related stimuli is an important component of this disease. If this is indeed the case, one would need to determine whether insula-specific interventions, such as sensitization or habituation of interoceptive sensitivity via real-time monitoring of the insular cortex activation, might help. Moreover, computational models such as those that have been proposed for addiction (Redish 2004) might provide a theoretical approach to better understand the complex pathology of this disorder.

Within the framework of the ventral and dorsal neurocircuits described above, there are also potential explanations for other core components of clinical dysfunction in AN. Negative affect – such as anxiety and harm avoidance – and anhedonia could be related to difficulties in accurately coding or integrating positive and negative emotions within ventral striatal circuits. There is considerable overlap between circuits that modulate emotionality and the rewarding aspects of food consumption (Volkow and Wise 2005). Food is pleasurable in healthy individuals but feeding is anxiogenic in AN patients, and starvation might serve to reduce dysphoric mood states. The neurobiologic mechanisms responsible for such behaviors remain to be elucidated, but it is possible that an enhancement of 5-HT-related aversive motivation and/or diminished DA-related appetitive drives (Daw et al. 2002; Cools et al. 2008) contribute to these behaviors.

Finally, it is possible that perfectionism and obsessional personality traits are related to exaggerated cognitive control by the DLPFC. The DLPFC might develop excessive inhibitory activity to dampen information processing through reward pathways (Chambers et al. 2003). Alternatively, increased activation of cognitive pathways might compensate for primary deficits in limbic function: when there are deficits in emotional regulation, overdependence upon cognitive rules is a reasonable strategy of self-management (Connan et al. 2003).

6 Conclusions and Future Directions

AN is thought to be a disorder of complex etiology, in which the genetic, biological, psychological, and sociocultural factors, and interactions between them, seem to contribute significantly to susceptibility (Connan et al. 2003; Jacobi et al. 2004; Lilenfeld et al. 2006; Stice 2002). Because no single factor has been shown to be either necessary or sufficient for causing AN, a multifactorial threshold model might be the most appropriate model (Connan et al. 2003). Typically, AN begins with a restrictive diet and weight loss during teenage years, which progresses to an out-of-control spiral. Thus, individuals might cross a threshold in which a premorbid temperament, interacting with stress and/or psychosocial factors, progresses to an illness with impaired insight and a powerful, obsessive preoccupation with dieting and weight loss. Adolescence is a time of profound biological, psychological, and sociocultural change, and it demands a considerable degree of flexibility to successfully manage the transition into adulthood. Psychologically, change might challenge the perfectionism, harm avoidance, and rigidity of those at risk for AN and thus fuel an underlying vulnerability.

We propose that somatic, autonomic, and visceral information is aberrantly processed in people who are vulnerable to developing AN. Brain changes associated with puberty might further challenge these processes. For example, orbital and DLPFC regions develop greatly during and after puberty (Huttenlocher and

Dabholkar 1997), and increased activity of these cortical areas might be a cause of the excessive worry, perfectionism, and strategizing in AN patients. It is possible that, in AN patients, hyperactivity of cognitive networks in the dorsal neurocircuit (e.g., DLPFC to dorsal striatum) directs motivated actions when the ability of the ventral striatal pathways to direct more "automatic" or intuitive motivated responses is impaired. Another possibility is that in AN patients (otherwise adequate) limbic–striatal information processing in the ventral circuit is too strongly inhibited by converging inputs from cognitive domains such as the DLPFC and the parietal cortex.

It is possible that such trait-related disturbances are related to altered monoamine neuronal modulation that predates the onset of AN and contributes to premorbid temperament and personality symptoms. Specifically, disturbances in the 5-HT system contribute to a vulnerability for restricted eating, behavioral inhibition, and a bias toward anxiety and error prediction, whereas disturbances in the DA system contribute to an altered response to reward. Several factors might act on these vulnerabilities to cause the onset of AN in adolescence. First, puberty-related female gonadal steroids or age-related changes might exacerbate 5-HT and DA system dysregulation. Second, stress and/or cultural and societal pressures might contribute by increasing anxious and obsessional temperament. Individuals find that restricting food intake is powerfully reinforcing because it provides a temporary respite from dysphoric mood. People with AN enter a vicious cycle – which could account for the chronicity of this disorder – because eating exaggerates, and food refusal reduces, an anxious mood.

AN has the highest mortality rate of any psychiatric disorder. It is expensive to treat and we have inadequate therapies. It is crucial to understand the neurobiologic contributions and their interactions with the environment, in order to develop more effective therapies. Thus, future imaging studies should focus on characterizing neural circuits, their functions, and their relationship to behavior in AN patients. Genetic studies might shed light on the complex interactions of molecules within these neural circuits. Finally, prospective and longitudinal studies should focus on identifying the neurobiologic traits and external factors that create a susceptibility for developing AN.

References

Abraham S, Beaumont P (1982) How patients describe bulimia or binge eating. Psychol Med 12:625–635

Anderluh MB, Tchanturia K, Rabe-Hesketh S, Treasure J (2003) Childhood obsessive-compulsive personality traits in adult women with eating disorders: defining a broader eating disorder phenotype. Am J Psychiatry 160:242–247

APA (2000) Diagnostic and statistical manual of mental disordes: DSM:VI-TR, 4th edn. American Psychological Association, Washington, DC.

Bergen A, Yeager M, Welch R, Haque K, Ganjei JK, Mazzanti C, Nardi I, Van Den Bree MBM, Fichter M, Halmi K, Kaplan A, Strober M, Treasure J, Woodside DB, Bulik C, Bacanu A,

Devlin B, Berrettini WH, Goldman D, Kaye W (2005) Association of multiple DRD2 polymorphisms with anorexia nervosa. Neuropsychopharmacology 30:1703–1710
Berns G, Mcclure S, Pagnoni G, Montague P (2001) Predictability modulates human brain response to reward. J Neurosci 21:2793–2798
Berridge K, Robinson T (1998) What is the role of dopamine in reward: hedonic impact, reward learning, or incentive salience? Brain Res 28:309–369
Bhagwagar Z, Rabiner E, Sargent P, Grasby P, Cowen P (2004) Persistent reduction in brain serotonin$_{1A}$ receptor binding in recovered depressed men mesured by positron emission tomography with [^{11}C]WAY-100635. Mol Psychiatry 9:386–392
Boyar RK, Finkelstein J, Kapen S, Weiner H, Weitzman E, Hellman L (1974) Anorexia nervosa. Immaturity of the 24-hour luteinizing hormone secretory pattern. NEJM 291:861–865
Breiter HC, Aharon I, Kahneman D, Dale A, Shizgal P (2001) Functional imaging of neural responses to expectancy and experience of monetary gains and losses. Neuron 30:619–639
Bruch H (1962) Perceptual and conceptual disturbances in anorexia nervosa. Psychosom Med 24:187–194
Bulik C, Hebebrand J, Keski-Rahkonen A, Klump K, Reichborn-Kjennerud KS, Mazzeo S, Wade T (2007) Genetic epidemiology, endophenotypes, and eating disorder classification. Int J Eat Disord Suppl:S52–S60
Carmichael S, Price J (1996) Connectional networks within the orbital and medial prefrontal cortex of macaque monkeys. J Comp Neurol 371:179–207
Carter C, Botvinick M, Cohan J (1999) 42: The contribution of the anterior cingulate cortex to executive processes in cognition. Rev Neurosci 10:49–57
Carter CS, Macdonald A, Botvinick M, Ross L, Stenger V, Noll D, Cohen B (2000) Parsing executive processes: strategic vs. evaluative functions of the anterior cingulate cortex. Proc Natl Acad Sci USA 97:1944–1948
Casper RC (1990) Personality features of women with good outcome from restricting anorexia nervosa. Psychosom Med 52:156–170
Cassin S, Von Ranson K (2005) Personality and eating disorders: a decade in review. Clin Psychol Rev 25:895–916
Chambers R, Taylor J, Potenza M (2003) Developmental neurocircuitry of motivation in adolescence: a critical period of addiction vulnerability. Am J Psychiatry 160:1041–1052
Chandraskekar J, Hoon M, Ryba N, Zuker C (2006) The receptors and cells for mammalian taste. Nature 444:288–294
Chikama M, Mcfarland N, Armaral D, Haber S (1997) Insular cortical projections to functional regions of the striatum correlate with cortical cytoarchitectonic organization in the primate. J Neurosci 17:9686–9705
Clarke H, Walker SD, Robbins T, Roberts A (2007) Cognitive inflexibility after prefrontal serotonin depletion is behaviorally and neurochemically specific. Cereb Cortex 17:18–27
Connan F, Campbell I, Katzman M, Lightman S, Treasure J (2003) A neurodevelopmental model for anorexia nervosa. Physiol Behav 79:13–24
Cools R, Roberts A, Robbins T (2008) Serotoninergic regulation of emotional and behavioural control processes. Trends Cogn Sci 12:31–40
Craig AD (2002) How do you feel? Interoception: the sense of the physiological condition of the body. Nat Rev Neurosci 3:655–666
Critchley H, Rolls E (1996) Hunger and satiety modify the responses of olfactory and visual neurons in the primate orbitofrontal cortex. J Neurophysiol 75:1673–1686
Daw ND, Kakade S, Dayan P (2002) Opponent interactions between serotonin and dopamine. Neural Netw 15:603–616
Delgado M, Nystrom L, Fissel C, Noll D, Fiez J (2000) Tracking the hemodynamic responses to reward and punishment in the striatum. J Neurophysiol 84:3072–3077
Drevets WC (2001) Neuroimaging and neuropathological studies of depression: implications for the cognitive-emotional features of mood disorders. Curr Opin Neurobiol 11:240–249

Drevets WC, Frank E, Price JC, Kupfer DJ, Holt D, Greer PJ, Huang Y, Gautier C, Mathis C (1999) PET imaging of serotonin 1A receptor binding in depression. Biol Psychiatry 46:1375–1387
Drevets W, Gautier C, Price J, Kupfer D, Kinahan P, Grace A, Price J, Mathis C (2001) Amphetamine-induced dopamine release in human ventral striatum correlates with euphoria. Biol Psychiatry 49:81–96
Drewnowski A, Pierce B, Halmi K (1988) Fat aversion in eating disorders. Appetite 10:119–131
Ellison AR, Fong J (1998) Neuroimaging in eating disorders. In: Hoek HW, Treasure JL, Katzman MA (eds) Neurobiology in the treatment of eating disorders. Wiley, Chichester
Ellison Z, Foong J, Howard R, Bullmore E, Williams S, Treasure J (1998) Functional anatomy of calorie fear in anorexia nervosa. Lancet 352:1192
Elman I, Borsook D, Lukas S (2006) Food intake and reward mechanisms in patients with schizophrenia: implications for metabolic disturbances and treatment with second-generation antipsychotic agents. Neuropsychopharmacology 31:2091–2120
Fairburn CG, Cooper JR, Doll HA, Welch SL (1999) Risk factors for anorexia nervosa: three integrated case-control comparisons. Arch Gen Psychiatry 56:468–476
Fassino S, Piero A, Gramaglia C, Abbate-Daga G (2004) Clinical, psychopathological and personality correlates of interoceptive awareness in anorexia nervosa, bulimia nervosa and obesity. Psychopathology 37:168–174
Faurion A, Cerf B, Van De Moortele PF, Lobel E, Mac Leod P, Le Bihan D (1999) Human taste cortical areas studied with functional magnetic resonance imaging: evidence of functional lateralization related to handedness. Neurosci Lett 277:189–192
Fernstrom MH, Weltzin TE, Neuberger S, Srinivasagam N, Kaye WH (1994) Twenty-four-hour food intake in patients with anorexia nervosa and in healthy control subjects. Biol Psychiatry 36:696–702
Frank G, Bailer UF, Henry S, Drevets W, Meltzer CC, Price JC, Mathis C, Wagner A, Hoge J, Ziolko SK, Barbarich N, Weissfeld L, Kaye W (2005) Increased dopamine D2/D3 receptor binding after recovery from anorexia nervosa measured by positron emission tomography and [^{11}C]raclopride. Biol Psychiatry 58:908–912
Frank G, Bailer UF, Meltzer CC, Price J, Mathis C, Wagner A, Becker C, Kaye WH (2007) Regional cerebral blood flow after recovery from anorexia and bulimia nervosa. Int J Eat Disord 40:488–492
Friederich HC, Kumari V, Uher R, Riga M, Schmidt U, Campbell IC, Heryog W, Treasure J (2006) Differential motivational responses to food and pleasurable cues in anorexia and bulimia nervosa: a startle reflex paradigm. Psychol Med 36:1327–1335
Fudge J, Breitbart M, Mcclain C (2004) Amygdaloid inputs define a caudal component of the ventral striatum in primates. J Comp Neurol 476:330–347
Fudge J, Breitbart M, Danish M, Pannoni V (2005) Insular and gustatory inputs to the caudal ventral striatum in primates. J Comp Neurol 490:101–118
Garfinkel P, Moldofsky H, Garner DM, Stancer HC, Coscina D (1978) Body awareness in anorexia nervosa: disturbances in "body image" and "satiety". Psychosom Med 40:487–498
Garfinkel P, Moldofsky H, Garner DM (1979) The stability of perceptual disturbances in anorexia nervosa. Psychol Med 9:703–708
Garner DM, Olmstead MP, Polivy J (1983) Development and validation of a multidimensional eating disorder inventory for anorexia and bulimia nervosa. Int J Eat Disord 2:15–34
Gehring W, Willoughby A (2002) The medial frontal cortex and the rapid processing of monetary gains and losses. Science 295:2279–2282
Godart N, Perdereau F, Rein Z, Berthoz S, Wallier J, Jeammet P, Flament M (2007) Comorbidity studies of eating disorders and mood disorders. Critical review of the literature. J Affect Disord 97:37–49
Gordon I, Lask B, Bryant-Waugh R, Christie D, Timimi S (1997) Childhood-onset anorexia nervosa: towards identifying a biological substrate. Int J Eat Disord 22:159–165

Gordon CM, Dougherty DD, Fischman AJ, Emans SJ, Grace E, Lamm R, Alpert NM, Majzoub JA, Rausch SL (2001) Neural substrates of anorexia nervosa: a behavioral challenge study with positron emission tomography. J Pediatr 139:51–57

Goto Y, Grace A (2005) Dopaminergic modulation of limbic and cortical drive of nucleus accumbens in goal-directed behavior. Nat Neurosci 386:14–17

Gottfried J, O'doherty J, Dolan R (2003) Encoding predictive reward value in human amygdala and orbitofrontal cortex. Science 301:1104–1107

Haber S, Kunishio K, Mizobuhi M, Lynd-Balta E (1995) The orbital and medial prefrontal circuit through the primate basal ganglia. J Neurosci 15:4851–4867

Haber SN, Kim K, Mailly P, Calzavara R (2006) Reward-related cortical inputs define a large striatal region in primates that interface with associative cortical connections, providing a substrate for incentive-based learning. J Neurosci 26:8368–8376

Halford J, Cooper G, Dovey T (2004) The pharmacology of human appetite expression. Curr Drug Targets 5:221–240

Halmi K, Agras WS, Crow S, Mitchell J, Wilson G, Bryson S, Kraemer HC (2005) Predictors of treatment acceptance and completion in anorexia nervosa. Arch Gen Psychiatry 62:776–781

Hikosaka K, Watanabe M (2000) Delay activity of orbital and lateral prefrontal neurons of the monkey varying with different rewards. Cereb Cortex 10:263–271

Hilbert A, Tuschen-Caffier B (2007) Maintenance of binge eating through negative mood: a naturalistic comparison of binge eating disorder and bulimia nervosa. Int J Eat Disord 40:521–530

Hinton E, Parkinson JA, Holland A, Arana F, Roberts A, Owen A (2004) Neural contributions to the motivational control of appetite in humans. Eur J Neurosci 20:1411–1418

Huttenlocher P, Dabholkar A (1997) Regional differences in synaptogenesis in human cerebral cortex. J Comp Neurol 387:167–178

Insel TR (1992) Toward a neuroanatomy of obsessive-compulsive disorder. Arch Gen Psychiatry 49:739–744

Izquierdo I, Cammarota M, Medina J, Bevilaqua L (2004) Pharmacological findings on the biochemical bases of memory processes: a general view. Neural Plast 11:159–189

Jacobi C, Hayward C, De Zwaan M, Kraemer H, Agras W (2004) Coming to terms with risk factors for eating disorders: application of risk terminology and suggestions for a general taxonomy. Psychol Bull 130:19–65

Jimerson D, Wolfe B (2006) Psychobiology of eating disorders. In: Wonderlich MJS, De Zwaan M, Steiger H (eds) Annual review of eating disorders: Part 2 – 2006. Radcliffe, Oxford

Jimerson DC, Lesem MD, Kaye WH, Brewerton TD (1992) Low serotonin and dopamine metabolite concentrations in cerebrospinal fluid from bulimic patients with frequent binge episodes. Arch Gen Psychiatry 49:132–138

Johnson C, Larson R (1982) Bulimia: an analysis of mood and behavior. Psychosom Med 44:341–351

Kaye W (2008) Neurobiology of anorexia and bulimia nervosa. Physiol Behav 94:121–135

Kaye WH, Gwirtsman HE, George DT, Weiss SR, Jimerson DC (1986) Relationship of mood alterations to bingeing behaviour in bulimia. Br J Psychiatry 149:479–485

Kaye WH, Ballenger JC, Lydiard RB, Stuart GW, Laraia MT, O'neil P, Fossey MD, Stevens V, Lesser S, Hsu G (1990) CSF monoamine levels in normal-weight bulimia: evidence for abnormal noradrenergic activity. Am J Psychiatry 147:225–229

Kaye WH, Weltzin TE, Mckee M, Mcconaha C, Hansen D, Hsu LK (1992) Laboratory assessment of feeding behavior in bulimia nervosa and healthy women: methods for developing a human-feeding laboratory. Am J Clin Nutr 55:372–380

Kaye WH, Frank GK, Mcconaha C (1999) Altered dopamine activity after recovery from restricting-type anorexia nervosa. Neuropsychopharmacology 21:503–506

Kaye WH, Barbarich NC, Putnam K, Gendall KA, Fernstrom J, Fernstrom M, Mcconaha CW, Kishore A (2003) Anxiolytic effects of acute tryptophan depletion in anorexia nervosa. Int J Eat Disord 33:257–267

Kaye W, Bulik C, Thornton L, Barbarich N, Masters K, Fichter M, Halmi K, Kaplan A, Strober M, Woodside DB, Bergen A, Crow S, Mitchell J, Rotondo A, Mauri M, Cassano G, Keel PK, Plotnicov K, Pollice C, Klump K, Lilenfeld LR, Devlin B, Quadflieg R, Berrettini WH (2004) Comorbidity of anxiety disorders with anorexia and bulimia nervosa. Am J Psychiatry 161:2215–2221

Kaye W, Wagner A, Frank G, Uf B (2006) Review of brain imaging in anorexia and bulimia nervosa. In: Mitchell J, Wonderlich S, Steiger H, Dezwaan M (eds) AED annual review of eating disorders, Part 2. Radcliffe, Abingdon, UK

Kaye W, Fudge J, Paulus M (2009) New insight into symptoms and neurocircuit function of anorexia nervosa. Nat Rev Neurosci 10:573–584

Kazama A, Bachevalier J (2006) Selective aspiration of neurotoxic lesions of the orbitofrontal areas 11 and 13 spared monkeys' performance on the object reversal discrimination task. Soc Neurosci Abstr 32:670.25

Kelley AE (2004) Ventral striatal control of appetite motivation: role in ingestive behavior and reward-related learning. Neurosci Biobehav Rev 27:765–776

Kerns J, Cohen J, Macdonald A, Cho R, Stenger V, Carter C (2004) Anterior cingulate conflict monitoring and adjustments in control. Science 303:1023–1026

Kringelbach ML, O'doherty J, Rolls E, Andrews C (2003) Activation of the human orbitofrontal cortex to a liquid food stimulus is correlated with its subjective pleasantness. Cereb Cortex 13:1064–1071

Lanzenberger R, Mitterhauser M, Spindelegger C, Wadsak W, Klein N, Mien L, Holik A, Attarbaschi T, Mossaheb N, Sacher J, Geiss-Granadia T, Keletter K, Kasper S, Tauscher J (2007) Reduced serotonin-1A receptor binding in social anxiety disorder. Biol Psychiatry 61:1081–1089

Lawrence A (2003) Impaired visual discrimination learning in anorexia nervosa. Appetite 20:85–89

Lilenfeld L, Wonderlich S, Riso LP, Crosby R, Mitchell J (2006) Eating disorders and personality: a methodological and empirical review. Clin Psychol Rev 26:299–320

Lopez C, Tchanturia K, Stahl D, Booth R, Holliday J, Treasure J (2008) An examination of central coherence in women with anorexia nervosa. Int J Eat Disord 41(4):340–347

Marsh R, Steinglass J, Gerber A, Graziano O'leary K, Wang Z, Murphy D, Walsh B, Bs P (2009) Deficient activity in the neural systems that mediate self-regulatory control in bulimia nervosa. Arch Gen Psychiatry 66:51–63

Mcclure S, Berns G, Montague P (2003) Temporal prediction errors in a passive learning task activate human striatum. Neuron 38:339–346

Montague R, Hyman S, Cohen J (2004) Computational roles for dopamine in behavioural control. Nature 431:760–767

Naruo T, Nakabeppu Y, Sagiyama K, Munemoto T, Homan N, Deguchi D, Nakajo M, Nozoe S (2000) Characteristic regional cerebral blood flow patterns in anorexia nervosa patients with binge/purge behavior. Am J Psychiatry 157:1520–1522

Neumeister A, Brain E, Nugent A, Carson R, Bonne O, Lucnekbaugh D, Eckelman W, Herscho-vitch P, Charney D, Drevets W (2004) Reduced serotinin type 1_A receptor binding in panic disorder. J Neurosci 24:589–591

O'doherty J (2004) Reward representations and reward related learning in the human brain: insights from neuroimaging. Science 14:769–776

O'doherty J, Rolls ET, Francis S, Bowtell R, Mcglone F, Kobal G, Renner B, Ahne G (2000) Sensory-specific satiety-related olfactory activation of the human orbitofrontal cortex. Neuroreport 11:893–897

O'doherty J, Kringelbach ML, Rolls ET, Hornak J, Andrews C (2001) Abstract reward and punishment representations in the human orbitofrontal cortex. Nat Neurosci 4:95–102

O'doherty J, Dayan P, Schultz J, Deichmann R, Friston KJ, Dolan RJ (2004) Dissociable roles of ventral and dorsal striatum in instrumental conditioning. Science 304:452–454

Ogawa H (1994) Gustatory cortex of primates: anatomy and physiology. Neurosci Res 20:1–13

Ongur D, Price JL (2000) Organization of networks within the orbital and medial prefrontal cortex of rats, monkeys, and humans. Cereb Cortex 10:206–219
Paulus M, Stein MB (2006) An insular view of anxiety. Biol Psychiatry 60:383–387
Phillips M, Drevets WR, Lane R (2003) Neurobiology of emotion perception I: the neural basis of normal emotion perception. Biol Psychiatry 54:504–514
Pollice C, Kaye WH, Greeno CG, Weltzin TE (1997) Relationship of depression, anxiety, and obsessionality to state of illness in anorexia nervosa. Int J Eat Disord 21:367–376
Preuschoff K, Quartz SBP (2008) Human insula activation reflects risk prediction errors as well as risk. J Neurosci 28:2745–2752
Protopopescu X, Pan H, Tuescher O, Cloitre M, Goldstein M, Engelien W, Epstein J, Yang Y, Gorman JM, Ledoux J, Silbersweig D, Stern E (2005) Differential time courses and specificity of amygdala activity in posttraumatic stress disorder subjects and normal control subjects. Biol Psychiatry 57:464–473
Rastam M, Bjure J, Vestergren E, Uvebrant P, Gillberg IC, Wentz E, Gillberg C (2001) Regional cerebral blood flow in weight-restored anorexia nervosa: a preliminary study. Dev Med Child Neurol 43:239–242
Redish A (2004) Addiction as a computational process gone awry. Science 306:1944–1947
Reynolds S, Zahm D (2005) Specificity in the projections of prefrontal and insular cortex to ventral striatopallidum and the extended amygdala. J Neurosci 25:11757–11767
Rolls ET (1996) The orbitofrontal cortex. Philos Trans R Soc Lond B Biol Sci 351:1433–1443
Rolls ET (1997) Taste and olfactory processing in the brain and its relation to the control of eating. Crit Rev Neurobiol 11:263–287
Rolls ET (2005) Taste, olfactory, and food texture processing in the brain, and the control of food intake. Physiol Behav 85:45–56
Santel S, Baving L, Krauel K, Munte T, Rotte M (2006) Hunger and satiety in anorexia nervosa: fMRI during cognitive processing of food pictures. Brain Res 1114:138–148
Sargent PA, Kjaer KH, Bench CJ, Rabiner EA, Messa C, Meyer J, Gunn RN, Grasby PM, Cowen PJ (2000) Brain serotonin$_{1A}$ receptor binding measured by positron emission tomography with [^{11}C]WAY-100635: effects of depression and antidepressant treatment. Arch Gen Psychiatry 57:174–180
Saxena S (2003) Neuroimaging and the pathophysiology of obsessive-compulsive disorder. In: Fu C, Senior C, Russell T, Weinberger D, Murray R (eds) Neuroimaging in psychiatry. Martin Dunitz, London
Schienle A, Schafer A, Hermann A, Vaitl D (2008) Binge-eating disorder: reward sensitivity and brain action to images of food. Biol Psychiatry 65:654–661
Schoenfeld M, Neuer G, Tempelmann C, Schussler K, Noesselt T, Hopf J, Heinze H (2004) Functional magnetic resonance tomography correlates of taste perception in the human primary taste cortex. Neuroscience 127:347–353
Schultz W (2004) Neural coding of basic reward terms of animal learning theory, game theory, microeconomics and behavioural ecology. Science 14:139–147
Schultz W, Tremblay L, Hollerman JR (2000) Reward processing in primate orbitofrontal cortex and basal ganglia. Cereb Cortex 10:272–284
Schwartz MW, Woods SC, Porte D Jr, Seeley RJ, Baskin DG (2000) Central nervous system control of food intake. Nature 404:661–671
Scott TR, Yaxley S, Sienkiewicz Z, Rolls E (1986) Gustatory responses in the frontal opercular cortex of the alert cynomolgus monkey. J Neurophysiol 56:876–890
Simon S, De Araujo I, Gutierrez R, Nicolelis M (2006) The neural mechanisms of gustation: a distributed processing code. Nat Rev Neurosci 7:890–901
Small D (2002) Toward an understanding of the brain substrates of reward in humans. Neuron 22:668–671
Small D, Zatorre R, Dagher A, Evans A, Jones-Gotman M (2001) Changes in brain activity related to eating chocolate: from pleasure to aversion. Brain 124:1720–1733

Smyth J, Wonderlich S, Heron K, Sliwinski M, Crosby R, Mitchell J, Engel S (2007) Daily and momentary mood and stress are associated with binge eating and vomiting in bulimia nervosa patients in the natural environment. J Consult Clin Psychol 75:629–638
Srinivasagam NM, Kaye WH, Plotnicov KH, Greeno C, Weltzin TE, Rao R (1995) Persistent perfectionism, symmetry, and exactness after long-term recovery from anorexia nervosa. Am J Psychiatry 152:1630–1634
Stein M, Simmons A, Feinsteim J, Paulus M (2007) Increased amygdala and insula activation during emotion processing in anxiety-prone subjects. Am J Psychiatry 164:318–327
Steinhausen HC (2002) The outcome of anorexia nervosa in the 20th century. Am J Psychiatry 159:1284–1293
Stice E (2002) Risk and maintenance factors for eating pathology: a meta-analytic review. Psychopharmacol Bull 128:825–848
Strober M (1995) Family-genetic perspectives on anorexia nervosa and bulimia nervosa. In: Brownell K, Fairburn C (eds) Eating disorders and obesity: a comprehensive handbook. Guilford, New York
Strober M, Freeman R, Morrell W (1997) The long-term course of severe anorexia nervosa in adolescents: survival analysis of recovery, relapse, and outcome predictors over 10–15 years in a prospective study. Int J Eat Disord 22:339–360
Strupp BJ, Weingartner H, Kaye W, Gwirtsman H (1986) Cognitive processing in anorexia nervosa. A disturbance in automatic information processing. Neuropsychobiology 15:89–94
Thut G, Schultz W, Roelcke U, Nienhusmeier M, Missimer J, Maguire RP, Leenders KL (1997) Activation of the human brain by monetary reward. Neuroreport 8:1225–1228
Tremblay LK, Naranjo CA, Graham SJ, Herrmann N, Mayberg HS, Hevenor SJ, Busto UE (2005) Functional neuroanatomical substrates of altered reward processing in major depressive diorder revealed by a dopaminergic probe. Arch Gen Psychiatry 62:1228–1236
Tricomi EM, Delgado MR, Fiez JA (2004) Modulation of caudate activity by action contingency. Neuron 41:281–292
Uher R, Brammer M, Murphy T, Campbell I, Ng V, Williams S, Treasure J (2003) Recovery and chronicity in anorexia nervosa: brain activity associated with differential outcomes. Biol Psychiatry 54:934–942
Uher R, Murphy T, Brammer M, Dalgleish T, Phillips M, Ng V, Andrew C, Williams S, Campbell I, Treasure J (2004) Medial prefrontal cortex activity associated with symptom provocation in eating disorders. Am J Psychiatry 161:1238–1246
Vitousek K, Manke F (1994) Personality variables and disorders in anorexia nervosa and bulimia nervosa. J Abnorm Psychol 103:137–147
Volkow ND, Wise RA (2005) How can drug addiction help us understand obesity? Nat Neurosci 8:555–560
Volkow ND, Wang G, Fowler J, Logan J, Jayne M, Franceschi D, Wong C, Gatley S, Gifford A, Ding Y, Pappas N (2002) "Nonhedonic" food motivation in humans involves dopamine in the dorsal striatum and methylephenidate amplifies this effect. Synapse 44:175–180
Wagner A, Barbarich N, Frank G, Bailer U, Weissfeld L, Henry S, Achenbach S, Vogel V, Plotnicov K, Mcconaha C, Kaye W, Wonderlich S (2006a) Personality traits after recovery from eating disorders: do subtypes differ? Int J Eat Disord 39:276–284
Wagner A, Greer P, Bailer U, Frank G, Henry S, Putnam K, Meltzer CC, Ziolko SK, Hoge J, Mcconaha C, Kaye WH (2006b) Normal brain tissue volumes after long-term recovery in anorexia and bulimia nervosa. Biol Psychiatry 59:291–293
Wagner A, Aizenstein H, Venkatraman M, Fudge J, May J, Mazurkewicz L, Frank G, Bailer UF, Fischer L, Nguyen V, Carter C, Putnam K, Kaye WH (2007) Altered reward processing in women recovered from anorexia nervosa. Am J Psychiatry 164:1842–1849
Wagner A, Aizenstein H, Frank GK, Figurski J, May JC, Putnam K, Bailer UF, Fischer L, Henry SE, Mcconaha C, Kaye WH (2008) Altered insula response to a taste stimulus in individuals recovered from restricting-type anorexia nervosa. Neuropsychopharmacology 33:513–523

Wagner A, Aizeinstein H, Venkatraman V, Bischoff-Grethe A, Fudge J, May J, Frank G, Bailer U, Fischer L, Putnam K, Kaye W (2009) Altered striatal response to reward in bulimia nervosa after recovery. Int J Eat Disord 43(4):289–294

Waters A, Hill A, Waller G (2001) Bulimics' responses to food cravings: is binge-eating a product of hunger or emotional state? Behav Res Ther 39:877–886

Weltzin TE, Hsu LK, Pollice C, Kaye WH (1991) Feeding patterns in bulimia nervosa. Biol Psychiatry 30:1093–1110

Wright CI, Martis B, Mcmullin K, Shin LM, Rauch SL (2003) Amygdala and insular responses to emotionally valenced human faces in small animal specific phobia. Biol Psychiatry 54:1067–1076

Yaxley S, Rolls E, Sienkiewicz Z (1990) Gustatory responses of single neurons in the insula of the macaque monkey. J Neurophysiol 63:689–700

Yin H, Knowlton B (2006) The role of the basal ganglia in habit formation. Nat Neurosci Rev 7:464–476

Zastrow A, Kaiser SS, Walthe S, Herzog W, Tchanturia K, Belger A, Weisbrod M, Treasure J, Friederich H (2009) Neural correlates of impaired cognitive-behavioral flexibility in anorexia nervosa. Am J Psychiatry 166:608–616

Serotonin: Imaging Findings in Eating Disorders

Ursula F. Bailer and Walter H. Kaye

Contents

Abstract Anorexia nervosa (AN) and bulimia nervosa (BN) are disorders characterized by aberrant patterns of feeding behavior, weight regulation, and disturbances in attitudes and perceptions toward body weight and shape. Several lines of evidence nominate disturbances of serotonin (5-HT) pathways as playing a role in the pathogenesis and pathophysiology of AN and BN. For example, 5-HT pathways are known to contribute to the modulation of a range of behaviors commonly seen in individuals with AN and BN. New technology using brain imaging with radioligands offers the potential for understanding previously inaccessible brain 5-HT neurotransmitter function and its dynamic relationship with human behaviors.

W.H. Kaye (✉)
Department of Psychiatry, Eating Disorder Treatment and Research Program, University of California, San Diego, La Jolla, CA, USA
e-mail: whkaye@gmail.com
U.F. Bailer
Department of Psychiatry and Psychotherapy, Division of Biological Psychiatry, Medical University of Vienna, Vienna, Austria
Department of Psychiatry, Eating Disorder Treatment and Research Program, University of California, San Diego, La Jolla, CA, USA

R.A.H. Adan and W.H. Kaye (eds.), *Behavioral Neurobiology of Eating Disorders*,
Current Topics in Behavioral Neurosciences 6, DOI 10.1007/7854_2010_78

Recent studies using positron emission tomography and single photon emission computed tomography with 5-HT-specific radioligands have consistently shown 5-HT_{1A} and 5-HT_{2A} receptor and 5-HT transporter alterations in AN and BN in cortical and limbic structures, which may be related to anxiety, behavioral inhibition, and body image distortions. These disturbances are present when subjects are ill and persist after recovery, suggesting that these may be traits that are independent of the state of the illness. Effective treatments for AN and BN have been elusive. A better understanding of neurobiology is likely to be important for developing specific and more powerful therapies for these often chronic and deadly disorders.

Keywords Anorexia nervosa · Bulimia nervosa · Serotonin · Receptor · Transporter · Brain imaging · PET · SPECT

1 Introduction

Anorexia nervosa (AN) and bulimia nervosa (BN) are disorders characterized by aberrant patterns of feeding behavior, weight regulation, and disturbances in attitudes and perceptions toward body weight and shape. In AN, there is an inexplicable fear of weight gain and unrelenting obsession with fatness even in the face of increasing cachexia. BN usually emerges after a period of dieting, which may or may not have been associated with weight loss. Binge eating is followed by either self-induced vomiting or some other means of compensation for the excess of food ingested. The majority of people with BN have irregular feeding patterns and satiety may be impaired. Although abnormally low body weight is an exclusion for the diagnosis of BN, some 25–30% of individuals with BN presenting to treatment centers have a prior history of AN; however, all individuals with BN have pathological concern with weight and shape.

It has been argued that AN and BN share some risk and liability factors because these disorders are often cross-transmitted in families and share many behavioral traits (Kendler et al. 1991; Lilenfeld et al. 1998; Strober et al. 2000; Walters and Kendler 1995). Similar to AN, individuals with BN commonly have anxiety, obsessionality, depression, a seemingly relentless drive to restrain food intake, an extreme fear of weight gain, and a distorted view of their body shape. AN tend to be over-controlled, whereas BN tend to have poor impulse control, greater novelty seeking (Strober et al. 1997; Wagner et al. 2006; Lilenfeld et al. 2006), and high rates of drug and substance abuse (Hudson et al. 2007). It is important to note that both AN and BN tend to restrict their eating and lose normal meal patterns (Mitchell et al. 1998). We hypothesize that AN can maintain this inhibition continuously, whereas BN have periodic disinhibition and overconsume food.

Although psychosocial factors are hypothesized to cause AN and BN, recent studies show that genetic heritability accounts for approximately 50–80% of the risk and creates neurobiological vulnerabilities (Kendler et al. 1991; Berrettini 2000; Bulik et al. 2006; Kaye et al. 2008; Steinglass and Walsh 2004; Treasure and Campbell 1994). It is important to note that considerable evidence suggests that childhood temperament and personality traits can create a vulnerability for developing AN and BN during adolescence. Recent studies (Lilenfeld et al. 2006; Stice 2002; Anderluh et al. 2003) describe negative emotionality, harm avoidance (HA), perfectionism, inhibition, drive for thinness, altered interoceptive awareness, and obsessive-compulsive personality traits as childhood predisposing factors that precede the onset of an ED, persist after recovery, and are elevated in unaffected family members (Bulik et al. 2007).

Several lines of evidence nominate disturbances of serotonin (5-HT) pathways as playing a role in the pathogenesis and pathophysiology of AN and BN. For example, 5-HT pathways are known to contribute to the modulation of a range of behaviors commonly seen in individuals with AN and BN. That is, 5-HT has been implicated in personality or temperament traits such as HA (Cloninger 1987) and behavioral inhibition (Soubrie 1986). Moreover, 5-HT has been implicated in psychiatric symptoms such as obsessionality (Barr et al. 1992), anxiety and fear (Charney and Deutch 1996), depression (Grahame-Smith 1992), as well as physiological traits such as satiety for food consumption. Second, many studies show disturbances of 5-HT activity in individuals who were ill or recovered from AN and BN. Third, medications that act on 5-HT pathways have some degree of efficacy in individuals with ED.

It is important to emphasize that brain neurotransmitter pathways do not work in isolation. For example (Tremblay and Blier 2006), the norepinephrine, dopamine, and 5-HT systems have reciprocal interactions so that it is virtually impossible to act on a specific neuronal element without a cascade effect on the other systems. In terms of clinical research in humans, we have perhaps more tools for investigating 5-HT activity and more understanding of its function than for other neurotransmitter systems. For that reason, this chapter will focus on 5-HT.

2 The Role of 5-HT Neurotransmitter Function in Eating Disorders

A role for biological determinants in the pathogenesis of eating disorders (EDs) has been proposed for the past 60 years (Treasure and Campbell 1994). In particular, an increased knowledge of the neurotransmitter modulation of feeding behavior has raised questions as to whether a disturbance in monoamine function may play a role in these disorders. 5-HT pathways play an important role in postprandial satiety. Treatments that increase intrasynaptic 5-HT, or directly activate 5-HT receptors, tend to reduce food consumption, whereas interventions that dampen 5-HT neurotransmission or block receptor activation reportedly increase food consumption and promote weight gain (Blundell 1984; Leibowitz and Shor-Posner 1986). Moreover, CNS 5-HT pathways have been implicated in the modulation of mood, impulse regulation and behavioral constraint, and obsessionality, and they affect a variety of neuroendocrine systems.

5-HT is synthesized from its precursor tryptophan, an essential amino acid that must be obtained through the diet. Following dietary consumption, tryptophan is taken up by the brain and hydroxylated by the enzyme tryptophan-5-hydroxylase (Petty et al. 1996). The product of this reaction, 5-hydroxytryptophan, is then decarboxylated by aromatic amino acid decarboxylase to the compound 5-hydroxy-tryptamine to the metabolite product known as 5-hydroxyindoleacetic (5-HIAA), which may be measured as a means of assessing 5-HT turnover or metabolism (Petty et al. 1996).

There has been considerable interest in the role that 5-HT may play in AN and BN (Jimerson et al. 1990; Kaye and Weltzin 1991; Treasure and Campbell 1994; Brewerton 1995; Kaye et al. 1998b). In part, this is related to the fact that studies, using hormonal response to 5-HT agents or other methods, have found that AN and BN have alterations in 5-HT metabolism.

When underweight, individuals with AN have a significant reduction in basal concentrations of the 5-HT metabolite 5-HIAA in the cerebral spinal fluid (CSF) compared to healthy controls (Kaye et al. 1984), as well as blunted plasma prolactin response to drugs with 5-HT activity and reduced ^{3}H-imipramine binding (Jimerson et al. 1990; Kaye and Weltzin 1991; Treasure and Campbell 1994; Brewerton 1995; Kaye et al. 1998a). Together, these findings suggest reduced serotonergic activity during the acute phase of illness, although this may arise secondarily from reductions in dietary supplies of the 5-HT synthesizing amino acid tryptophan. By contrast, CSF concentrations of 5-HIAA are reported to be elevated in long-term weight-recovered AN individuals (Kaye et al. 1991a). These contrasting findings of reduced and heightened serotonergic activity in acutely ill and long-term recovered AN individuals, respectively, may seem counterintuitive; however, since dieting lowers plasma tryptophan levels in otherwise healthy women (Anderson et al. 1990), resumption of normal eating in individuals with AN may unmask intrinsic abnormalities in serotonergic systems that mediate certain core behavioral or temperamental underpinnings of risk and vulnerability.

Given that food restriction is not an inherently reinforcing behavior in healthy individuals, persistent dieting to the point of starvation suggests that food restriction may have some intrinsic benefit for individuals with AN. The ratio of tryptophan to other large neutral amino acids has been found to be significantly reduced in AN (Favaro et al. 2000). This reduction is likely to be a consequence of starvation since food restriction results in a decrease in dietary tryptophan consumption and thereby a decrease in the concentration of tryptophan available for 5-HT synthesis.

Premorbidly, individuals with AN report high levels of anxiety and obsessionality. Evidence suggests that individuals with AN may have an intrinsic defect in the 5-HT system. These individuals may have high levels of 5-HT in the synapse premorbidly resulting in a dysphoric state. Dieting may serve as a means of regulating this overactivity of 5-HT by decreasing the amount of tryptophan available for 5-HT synthesis, as evidence of reduced 5-HT activity is found during the acute phase. A study of acute tryptophan depletion found that a reduction of dietary tryptophan was associated with decreased anxiety and an elevation of mood in individuals with AN during the acute phase of illness and following long-term recovery (Kaye et al. 2003). Acute tryptophan depletion did not have significant anxiolytic effects for control women.

Considerable evidence also exists for a dysregulation of serotonergic processes in BN. Examples include blunted prolactin response to the 5-HT receptor agonists *m*-chlorophenylpiperazine, 5-hydroxytrytophan, and DL-fenfluramine, and enhanced migraine-like headache response to *m*-CPP challenge (Jimerson et al. 1990; Kaye and Weltzin 1991; Treasure and Campbell 1994; Brewerton 1995; Kaye et al. 1998a; Steiger et al. 2001a, b, c). Acute perturbation of serotonergic tone by dietary depletion of tryptophan has also been linked to increased food intake and mood irritability in individuals with BN compared to healthy controls. And, like AN, women with long-term recovery from BN have been shown to have elevated concentrations of 5-HIAA in the CSF, whereas CSF 5-HIAA levels are normal in ill BN (Kaye et al. 1984, 1988, 1990, 1991a, 1998b; Jimerson et al. 1992). Furthermore, Steiger and colleagues (Steiger et al. 2005b) found that individuals recovered from BN have reduced platelet [3H-] paroxetine binding, which is thought to be a marker of 5-HTT activity.

It has been found that *low* levels of CSF 5-HIAA are associated with impulsive and non-premeditated aggressive behaviors (Stein et al. 1993), which cut across traditional diagnostic boundaries. Thus, it is of interest that recovered AN and BN women had elevated CSF 5-HIAA concentrations. Behaviors found after recovery from AN and BN, such as obsessionality with symmetry and exactness, anxiety, and perfectionism, tend to be opposite in character to behaviors displayed by people with low 5-HIAA levels. Together, these studies contribute to a growing literature suggesting that CSF 5-HIAA concentrations may correlate with a spectrum of behavior. Reduced CSF 5-HIAA levels appear to be related to behavioral under-control, whereas increased CSF 5-HIAA concentrations may be related to behavioral over-control.

3 Brain Imaging Studies

New technology using brain imaging with radioligands offers the potential for understanding previously inaccessible brain 5-HT neurotransmitter function and its dynamic relationship with human behaviors. Technologies that are used to date include single photon emission computed tomography (SPECT) and positron emission tomography (PET). Studies that have used these imaging techniques in EDs are summarized in Table 1.

The marriage of PET imaging with *selective* neurotransmitter *radioligands* has resulted in a technology permitting new insights into regional binding and specificity of 5-HT neurotransmission in vivo in humans and their relationship to behaviors.

Table 1 SPECT and PET studies assessing 5-HT function in anorexia and bulimia nervosa compared to healthy controls [modified from (Kaye et al. 2005)]

Year	Author	Method	ILL	REC	*N*	Frontal cortex	Temporal/ amygdala	Cingulate cortex	Parietal cortex
$5\text{-}HT_{2A}$ receptor									
2001	Kaye	PET5HT_{2A}		BN	9	↓	nl	nl	nl
2002	Frank	PET 5-HT_{2A}		AN	16	nl	↓	↓	nl
2003	Audenaert	SPECT 5-HT_{2A}	AN*		15	↓	nl		
2004	Bailer	PET 5-HT_{2A}		AN–BN	10	nl	↓	↓	↓
2004	Goethals	SPECT 5-HT_{2A}	BN		10	nl	nl	nl	nl
2007	Bailer	PET 5-HT_{2A}	AN + AN–BN		15	nl	nl	nl	nl
$5\text{-}HT_{1A}$ receptor									
2004	Tiihonen	PET 5HT_{1A}	BN		8	↑	nl	↑	
2005	Bailer	PET 5HT_{1A}		AN	13	nl	nl	nl	nl
2005	Bailer	PET 5HT_{1A}		AN–BN	12	↑	↑	↑	↑
2007	Bailer	PET 5HT_{1A}	AN + AN–BN		15	↑	↑	↑	↑
2008	Galusca	PET 5HT_{1A}	AN		8	↑	↑		↑
2008	Galusca	PET 5HT_{1A}		AN	9	↑	↑		↑
In press	Bailer	PET 5HT_{1A}		BN	13		↑	↑	↑
5-HT transporter									
2001	Tauscher	SPECT 5-HTT	BN		10	Decreased subcortical			
2007	Bailer	PET 5-HTT		AN	11			nl	
2007	Bailer	PET 5-HTT		AN–BN	7			nl	
2007	Bailer	PET 5-HTT		BN	9			nl	
2007	Koskela	SPECT 5-HTT	BN		13	Increased in midbrain in purging type only			

nl normal, ↓ decreased compared to controls, ↑ increased compared to controls, *AN* anorexia nervosa, restricting type, *AN–BN* anorexia nervosa, binging–purging type, *AN** diagnostic subgroup not specified, *BN* bulimia nervosa, *REC* recovered, *PET* positron emission tomography, *SPECT* single photon emission computed tomography, *5-HT* serotonin, *5-HTT* 5-HT transporter

It is important to note that the 5-HT system involves 14 or more receptors and interacts with many other neurotransmitters and molecules. Only a few of these components can currently be measured in vivo in humans. Although the complexity of 5-HT circuits cannot be fully elucidated in humans, such imaging studies of 5-HT functional activity are useful in that they can characterize potential state and trait differences between AN and BN patients and healthy controls, be used to model relationships of 5-HT activity to behavior, and provide new insights into targets for more effective treatment.

3.1 5-HT$_{2A}$ Receptor

Postsynaptic 5-HT$_{2A}$ receptors are in high densities in the cerebral cortex and other brain regions of rodents and humans (Burnet et al. 1997; Saudou and Hen 1994). The 5-HT$_{2A}$ receptor is of interest in ED because it has been implicated in the modulation of feeding and mood, as well as SSRI response (Bonhomme and Esposito 1998; De Vry and Schreiber 2000; Simansky 1996; Stockmeier 1997).

Our group has used this technology to study women ill with AN and after recovery from AN and BN (>1 year no binging or purging, normal weight, and regular menstrual cycles) to confirm 5-HT disturbances and provide new insights into the disorder. Although women ill with BN have been found to have normal 5-HT$_{2A}$ receptor binding (Goethals et al. 2004), studies of recovered BN women (Kaye et al. 2001a), using PET with [^{18}F]altanserin, a specific 5-HT$_{2A}$ receptor antagonist, found a significant reduction in bilateral medial orbital frontal cortex 5-HT$_{2A}$ binding. These data lend further support to the possibility that vulnerabilities for impulse dyscontrol and mood disturbances cut across diagnostic boundaries and involve 5-HT and frontal lobe activity. REC BN women did not show the age-related decrease in 5-HT$_{2A}$ binding found in control women. This is further evidence of persistent 5-HT alterations after recovery from this illness. Moreover, studies using PET and [^{18}F]altanserin (Frank et al. 2002) investigated women who were recovered from restricting-type AN (REC AN). REC AN had reduced 5-HT$_{2A}$ activity, relative to CW, in mesial temporal (amygdala and hippocampus) regions, as well as cingulate, sensorimotor, and occipital/parietal cortical regions. Bailer and colleagues (Bailer et al. 2004) found that women who were recovered from binging–purging type AN (REC AN–BN) had significantly reduced 5-HT$_{2A}$ receptor binding in the left subgenual cingulate, left parietal cortex, and right occipital cortex compared to CW. Audenaert et al. (2003) used SPECT and 123I-5-I-R91159 and found that ILL AN subjects had reduced binding of postsynaptic 5-HT$_{2A}$ receptors in the left frontal, bilateral parietal, and occipital cortex, while bulimic type AN had reduced 5-HT$_{2A}$ binding in the parietal cortex in comparison to restricting-type AN (Goethals et al. 2007). However, using PET and [^{18}F]altanserin we found similar 5-HT$_{2A}$ receptor binding in a mixed group of ILL AN and AN–BN compared to CW (Bailer et al. 2007a). However, the SPECT study did not account for possible brain volume loss in ILL AN, so that the reduced binding may be the result of partial volume averaging, leading to an underestimation of binding per unit brain volume in the ILL AN group. Different imaging techniques also vary in terms of resolution; thus it makes it difficult to directly compare studies.

While REC AN showed significantly negative relationships between age and 5-HT_{2A} receptor binding for most cortical regions (Frank et al. 2002), REC AN–BN (Bailer et al. 2004) and REC BN (Kaye et al. 2001a) did not show any significant relations to age. These data raise the question of whether 5-HT activity in BN is dissociated from normal age-associated changes, a finding that may offer new clues into the pathophysiologic mechanisms contributing to EDs. Whether the 5-HT system becomes free-running and insensitive to normal developmental mechanisms remains to be explored.

In summary, when 5-HT_{2A} receptor binding is compared between subgroups, both REC AN and AN–BN have reductions in the subgenual cingulate, parietal, and occipital cortex. In comparison, only REC AN have reduced 5-HT_{2A} receptor binding of the mesial temporal region and pregenual cingulate (Frank et al. 2002).

The PET imaging studies in ill and REC AN subjects described above found significant correlations between HA and binding for the 5-HT_{2A} receptors in mesial temporal and other limbic regions. Bailer and colleagues (Bailer et al. 2004) found that REC AN–BN subjects showed a positive relationship between [^{18}F]altanserin binding in the left subgenual cingulate and mesial temporal cortex and HA. For ill AN subjects, [^{18}F]altanserin binding was positively related to HA in the suprapragenual cingulate, frontal, and parietal regions (Bailer et al. 2007a).

The anterior cingulate cortex has an executive function (Devinsky et al. 1995). The subcaudal cingulate regions play a role in emotion (affect component) and have extensive connections with the amygdala, periaqueductal gray, frontal lobes, ventral striatum, etc. It is involved in conditioned emotional learning, vocalizations associated with expressing internal states, and assigning emotional valence to internal and external stimuli (Devinsky et al. 1995; Takenouchi et al. 1999; Bush et al. 1999). Mesial temporal regions include the amygdala and related regions which play a pivotal role in anxiety and fear (Charney and Deutch 1996) as well the modulation and integration of cognition and mood. The amygdala may enable the individual to initiate adaptive behaviors to threat based on the nature of the threat and prior experience (Charney and Deutch 1996). Most other brain imaging studies also show that ill and REC AN have cingulate and temporal alterations.

Do people with AN have symptoms that might be related to cingulate–temporal dysfunction? HA is common in AN (Brewerton et al. 1993; Kleifield et al. 1993; Bulik et al. 1995; O'Dwyer et al. 1996; Ward et al. 1998; Klump et al. 2000). Most ED subjects have one or more anxiety disorder diagnosis, who usually have an onset *before* the onset of their ED. Moreover, AN have an obsessive, perseverative, and rigid personality style and have difficulty shifting sets. While AN do well on goal-directed behavior, they have difficulties incorporating feedback and modifying their behavior. For example, they often feel that they should be able to do things perfectly and not make mistakes, and have little appreciation for the fact that mistakes are a normal learning experience. Moreover, they often fail to accurately recognize and incorporate affective and social stimuli in the environment, as confirmed by laboratory tests (Strupp et al. 1986; Kingston et al. 1996). We hypothesize that people with AN do not seem to be able to access and use conventional strategies for problem solving, such as learning from making mistakes. Rather, they obsessively repeat the same strategies despite the fact that such strategies are maladaptive and are not productive. These characteristic styles raise the question of whether there is some physiologic disturbance of executive brain mechanisms that detect errors or plan and verify actions. Perhaps a disturbance in cingulate–temporal pathways results in an obsessive focus on certain events and excludes the comprehension and incorporation of other stimuli.

Moreover, a most puzzling symptom in AN is their severe and intense body image distortion in which emaciated subjects perceive themselves as fat. We have previously shown that REC AN–BN had a negative relationship between the Eating Disorder Inventory – 2 Drive for Thinness (Garner 1991) (EDI-DT) subscale and [^{18}F]altanserin binding in the right subgenual cingulate, right pregenual cingulate, the lateral temporal cortex, the left parietal cortex, and the prefrontal cortex (Bailer et al. 2004). Furthermore, the AN studies described above (Frank et al. 2002; Bailer et al. 2004; Audenaert et al. 2003) all found alterations in 5-HT$_{2A}$ activity in the left parietal region. These findings raise the speculation that left parietal alterations in REC AN and AN–BN might contribute to body image distortions. It is well known that lesions in the right parietal cortex may not only result in denial of illness, but may also produce experiences of disorientation of body parts and body image distortion (Critchley 1953). Theoretically, body image distortion might be related to the syndrome of neglect (Mesulam 1981), which may be coded in parietal, frontal, and cingulate regions that assign motivational relevance to sensory events. The refractory body image distortion in patients suffering from AN is a central feature of the illness. Other studies, using functional magnetic resonance imaging, support the speculation that left parietal disturbances may contribute to body image distortion (Wagner et al. 2003).

Only REC BN have reductions of the medial orbital frontal cortex (Kaye et al. 2001a). It is well recognized that BN subjects have extremes of self-control, such as alternating between undereating and overeating. Both 5-HT activity and frontal lobe function have been associated with behavioral disinhibition and extremes of self-control, such as obsessionality and impulsive aggressive behaviors (Tucker et al. 1995). We postulate that inherent disturbance of orbital frontal 5-HT circuits in BN contributes to a vulnerability for imprecise and poorly modulated behavioral control, which is reflected in reduced 5-HT$_{2A}$ receptor binding.

3.2 5-HT$_{1A}$ Receptor

Our group used PET imaging with the radioligand [^{11}C]WAY100635 to assess the BP of the 5-HT$_{1A}$ receptor. The 5-HT$_{1A}$ autoreceptor is located presynaptically on serotonergic somatodendritic cells in the raphe nucleus, where it functions to decrease 5-HT neurotransmission (Staley et al. 1998). High densities of postsynaptic 5-HT_{1A} exist in the hippocampus, septum, amygdala, and entorhinal and frontal cortex, where they serve to mediate the effects of released 5-HT. Although the molecular organization for the receptor transduction seems to be identical in all of the areas where 5-HT_{1A} receptors are expressed, some differences in both functional and regulatory properties have been reported from area to area (Lanfumey and Hamon 2000). Studies in animals and humans implicate the 5-HT_{1A} receptor in anxiety (Cervo et al. 2000; File et al. 2000; Olivier et al. 2001) and depression and/or suicide (Matsubara et al. 1991; Arango et al. 1995; Mann 1999).

Bailer et al. (2007a) have reported that ill AN individuals have a 50–70% increase in $[^{11}C]$WAY100635 BP in subgenual, mesial temporal, orbital frontal, and raphe brain regions as well as prefrontal, lateral temporal, anterior cingulate, and parietal regions. Similarly, REC AN–BN and REC BN subjects (Bailer et al. 2005, 2010) have a significant 20–40% increase in $[^{11}C]$WAY100635 BP in these same regions, compared to CW. While women recovered from restrictive-type AN had normal $[^{11}C]$WAY 100635 BP (Bailer et al. 2005), $[^{11}C]$WAY 100635 BP values were markedly elevated in some subjects and were recently found to be significantly increased in lean and recovered restricting-type AN individuals (using the radioligand $[^{18}F]$MPPF) (Galusca et al. 2008). Increased 5-HT_{1A} BP was positively associated with HA in REC restricting-type AN individuals (Bailer et al. 2005). Increased 5-HT_{1A} postsynaptic activity has also been reported in ill BN subjects (Tiihonen et al. 2004). Several interpretations are possible, which will require further testing to confirm. First, in recovered state, increased binding of the 5-HT_{1A} receptor may be associated specifically with REC BN, whether they have had a history of AN. Second, elevated 5-HT_{1A} receptor binding may be further exaggerated in the ill state of both AN and BN individuals, suggesting a possible trait phenomenon that is exacerbated by nutritional abnormalities. These data also may provide insight into possible new pharmaceutical treatments for AN and BN. Although numerous controlled trials have shown some efficacy for a variety of antidepressant medications in BN, relatively few individuals achieve abstinence on medication, as most continue to binge and purge. For example, a large-scale controlled trial of fluoxetine, which showed that a relatively high dose of 60 mg/day was superior to 20 mg/day for BN (Romano et al. 2002), had a 1-year remission rate of only 17.7%. Many subjects remained symptomatic on medication, and there was a worsening on all measures of efficacy over time. This result is consistent with other clinical observations (Walsh et al. 1991) that suggest limited improvement and considerable relapse with long-term antidepressant treatment in BN. The efficacy of SSRIs is dependent on neuronal release of 5-HT (Tollefson 1995), and 5-HT release in turn results in desensitization of the 5-HT_{1A} receptor (Blier and de Montigny 1999). Highly elevated 5-HT_{1A} receptor activity in BN raises the question of whether BN individuals have difficulty in achieving SSRI-induced 5-HT_{1A} autoreceptor desensitization. Such a difficulty could explain the need for higher doses of fluoxetine as well as partial response to drugs. Perhaps, higher doses of SSRIs or the addition of 5-HT_{1A} specific agents may prove useful in BN. With regard to AN, despite considerable evidence of 5-HT abnormalities, ill AN patients show little response to SSRI administration (Attia and Schroeder 2005), in terms of improvement of mood or reduction of core ED symptoms. It is possible that elevated activity of 5-HT$_{1A}$ receptors in the raphe nucleus in ill AN patients results in reduced 5-HT neuronal firing, and thus decreased extracellular 5-HT levels (Kaye et al. 1988), consistent with the reduced CSF 5-HIAA levels found in these patients. Thus, it is possible that SSRIs are not effective in ill AN patients because SSRIs would not have much effect if synaptic 5-HT levels are depleted by malnutrition.

As noted above, EDs are frequently comorbid with depression and anxiety disorders. However, while individuals with EDs tend to have elevated [^{11}C]WAY 100635 BP, reduced binding of 5-HT$_{1A}$ receptor ligands was found in most [for review, see (Drevets et al. 2007; Savitz et al. 2009)], but not all, studies of major depression (Miller et al. 2009). In addition, reduced binding of 5-HT$_{1A}$ receptor ligands was found in social phobia (Lanzenberger et al. 2007) and panic disorder (Neumeister et al. 2004; Nash et al. 2008). Thus, it can be argued that these disorders may be etiologically different.

3.3 Brain Regions/Pathways Enervated by 5HT$_{1A/2A}$ Receptors

In REC subjects, altered 5-HT$_{1A}$ and 5-HT$_{2A}$ receptor BP shows persistent alterations in frontal, subgenual cingulate, and mesial temporal regions that are part of the ventral limbic system.

Several lines of evidence show that 5-HT$_{1A}$ and 5-HT$_{2A}$ receptors interact in the brain. In rats, 5-HT$_{1A}$ and 5-HT$_{2A}$ receptors interact robustly to regulate the inhibition of exploration of novel environments produced by either 5-HT$_{1A}$ or 5-HT$_{2A}$ receptor agonists (Krebs-Thomson and Geyer 1998). 5-HT$_{2A}$ and 5-HT$_{1A}$ receptors are highly co-localized in rodent frontal cortex (Amargos-Bosch et al. 2004). Postsynaptic 5-HT$_{1A}$ and 5-HT$_{2A}$ receptors mediate, respectively, the direct hyperpolarizing and depolarizing actions of 5-HT on prefrontal neurons (Santana et al. 2004), which in turn project to numerous cortical and subcortical areas. Thus, a balance between postsynaptic 5-HT$_{1A}$ and 5-HT$_{2A}$ receptor activity on neurons may modulate the descending excitatory input into limbic and motor structures. These data raise the speculation that postsynaptic 5-HT$_{1A}$ and 5-HT$_{2A}$ receptors fine-tune cortical systems that modulate behavioral inhibition and self-control. Mixed 5-HT$_{2A/1A}$ agonists, e.g., psilocybin, seem to disrupt the 5-HT$_{1A/2A}$ balance (Vollenweider et al. 1999) by driving 5-HT$_{2A}$ activity, resulting in excessive neuronal output contributing to extremes of disinhibition, disorganization, and loss of self-control. In our studies, REC ED subjects had a relative increase in 5-HT$_{1A}$ receptor activity compared to 5-HT$_{2A}$ receptor binding. While speculative, this possible imbalance could contribute to behavioral inhibition and over-control commonly seen in ED. As discussed before, we found considerable correlations between binding of these two receptors and HA. Taken together, these findings raise the possibility that mesial temporal (amygdala)–cingulate 5HT$_{1A/2A}$

imbalance may also be a trait shared by AN subgroups related to behavioral inhibition, anticipatory anxiety, or integration of cognition and mood.

3.4 5-HT Transporter

Studies with SPECT and [123I] beta-CIT showed that ill BN subjects had a 17% reduced 5-HT transporter (5-HTT) availability in the hypothalamus and thalamus (Tauscher et al. 2001). Two of those individuals had a history of AN, but specific imaging data for those two individuals were not identified. Our group (Bailer et al. 2007b) used PET imaging with [^{11}C]McN5652 comparing 11 subjects recovered (>1 year normal weight, regular menstrual cycles, no binging or purging) REC AN, 7 REC AN–BN, 9 REC BN, and 10 healthy CW. After correcting for multiple comparisons, we found that the REC AN had significantly increased [^{11}C] McN5652 BP compared to REC AN–BN for the dorsal raphe and antero-ventral striatum. However, neither group was different from healthy CW. In addition, REC BN were similar to CW and REC AN. No other studies have been done in AN. However, other imaging and peripheral platelet studies have found evidence of reduced 5-HTT in BN (Tauscher et al. 2001; Steiger et al. 2005b) and binge eating disorder individuals (Kuikka et al. 2001). A SPECT study compared 5-HTT availability in the midbrain and thalamus in 13 female twins with BN (nine with purging and four with nonpurging) versus 25 CW using a different radioligand for 5-HTT, [123I]ADAM (Koskela et al. 2007). They found that purging type BN had increased midbrain [123I]ADAM binding compared to CW, supporting a 5-HT-based distinction between those with purging and nonpurging behaviors, across both studies. While BN individuals show a response to higher doses of fluoxetine (Fluoxetine Bulimia Nervosa Collaborative Study Group 1992), the efficacy of such medication has been questioned, as relatively few individuals abstain from binge and purge behaviors, and relapse during treatment is common (Walsh 1991). It remains controversial whether SSRIs are effective in AN individuals. Our clinical experience and data (Kaye et al. 1991b, 2001b; Walsh et al. 2006) suggest that individuals with AN respond better to fluoxetine than do those with AN–BN. While highly speculative, our findings raise the provocative possibility that decreased 5-HTT function may be related to poor response to SSRI medication, whereas individuals with increased 5-HTT activity may respond to higher SSRI doses. In general, REC AN individuals had elevated 5-HTT binding, suggesting they have relatively greater 5-HT uptake, and reduced extracellular 5-HT, compared to REC AN–BN. In support of this possibility, the REC AN–BN individuals tend to have higher binding of 5-HT$_{1A}$ postsynaptic receptors and autoreceptors (Bailer et al. 2005), which may be a compensatory means of downregulating raphe activity (Cooper 1996; Hajos et al. 2003). Moreover, reduced 5-HTT activity, resulting from functional polymorphisms (Steiger et al. 2005a), has been associated with affect dysregulation, which tends to be more common in the BN subgroups. Traits such as sensation seeking and insecure attachment are elevated in BN syndromes carrying low function alleles of the 5-HTT promoter polymorphism, who report prior physical or sexual maltreatment (Steiger et al. 2007). Furthermore, in people with impulsive aggression, reduced 5-HTT binding was found in the anterior cingulate cortex, a region involved in affecting regulation (Frankle et al. 2005).

4 Conclusion

SPECT and PET-radioligand studies confirm that altered 5-HT neuronal pathway activity persists after recovery from AN and BN and support the possibility that these psychobiological alterations might contribute to traits, such as increased anxiety, that may contribute to a vulnerability to develop an ED. Clinical and epidemiological studies have consistently shown that one or more anxiety disorders occur in the majority of people with AN or BN (Godart et al. 2002; Walters and Kendler 1995; Kendler et al. 1995; Kaye et al. 2004). Silberg and Bulik (2005), using twins, found a unique genetic effect that influences liability to early anxiety and ED symptoms. When a lifetime anxiety disorder is present, the anxiety most commonly occurs first in childhood, preceding the onset of AN or BN (Deep et al. 1995; Bulik et al. 1997; Godart et al. 2000). Anxiety and HA remain elevated after recovery from AN, AN–BN, and BN (Wagner et al. 2006), even if individuals never had a lifetime anxiety disorder diagnosis (Kaye et al. 2004). The PET imaging data suggest that such behaviors are related to disturbances of 5-HT neurotransmitter function in limbic and executive pathways (Kaye et al. 2009). It is thought that in individuals with AN, dietary restraint reduces anxiety, whereas eating stimulates dysphoric mood (Kaye et al. 2003; Strober 1995; Vitousek and Manke 1994). Is altered 5-HT function the link between restricted feeding behavior and anxiety in AN patients? It is well known that carbohydrate intake increases extracellular 5-HT concentrations in the brain through complex metabolic effects on tryptophan, the amino acid precursor of 5-HT (Fernstrom and Wurtman 1972; Kaye et al. 2003). We hypothesize that both premorbidly and after recovery from AN, a normal level of food ingestion is associated with exaggerated extracellular brain 5-HT secretion (Kaye et al. 1991a). This is consistent with increased CSF 5-HIAA levels in recovered AN individuals (Kaye et al. 1991a). Increased 5-HT concentrations inhibit appetite, perhaps through activation of the 5-HT_{2C} receptor (Simansky et al. 2004); however, 5-HT_{2C} receptor binding has not been measured by imaging studies in individuals with AN. Increased 5-HT_{1A} BP is positively associated with HA in REC AN individuals (Bailer et al. 2005), and enhanced anxiety and HA are traits that are present premorbidly and persist after recovery from AN (Wagner et al. 2006). Thus, it is possible that carbohydrate-induced increases in extracellular 5-HT levels drive anxiety and HA through stimulation of the 5-HT_{1A} receptor, offering a potential explanation for feeding-related dysphoric mood in AN. By contrast, when individuals with AN starve, extracellular 5-HT concentrations might reduce, resulting in a brief respite from dysphoric mood. Studies in animals and healthy humans show that both a restricted diet (which significantly lowers plasma tryptophan) and experimentally reduced tryptophan depletion decrease brain 5-HT synthesis (Fernstrom and Wurtman 1972; Young and Gauthier 1981; Anderson et al. 1990). Indeed, malnourished and emaciated AN patients have reduced plasma tryptophan availability (Schweiger et al. 1986; Attia et al. 2005) and reduced CSF 5-HIAA (Kaye et al. 1988). Importantly, experimental manipulations that reduce brain tryptophan levels decrease anxiety in both ill and REC AN subjects (Kaye et al. 2003). However, starvation in AN seems to be associated with a compensatory increase in postsynaptic 5-HT_{1A} receptor BP (Bailer et al. 2007a). Moreover, 5-HT_{2A} receptor binding is also positively related to HA in ill AN patients (Bailer et al. 2007a). Thus, when AN patients are forced to eat, it is likely that they have a relative increase in extracellular 5-HT concentrations in the brain, leading to an exaggeration of dysphoric mood. Thus, AN patients might pursue starvation in an attempt to avoid the dysphoric consequences of eating and spiral out of control.

There is an extensive literature associating the serotonergic systems and fundamental aspects of behavioral inhibition (Geyer 1996; Soubrie 1986). Reduced CSF 5-HIAA levels are associated with increased impulsivity and aggression in humans and nonhuman primates, whereas increased CSF 5-HIAA levels are related to behavioral inhibition (Fairbanks et al. 2001; Westergaard et al. 2003). Brainstem 5-HT_{1A} receptors inhibit stress-induced sympathetic activity and inhibit fight-or-flight behavioral responses, supporting a role for this receptor in behavioral inhibition and self-control (Johnson et al. 2004). Furthermore, recent animal studies also support modulation of impulse control via 5-HT_{1A} receptors through effects on catecholamine systems (Winstanley et al. 2005). Other studies have shown that blunted 5-HT_{1A} receptor number or function is associated with increased aggression (Cleare and Bond 2000; Coccaro et al. 1990). A recent study (Fischer et al. 2007) found a significant inverse relationship between dorsal raphe 5-HT_{1A} autoreceptor BP and bilateral amygdala reactivity. 5-HT_{1A} receptor function could contribute to behavioral inhibition in BN. In support of this, we found that for CW [^{11}C]WAY 100635 binding was related negatively to novelty seeking, whereas for REC BN [^{11}C]WAY 100635 binding was related positively to HA and negatively to sensation seeking. Moreover, novelty seeking and HA accounted for approximately 30% of the variance for [^{11}C]WAY 100635 binding in CW and REC BN, respectively (Bailer et al. 2010). The instruments used to assess behavior in humans tend to assess complex phenomena that are likely to be a composite of many traits, therefore confounding the understanding of how behaviors might be associated with a 5-HT receptor. For example, HA measures anxiety and behavioral inhibition, whereas novelty seeking measures exploration and impulsivity (Cloninger et al. 1994). Similarly, assessment of behavior in animals is complex. Thus, while considerable studies in animals associate 5-HT_{1A} receptor function with anxiety, most tests of anxiety in rodents are based in part on the approach/avoidance conflict between the innate tendency of an animal to explore a novel place and the tendency to avoid novel stimuli or environments (Groenink et al. 2003). Still, other studies show that various measures of 5-HT activity are related to measures of affective instability and impulsivity in ill BN subjects (Steiger et al. 2001a, b, c). Taken together, these data raise the possibility that 5-HT_{1A} receptor may contribute to the emergent ability to inhibit or self-control the expression of a number of behaviors related to stimulus seeking, anxiety, aggression, and impulsivity in BN.

A better understanding of how 5-HT contributes to symptoms in ED may help us advance beyond the trial and error system and develop new methods for identifying effective medications and psychological treatments in these often chronic and devastating disorders.

References

Amargos-Bosch M, Bortolozzi A, Puig MJS, Adell A, Celada P, Toth M, Mengod G, Artigas F (2004) Co-expression and in vivo interaction of serotonin$_{1A}$ and serotonin$_{2A}$ receptors in pyramidal neurons of prefrontal cortex. Cerebral Cortex 14:281–299

Anderluh MB, Tchanturia K, Rabe-Hesketh S, Treasure J (2003) Childhood obsessive-compulsive personality traits in adult women with eating disorders: defining a broader eating disorder phenotype. Am J Psychiatry 160:242–247

Anderson IM, Parry-Billings M, Newsholme EA, Fairburn CG, Cowen PJ (1990) Dieting reduces plasma tryptophan and alters brain 5-HT function in women. Psychol Med 20:785–91

Arango V, Underwood MD, Gubbi AV, Mann JJ (1995) Localized alterations in pre- and postsynaptic serotonin binding sites in the ventrolateral prefrontal cortex of suicide victims. Brain Res 688:121–33

Attia E, Schroeder L (2005) Pharmacologic treatment of anorexia nervosa: where do we go from here? Int J Eat Disord 37:S60–S63; discussion S87–S89

Attia E, Wolk S, Cooper T, Glasofer D, Walsh B (2005) Plasma tryptophan during weight restoration in patients with anorexia nervosa. Biol Psychiatry 57:674–678

Audenaert K, Van Laere K, Dumont F, Vervaet M, Goethals I, Slegers G, Mertens J, Van Heeringen C, Dierckx R (2003) Decreased 5-HT2a receptor binding in patients with anorexia nervosa. J Nucl Med 44:163–169

Bailer UF, Price JC, Meltzer CC, Mathis CA, Frank GK, Weissfeld L, Mcconaha CW, Henry SE, Brooks-Achenbach S, Barbarich NC, Kaye WH (2004) Altered 5-HT$_{2A}$ receptor binding after recovery from bulimia-type anorexia nervosa: relationships to harm avoidance and drive for thinness. Neuropsychopharmacology 29:1143–1155

Bailer UF, Frank GK, Henry SE, Price JC, Meltzer CC, Weissfeld L, Mathis CA, Drevets WC, Wagner A, Hoge J, Ziolko SK, Mcconana CW, Kaye WH (2005) Altered brain serotonin 5-HT$_{1A}$ receptor binding after recovery from anorexia nervosa measured by positron emission tomography and [^{11}C]WAY100635. Arch Gen Psychiatry 62:1032–1041

Bailer UF, Frank G, Henry S, Price J, Meltzer C, Mathis C, Wagner A, Thornton L, Hoge J, Ziolko SK, Becker C, Mcconaha C, Kaye WH (2007a) Exaggerated 5-HT$_{1A}$ but normal 5-HT$_{2A}$ receptor activity in individuals ill with anorexia nervosa. Biol Psychiatry 61: 1090–1099

Bailer UF, Frank G, Henry S, Price J, Meltzer CC, Becker C, Ziolko SK, Mathis C, Wagner A, Barbarich-Marsteller N, Putman K, Kaye WH (2007b) Serotonin transporter binding after recovery from eating disorders. Psychopharmacology 195:315–324

Bailer U, Bloss C, Frank G, Price J, Meltzer C, Mathis C, Geyer M, Wagner A, Becker C, Schork N, Kaye W (2010) 5-HT$_{1A}$ receptor binding is increased after recovery from bulimia nervosa compared to control women and is associated with behavioral inhibition in both groups Int J Eat Disord (in press)

Barr LC, Goodman WK, Price LH, Mcdougle CJ, Charney DS (1992) The serotonin hypothesis of obsessive compulsive disorder: implications of pharmacologic challenge studies. J Clin Psychiatry 53(Suppl):17–28

Berrettini W (2000) Genetics of psychiatric disease. Annu Rev Med 51:465–479
Blier P, De Montigny C (1999) Serotonin and drug-induced therapeutic responses in major depression, obsessive-compulsive and panic disorders. Neuropsychopharmacology 21:91S–98S
Blundell JE (1984) Serotonin and appetite. Neuropharmacology 23:1537–51
Bonhomme N, Esposito E (1998) Involvement of serotonin and dopamine in the mechanism of action of novel antidepressant drugs: a review. J Clin Psychopharmacol 18:447–454
Brewerton TD (1995) Toward a unified theory of serotonin dysregulation in eating and related disorders. Psychoneuroendocrinology 20:561–90
Brewerton TD, Hand LD, Bishop ER Jr (1993) The tridimensional personality questionnaire in eating disorder patients. Int J Eat Disord 14:213–218
Bulik CM, Sullivan PF, Joyce PR, Carter FA (1995) Temperament, character, and personality disorder in bulimia nervosa. J Nerv Ment Dis 183:593–8
Bulik CM, Sullivan PF, Fear JL, Joyce PR (1997) Eating disorders and antecedent anxiety disorders: a controlled study. Acta Psychiatr Scand 96:101–7
Bulik C, Sullivan PF, Tozzi F, Furberg H, Lichtenstein P, Pedersen NL (2006) Prevalence, heritability and prospective risk factors for anorexia nervosa. Arch Gen Psychiatry 63:305–312
Bulik C, Hebebrand J, Keski-Rahkonen A, Klump K, Reichborn-Kjennerud KS, Mazzeo S, Wade T (2007) Genetic epidemiology, endophenotypes, and eating disorder classification. Int J Eat Disord Suppl:S52–S60
Burnet PW, Eastwood SL, Harrison PJ (1997) [3H]WAY-100635 for 5-HT1A receptor autoradiography in human brain: a comparison with [3H]8-OH-DPAT and demonstration of increased binding in the frontal cortex in schizophrenia. Neurochem Int 30:565–74
Bush G, Frazier JA, Rauch SL, Seidman LJ, Whalen PJ, Jenike MA, Rosen BR, Biederman J (1999) Anterior cingulate cortex dysfunction in attention-deficit/hyperactivity disorder revealed by fMRI and the Counting Stroop. Biol Psychiatry 45:1542–1552
Cervo L, Mocaer E, Bertaglia A, Samanin R (2000) Roles of 5-HT_{1A} receptors in the dorsal raphe and dorsal hippocampus in anxiety assessed by the behavioral effects of 8-OH-DPAT and S 15535 in a modified Geller-Seifter conflict model. Neuropharmacology 39:1037–43
Charney DS, Deutch A (1996) A functional neuroanatomy of anxiety and fear: implications for the pathophysiology and treatment of anxiety disorders. Crit Rev Neurobiol 10:419–46
Cleare A, Bond AJ (2000) Ipaspirone challenge in aggressive men shows an inverse correlation between 5HT1A receptor function and aggression. Psychopharmacology (Berl) 148:344
Cloninger CR (1987) A systematic method for clinical description and classification of personality variants. A proposal. Arch Gen Psychiatry 44:573–88
Cloninger CR, Przybeck TR, Svrakic DM, Wetzel RD (1994) The temperament and character inventory (TCI): a guide to its development and use. Center for Psychobiology of Personality, Washington University, St. Louis, MO, USA
Coccaro EF, Gabriel S, Siever LJ (1990) Buspirone challenge: preliminary evidence for a role of 5HT1A receptors in behavioral irritability in personality disorderd patients. Psychopharmacol Bull 26:285
Cooper SJ (1996) Cholecystokinin modulation of serotonergic control of feeding behavior. Ann NY Acad Sci 780:213–222
Critchley M (1953) The parietal lobes. Hafner, New York
De Vry J, Schreiber R (2000) Effects of selected serotonin 5-HT(1) and 5-HT(2) receptor agonists on feeding behavior: possible mechanisms of action. Neurosci Biobehav Rev 24:341–53
Deep AL, Nagy LM, Weltzin TE, Rao R, Kaye WH (1995) Premorbid onset of psychopathology in long-term recovered anorexia nervosa. Int J Eat Disord 17:291–297
Devinsky O, Morrell MJ, Vogt BA (1995) Contributions of anterior cingulate cortex to behaviour. Brain 118:279–306
Drevets W, Thase M, Moses-Kolko E, Price J, Frank E, Kupfer D, Mathis C (2007) Serotonin-1A receptor imaging in recurrent depression: replication and literature review. Nucl Med Biol 34:865–877

Fairbanks L, Melega W, Jorgensen M, Kaplan J, Mcguire M (2001) Social impulsivity inversely associated with CSF 5-HIAA and fluoxetine exposure in vermet monkeys. Neuropschopharmacology 24:370–378

Favaro A, Caregaro L, Burlina AB, Santonastaso P (2000) Tryptophan levels, excessive exercise, and nutritional status in anorexia nervosa. Psychosom Med 62:535–8

Fernstrom JD, Wurtman RJ (1972) Brain serotonin content: physiological regulation by plasma neutral amino acids. Science 178:414–6

File SE, Kenny PJ, Cheeta S (2000) The role of the dorsal hippocampal serotonergic and cholinergic systems in the modulation of anxiety. Pharmacol Biochem Behav 66:65–72

Fisher P, Meltzer C, Ziolko S, Price J, Hariri AR (2007) Capacity for 5-HT1A-mediated autoregulation predicts amygdala reactivity. Nat Neurosci 9:1362–1363

Fluoxetine Bulimia Nervosa Collaborative Study Group (1992) Fluoxetine in the treatment of bulimia nervosa. A multicenter, placebo-controlled, double-blind trial. Arch Gen Psychiatry 49:139–147

Frank GK, Kaye WH, Meltzer CC, Price JC, Greer P, Mcconaha C, Skovira K (2002) Reduced 5-HT2A receptor binding after recovery from anorexia nervosa. Biol Psychiatry 52:896–906

Frankle W, Lombardo I, New AS, Goodman M, Talbot P, Huang Y, Hwang D, Slifstein M, Curry S, Abi-Dargham A, Lauruelle M, Siever L (2005) Brain serotonin transporter distribution in subjects with impulsive aggressivity: a positron emission study with [11C]McN 5652. Am J Psychiatry 162:915–923

Galusca B, Costes N, Zito N, Peyron R, Bossu C, Lang F, Le Bars D, Estour B (2008) Organic background of restrictive-type anorexia nervosa suggested by increased serotonin(1A) receptor binding in right frontotemporal cortex of both lean and recovered patients: [(^{18})F]MPPF PET scan study. Biol Psychiatry 64:1009–1013

Garner DM (1991) The eating disorders inventory-2 manual. In: Psychological Assessment Resources. Psychological Assessment Resources, Odessa, FL, USA

Geyer MA (1996) Serotonergic functions in arousal and motor activity. Behav Brain Res 73:31–35

Godart NT, Flament MF, Lecrubier Y, Jeammet P (2000) Anxiety disorders in anorexia nervosa and bulimia nervosa: co-morbidity and chronology of appearance. Eur Psychiatry 15:38–45

Godart NT, Flament MF, Perdereau F, Jeammet P (2002) Comorbidity between eating disorders and anxiety disorders: a review. Int J Eat Disord 32:253–270

Goethals I, Vervaet M, Audenaert K, Van De Wiele C, Ham H, Vandecapelle M, Slegers G, Dierckx RA, Van Heeringen C (2004) Comparison of cortical 5-HT2A receptor binding in bulimia nervosa patients and healthy volunteers. Am J Psychiatry 161:1916–1918

Goethals I, Vervaet M, Audenaert K, Jacobs F, Ham H, Van Heeringen C (2007) Does regional brain perfusion correlate with eating disorder symptoms in anorexia and bulimia nervosa. J Psychiatr Res 41:1005–1011

Grahame-Smith DG (1992) Serotonin in affective disorders. Int Clin Psychopharmacol 6(Suppl 4):5–13

Groenink L, Van Bogaert M, Van Der Gugten J, Oosting R (2003) 5-HT$_{1A}$ receptor and 5-HT$_{1B}$ receptor knockout mice in stress and anxiety paradigms. Behav Pharmacol 14:369–383

Hajos M, Gartside SE, Varga V, Sharp T (2003) In vivo inhibition of neuronal activity in the rat ventromedial prefrontal cortex by midbrain-raphe nuclei: role of 5-HT$_{1A}$ receptors. Neuropharmacology 45:72–81

Hudson J, Hiripi E, Pope H, Kessler R (2007) The prevalence and correlates of eating disorders in the National Comorbidity Survey Replication. Biol Psychiatry 61:348–358

Jimerson DC, Lesem MD, Hegg AP, Brewerton TD (1990) Serotonin in human eating disorders. Ann NY Acad Sci 600:532–44

Jimerson DC, Lesem MD, Kaye WH, Brewerton TD (1992) Low serotonin and dopamine metabolite concentrations in cerebrospinal fluid from bulimic patients with frequent binge episodes. Arch Gen Psychiatry 49:132–138

Johnson P, Lightman S, Lowry C (2004) A functional subset of serotonergic neruons in the rat ventrolateral periaqueductal gray implicated in the inhibition of sympathoexcitation and panic. Ann NY Acad Sci 1018:58–64
Kaye WH, Weltzin TE (1991) Serotonin activity in anorexia and bulimia nervosa: relationship to the modulation of feeding and mood. J Clin Psychiatry 52(Suppl):41–48
Kaye WH, Ebert MH, Raleigh M, Lake R (1984) Abnormalities in CNS monoamine metabolism in anorexia nervosa. Arch Gen Psychiatry 41:350–5
Kaye WH, Gwirtsman HE, George DT, Jimerson DC, Ebert MH (1988) CSF 5-HIAA concentrations in anorexia nervosa: reduced values in underweight subjects normalize after weight gain. Biol Psychiatry 23:102–5
Kaye WH, Gwirtsman HE, George DT, Jimerson DC, Ebert MH, Lake CR (1990) Disturbances of noradrenergic systems in normal weight bulimia: relationship to diet and menses. Biol Psychiatry 27:4–21
Kaye WH, Gwirtsman HE, George DT, Ebert MH (1991a) Altered serotonin activity in anorexia nervosa after long-term weight restoration. Does elevated cerebrospinal fluid 5-hydroxyindoleacetic acid level correlate with rigid and obsessive behavior? Arch Gen Psychiatry 48:556–62
Kaye WH, Weltzin TE, Hsu LK, Bulik CM (1991b) An open trial of fluoxetine in patients with anorexia nervosa. J Clin Psychiatry 52:464–71
Kaye W, Gendall K, Strober M (1998a) Serotonin neuronal function and selective serotonin reuptake inhibitor treatment in anorexia and bulimia nervosa. Biol Psychiatry 44:825–38
Kaye WH, Greeno CG, Moss H, Fernstrom J, Fernstrom M, Lilenfeld LR, Weltzin TE, Mann JJ (1998b) Alterations in serotonin activity and psychiatric symptomatology after recovery from bulimia nervosa. Arch Gen Psychiatry 55:927–935
Kaye WH, Frank GK, Meltzer CC, Price JC, Mcconaha CW, Crossan PJ, Klump KL, Rhodes L (2001a) Altered serotonin 2A receptor activity in women who have recovered from bulimia nervosa. Am J Psychiatry 158:1152–1155
Kaye WH, Nagata T, Weltzin TE, Hsu LK, Sokol MS, Mcconaha C, Plotnicov KH, Weise J, Deep D (2001b) Double-blind placebo-controlled administration of fluoxetine in restricting- and restricting-purging-type anorexia nervosa. Biol Psychiatry 49:644–52
Kaye WH, Barbarich NC, Putnam K, Gendall KA, Fernstrom J, Fernstrom M, Mcconaha CW, Kishore A (2003) Anxiolytic effects of acute tryptophan depletion in anorexia nervosa. Int J Eat Disord 33:257–267
Kaye W, Bulik C, Thornton L, Barbarich N, Masters K, Fichter M, Halmi K, Kaplan A, Strober M, Woodside DB, Bergen A, Crow S, Mitchell J, Rotondo A, Mauri M, Cassano G, Keel PK, Plotnicov K, Pollice C, Klump K, Lilenfeld LR, Devlin B, Quadflieg R, Berrettini WH (2004) Comorbidity of anxiety disorders with anorexia and bulimia nervosa. Am J Psychiatry 161:2215–2221
Kaye W, Frank G, Bailer UF, Henry S, Meltzer CC, Price J, Mathis C, Wagner A (2005) Serotonin alterations in anorexia and bulimia nervosa: new insights from imaging studies. Physiol Behav 85:73–81
Kaye W, Strober M, Jimerson D (2008) The neurobiology of eating disorders. In: Charney D, Nestler E (eds) The neurobiology of mental illness, 3rd edn. Oxford Press, New York
Kaye W, Fudge J, Paulus M (2009) New insight into symptoms and neurocircuit function of anorexia nervosa. Nat Rev Neurosci 10:573–584
Kendler KS, Maclean C, Neale M, Kessler R, Heath A, Eaves L (1991) The genetic epidemiology of bulimia nervosa. Am J Psychiatry 148:1627–37
Kendler KS, Walters EE, Neale MC, Kessler RC, Heath AC, Eaves LJ (1995) The structure of the genetic and environmental risk factors for six major psychiatric disorders in women. Phobia, generalized anxiety disorder, panic disorder, bulimia, major depression, and alcoholism. Arch Gen Psychiatry 52:374–83
Kingston K, Szmukler G, Andrewes D, Tress B, Desmond P (1996) Neuropsychological and structural brain changes in anorexia nervosa before and after refeeding. Psychol Med 26:15–28
Kleifield EI, Sunday S, Hurt S (1993) Psychometric validation of the Tridimensional Personality Questionnaire: application to subgroups of eating disorders. Compr Psychiatry 34:249–53

Klump KL, Bulik CM, Pollice C, Halmi KA, Fichter MM, Berrettini WH, Devlin B, Strober M, Kaplan A, Woodside DB, Treasure J, Shabbout M, Lilenfeld LR, Plotnicov KH, Kaye WH (2000) Temperament and character in women with anorexia nervosa. J Nerv Ment Dis 188:559–67

Koskela AK-R, Sihvola E, Kauppinen T, Kaprio JA, Rissanen A (2007) Serotonin transporter binding of [123I]ADAM in bulimic women, their healthy twin sisters, and healthy women: a SPET study. BMC Psychiatry 21:7–9

Krebs-Thomson K, Geyer MA (1998) Evidence for a functional interaction between 5-HT_{1A} and 5-HT_{2A} receptors in rats. Psychopharmacology 140:69–74

Kuikka JT, Tammela L, Karhunen L, Rissanen A, Bergstrom KA, Naukkarinen H, Vanninen E, Karhu J, Lappalainen R, Repo-Tiihonen E, Tiihonen J, Uusitupa M (2001) Reduced serotonin transporter binding in binge eating women. Psychopharmacology 155:310–314

Lanfumey L, Hamon M (2000) Central 5-HT_{1A} receptors: regional distribution and functional characteristics. Nucl Med Biol 27:429–435

Lanzenberger R, Mitterhauser M, Spindelegger C, Wadsak W, Klein N, Mien L, Holik A, Attarbaschi T, Mossaheb N, Sacher J, Geiss-Granadia T, Keletter K, Kasper S, Tauscher J (2007) Reduced serotonin-1A receptor binding in social anxiety disorder. Biol Psychiatry 61:1081–1089

Leibowitz SF, Shor-Posner G (1986) Brain serotonin and eating behavior. Appetite 7:1–14

Lilenfeld LR, Kaye WH, Greeno CG, Merikangas KR, Plotnicov K, Pollice C, Rao R, Strober M, Bulik CM, Nagy L (1998) A controlled family study of anorexia nervosa and bulimia nervosa: psychiatric disorders in first-degree relatives and effects of proband comorbidity. Arch Gen Psychiatry 55:603–610

Lilenfeld L, Wonderlich S, Riso LP, Crosby R, Mitchell J (2006) Eating disorders and personality: a methodological and empirical review. Clin Psychol Rev 26:299–320

Mann JJ (1999) Role of the serotonergic system in the pathogenesis of major depression and suicidal behavior. Neuropsychopharmacology 21:99S–105S

Matsubara S, Arora RC, Meltzer HY (1991) Serotonergic measures in suicide brain: 5-HT1A binding sites in frontal cortex of suicide victims. J Neural Transm Gen Sect 85:181–94

Mesulam M (1981) A cortical network for directed attention and unilateral neglect. Ann Neurol 10:309–325

Miller JB, Ogden T, Oquendo M, Sullivan G, Mann J, Parsey R (2009) Elevated serotonin 1A binding in remitted major depressive disorder: evidence for a trait biological abnormality. Neuropsychopharmacology 34:2275–2284

Mitchell J, Crow S, Peterson CW, Crosby R (1998) Feeding laboratory studies in patients with eating disorders: A review. Int J Eat Disord 24:115–124

Nash J, Sargent P, Rabiner E, Hood SA, Potokar J, Grasby PN (2008) Serotonin 5-HT1A receptor binding in people with panic disorder: positron emission tomography study. Br J Psychiatry 193:229–234

Neumeister A, Brain E, Nugent A, Carson R, Bonne O, Lucnekbaugh D, Eckelman W, Herschovitch P, Charney D, Drevets W (2004) Reduced serotinin type 1_A receptor binding in panic disorder. J Neurosci 24:589–591

O'dwyer AM, Lucey JV, Russell GF (1996) Serotonin activity in anorexia nervosa after long-term weight restoration: response to D-fenfluramine challenge. Psychol Med 26:353–359

Olivier B, Pattij T, Wood S, Oosting R, Sarnyai Z, Toth M (2001) The 5-HT_{1A} receptor knockout mouse and anxiety. Behav Pharmacol 12:439–450

Petty F, Davis LL, Kabel D, Kramer GL (1996) Serotonin dysfunction disorders: a behavioral neurochemistry perspective. J Clin Psychiatry 57:11–6

Romano SJ, Halmi KA, Sarkar NP, Koke SC, Lee JS (2002) A placebo-controlled study of fluoxetine in continued treatment of bulimia nervosa after successful acute fluoxetine treatment. Am J Psychiatry 159:96–102

Santana N, Bortolozzi A, Serrats J, Mengod G, Artigas F (2004) Expression of $serotoinin_{1A}$ and $serotonin_{2A}$ receptor in pyramidal and GABAergic neurons of the rat prefrontal cortex. Cereb Cortex 14:1100–1109

Saudou F, Hen R (1994) 5-Hydroxytryptamine receptor subtypes in vertebrates and invertebrates. Neurochem Int 25:503–32
Savitz J, Lucki I, Drevets W (2009) 5-HT(1A) receptor function in major depressive disorder. Prog Neurobiol 88:17–31
Schweiger U, Warnhoff M, Pahl J, Pirke KM (1986) Effects of carbohydrate and protein meals on plasma large neutral amino acids, glucose, and insulin plasma levels of anorectic patients. Metabolism 35:938–43
Silberg J, Bulik C (2005) Developmental association between eating disorders symptoms and symptoms of depression and anxiety in juvenile twin girls. J Child Psychol Psychiatr 46:1317–1326
Simansky KJ (1996) Serotonergic control of the organization of feeding and satiety. Behav Brain Res 73:37–42
Simansky K, Dave K, Inemer B, Nicklous D, Padron J, Aloyo V, Romano A (2004) A 5-HT2C agonist elicits hyperactivity and oral dyskinesia with hypophagia in rabbits. Physiol Behav 82:97–107
Soubrie P (1986) Reconciling the role of central serotonin neurons in human and animal behavior. Behav Brain Sci 9:319
Staley J, Malison R, Innis R (1998) Imaging of the serotonergic system: interactions of neuroanatomical and functional abnormalities of depression. Biol Psychiatry 44:534–549
Steiger H, Gauvin L, Israel M, Lkoerner N, Ng Ying Kin N, Paris J, Young SN (2001a) Association of serotonin and cortisol indices with childhood abuse in bulimia nervosa. Arch Gen Psychiatry 58:837–843
Steiger H, Koerner N, Engelberg MJ, Israel M, Ng Ying Kin NM, Young SN (2001b) Self-destructiveness and serotonin function in bulimia nervosa. Psychiatry Res 103:15–26
Steiger H, Young SN, Kin NM, Koerner N, Israel M, Lageix P, Paris J (2001c) Implications of impulsive and affective symptoms for serotonin function in bulimia nervosa. Psychol Med 31:85–95
Steiger H, Joober R, Israel M, Young S, Ng Ying Kin N, Gauvin L, Bruce K, Joncas J, Torkaman-Zehi A (2005a) The 5HTTLPR polymorphism, psychopathological symptoms, and platelet [3H-] paroxetine binding in bulimic syndromes. Int J Eat Disord 37:57–60
Steiger H, Richardson J, Israel M, Ng Ying Kin N, Bruce K, Mansour S, Marie Parent A (2005b) Reduced density of platelet-binding sites for [3H]paroxetine in remitted bulimic women. Neuropsychopharmacology 30:1028–1032
Steiger H, Richardson J, Joober R, Gauvin L, Israel M, Bruce K, Ying Kin N, Howard H, Young S (2007) The 5HTTLPR polymorphism, prior maltreatment and dramatic-erratic personality manifestations in women with blimic syndromes. J Psychiatr Res 32:354–362
Stein DJ, Hollander E, Liebowitz MR (1993) Neurobiology of impulsivity and the impulse control disorders. J Neuropsychiatry Clin Neurosci 5:9–17
Steinglass JE, Walsh T (2004) Psychopharmacology of anorexia nervosa, bulimia nervosa, and binge eating disorder. In: Brewerton TD (ed) Clinical handbook of eating disorders: an integrated approach. Marcel Dekker, New York
Stice E (2002) Risk and maintenance factors for eating pathology: a meta-analytic review. Pychopharm Bull 128:825–848
Stockmeier CA (1997) Neurobiology of serotonin in depression and suicide. Ann NY Acad Sci 836:220–32
Strober M (1995) Family-genetic perspectives on anorexia nervosa and bulimia nervosa. In: Brownell K, Fairburn C (eds) Eating disorders and obesity-a comprehensive handbook. Guilford, New York
Strober M, Freeman R, Morrell W (1997) The long-term course of severe anorexia nervosa in adolescents: survival analysis of recovery, relapse, and outcome predictors over 10–15 years in a prospective study. Int J Eat Disord 22:339–360
Strober M, Freeman R, Lampert C, Diamond J, Kaye W (2000) Controlled family study of anorexia nervosa and bulimia nervosa: evidence of shared liability and transmission of partial syndromes. Am J Psychiatry 157:393–401

Strupp BJ, Weingartner H, Kaye W, Gwirtsman H (1986) Cognitive processing in anorexia nervosa. A disturbance in automatic information processing. Neuropsychobiology 15:89–94

Takenouchi K, Nishijo H, Uwano T, Tamura R, Takigawa M, Ono T (1999) Emotional and behavioral correlates of the anterior cingulate cortex during associative learning in rats. Neuroscience 93:1271–1287

Tauscher J, Pirker W, Willeit M, De Zwaan M, Bailer U, Neumeister A, Asenbaum S, Lennkh C, Praschak-Rieder N, Brücke T, Kasper S (2001) [^{123}I]beta-CIT and single photon emission computed tomography reveal reduced brain serotonin transporter availability in bulimia nervosa. Biol Psychiatry 49:326–332

Tiihonen J, Keski-Rahkonen A, Lopponen M, Muhonen M, Kajander J, Allonen T, Nagren K, Hietala J, Rissanen A (2004) Brain serotonin 1A receptor binding in bulimia nervosa. Biol Psychiatry 55:871–873

Tollefson GD (1995) Selective serotonin reuptake inhibitors. In: Schatzberg AF, Nemeroff CB (eds) Textbook of psychopharmacology. American Psychiatric Press, Washington DC, USA

Treasure J, Campbell I (1994) The case for biology in the aetiology of anorexia nervosa. Psychol Med 24:3–8

Tremblay P, Blier P (2006) Catecholaminergic strategies for the treatment of major depression. Curr Drug Targets 7:149–158

Tucker DM, Luu P, Pribram KH (1995) Social and emotional self-regulation. Ann NY Acad Sci 769:213–39

Vitousek K, Manke F (1994) Personality variables and disorders in anorexia nervosa and bulimia nervosa. J Abnorm Psychol 103:137–147

Vollenweider F, Vontobel P, Hell D, Leenders K (1999) 5-HT modulation of dopamine release in basal ganglia in psilocybin-induced psychosis in man–a PET study with [11C]raclopride. Neuropsychopharmacology 20:424–433

Wagner A, Ruf M, Braus DF, Schmidt MH (2003) Neuronal activity changes and body image distortion in anorexia nervosa. NeuroReport 14:2193–2197

Wagner A, Barbarich N, Frank G, Bailer U, Weissfeld L, Henry S, Achenbach S, Vogel V, Plotnicov K, Mcconaha C, Kaye W, Wonderlich S (2006) Personality traits after recovery from eating disorders: Do subtypes differ? Int J Eat Disord 39:276–284

Walsh BT (1991) Psychopharmacologic treatment of bulimia nervosa. J Clin Psychiatry 52 (Suppl):34–38

Walsh BT, Hadigan CM, Devlin MJ, Gladis M, Roose SP (1991) Long-term outcome of antidepressant treatment for bulimia nervosa. Am J Psychiatry 148:1206–1212

Walsh B, Kaplan A, Attia E, Olmsted M, Parides M, Carter J, Pike K, Devlin M, Woodside B, Roberto C, Rocket W (2006) Fluoxetine after weight restoration in anorexia nervosa: a randomized controlled trial. JAMA 295:2605–2612

Walters EE, Kendler KS (1995) Anorexia nervosa and anorexic-like syndromes in a population-based female twin sample. Am J Psychiatry 152:64–71

Ward A, Brown N, Lightman S, Campbell IC, Treasure J (1998) Neuroendocrine, appetitive and behavioural responses to d-fenfluramine in women recovered from anorexia nervosa. Br J Psychiatry 172:351–358

Westergaard G, Suomi S, Chavanne T, Houser L, Hurley A, Cleveland A, Snoy P, Higley J (2003) Physiological correlates of aggression and impulsivity in free-ranging female primates. Neuropsychopharmacology 28:1045–1055

Winstanley CA, Theobald DE, Dalley JW, Robbins TW (2005) Interactions between serotonin and dopamine in the control of impulsive choice in rats: Therapeutic implications for impulse control disorders. Neuropschopharmacology 30:669–682

Young SN, Gauthier S (1981) Effect of tryptophan administration on tryptophan, 5-hydroxyindoleacetic acid and indoleacetic acid in human lumbar and cisternal cerebrospinal fluid. J Neurol Neurosurg Psychiatry 44:323–327

Dopamine-Based Reward Circuitry Responsivity, Genetics, and Overeating

Eric Stice, Sonja Yokum, David Zald, and Alain Dagher

Contents

Abstract Data suggest that low levels of dopamine D2 receptors and attenuated responsivity of dopamine-target regions to food intake is associated with increased eating and elevated weight. There is also growing (although mixed) evidence that genotypes that appear to lead to reduced dopamine signaling (e.g., DRD2, DRD4, and DAT) and certain appetite-related hormones and peptides (e.g., ghrelin, orexin A, leptin) moderate the relation between dopamine signaling, overeating, and obesity. This chapter reviews findings from studies that have investigated the relation between dopamine functioning and food intake and how certain genotypes and appetite-related hormones and peptides affect this relation.

E. Stice (✉) and S. Yokum
Oregon Research Institute, Eugene, OR, USA
e-mail: estice@ori.org
D. Zald
Vanderbilt University, Nashville, TN, USA
A. Dagher
McGill University, Montreal, QC, Canada

R.A.H. Adan and W.H. Kaye (eds.), *Behavioral Neurobiology of Eating Disorders*,
Current Topics in Behavioral Neurosciences 6, DOI 10.1007/7854_2010_89

Keywords Dopamine · Food reward · Genetics · Overeating · Obesity

1 Introduction: Dopamine Reward and Overeating

Considerable research has implicated dopamine-based meso-limbic and meso-cortical circuitries in eating behavior. Microdialysis studies with animals reveal that consumption of high-sugar or high-fat food results in dopamine release in the nucleus accumbens (Avena et al. 2008). Consumption of a pleasant meal in humans results in dopamine release in the dorsal striatum and the magnitude of release correlates with ratings of meal pleasantness (Small et al. 2003). Furthermore, brain dopamine release increases during the anticipation of food intake (Volkow et al. 2002).

Incentive salience theory posits that consummatory and anticipatory reward operates in tandem in the development of reinforcing value of food, but that with repeated intake of a food, hedonics decrease, while anticipatory reward increases (Robinson and Berridge 2000). By conditioning, food images and cues come to activate reward circuitry, leading to food cravings, and possibly to overeating and weight gain (Ritchie and Noble 2003). This motivational state, mediated by increased dopamine in reward circuitry, draws attention to food and results in attentional bias for food stimuli (Berridge and Robinson 1998; Bowirrat and Oscar-Berman 2005). After conditioning, the reward value of food shifts from food intake to anticipated food intake. For example, naive monkeys that had not experienced rewards in a setting showed increased firing of dopamine neurons in midbrain only in response to food taste; however, after conditioning, increased firing began to precede reward delivery and eventually maximal firing was elicited by the conditioned stimuli that predicted food reward versus actual food receipt (Schultz et al. 1993). Kiyatkin and Gratton (1994) found that the greatest dopamine neuronal activity occurred in an anticipatory fashion as rats approached and pressed a bar that produced food reward, with activity decreasing as the rat received and ate the food. Indeed, although DA release occurs in response to unexpected palatable food, Blackburn et al. (1989) found that dopamine release was greater in the nucleus accumbens of rats after presentation of a conditioned stimulus that usually signaled food receipt than after delivery of an unexpected meal.

2 Relation of Dopamine Functioning to Overeating and Obesity

Increasing data suggests that variations in dopamine receptors and dopamine release play a role in overeating and obesity. Animal research implicates that when exposed to the same high-fat diet, mice with lower dopamine D2 receptor density in the putamen show more weight gain than mice with higher D2 receptor density in this region (Huang et al. 2006). A transgenic mouse with persistently elevated striatal dopamine displayed increased food intake and increased incentive motivation to consume food (Orosco et al. 1996). Compared to lean Zucker rats, obese Zucker rats, which have defective leptin receptor function, have fewer D2 receptors and reduced hypothalamic dopamine activity when fasting, but release more phasic dopamine when eating and do not stop eating in response to insulin and glucose administration (Orosco et al. 1996). In addition, D2 receptor blockade causes obese, but not lean, Zucker rats to overeat (Fetissov et al. 2002), implying that blockade of already low D2 receptor availability may increase the incentive value of food in obese rats. Obesity-prone (DIO) Sprague-Dawley rats, compared to the obesity-resistant strain, also have reduced dopamine turnover in the hypothalamus compared to the obesity-resistant strain before they become obese, and develop obesity only when given a palatable high-energy diet (Levin and Meynell 2002). When exposed to the same high-fat diet, mice with lower D2 receptor density in the putamen show more weight gain than mice with higher D2 receptor density in this region (Huang et al. 2006).

Positron emission tomography (PET) studies suggest that obese versus lean adults show less striatal D2 receptor binding (Volkow et al. 2008), and that D2 receptors are reduced in the striatum in morbidly obese individuals in proportion to their body mass (Wang et al. 2001). Lower D2 striatal receptor density is also correlated with lower resting metabolism in the prefrontal cortex, which may contribute to overeating because of reduced inhibitory control (Volkow et al. 2008). These results echo evidence that substance abuse is associated with low D2 receptor density and less sensitivity of corticolimbic reward circuitry in response to general reinforcers, such as money (Goldstein et al. 2007), and also evidence that experimentally increased expression of D2 receptors reduces alcohol self-administration in rats (Thanos et al. 2001).

Using functional magnetic resonance imaging (fMRI), we found that obese relative to lean adolescents show weaker activation in dopamine-target regions (caudate and putamen) in response to food intake, and that those showing reduced striatal response to food intake who had an A1 allele of the TaqIA (DRD2/ANKK1) gene, which is associated with lower striatal D2 receptor density and attenuated striatal dopamine signaling, were at elevated risk for weight gain over a 1-year follow-up (Stice et al. 2008a,b). Because BOLD responses were measured, we can only speculate that the effects reflect lower D2 receptor density. This interpretation seems reasonable because the presence of the Taq1A A1 allele, which has been associated with reduced dopaminergic signaling in several postmortem and PET studies (Ritchie and Noble 2003; Tupala et al. 2003), significantly moderated the observed BOLD effects. Yet, the blunted striatal activation may also implicate altered dopamine release from food intake rather than a lower D2 receptor density (Knutson and Gibbs 2007). Research has also found that obese rats (due to a cafeteria diet) had lower extracellular dopamine and lower amphetamine-induced dopamine release (Geiger et al. 2009). Furthermore, obese individuals show greater activation in the insula and anterior cingulate cortex in response to food receipt relative to healthy controls (Stice et al. 2008b). Although the above-mentioned data suggest that obese versus lean individuals have fewer striatal D2 receptors available and process food reward differently, there are key gaps in the literature. First, existing studies in humans have used either fMRI measures of responses to food or PET measures of dopamine binding, but have never measured both in the same participants. Thus, it is unclear to what extent the fMRI results are dependent on dopamine mechanisms. It is possible that the blunted striatal activation may also implicate altered dopamine release from food intake rather than a lower D2 receptor density. It is theorized that consumption of high-fat, high-sugar diet leads to downregulation of D2 receptors (Small et al. 2003) and reduced dopamine turnover (Davis et al. 2008). In support, animal studies suggest that repeated intake of sweet and fatty foods results in downregulation of postsynaptic D2 receptors, increased D1 receptor binding, and decreased D2 sensitivity (Bello et al. 2002; Kelley et al. 2003), changes that also occur in response to chronic substance abuse. As well, experimental exposure to a high-fat diet results in reduced response to psychostimulants and meso-limbic dopamine turnover (Davis et al. 2008; Wellman et al. 2007). Paradoxically, regular exposure to high-sugar food results in subsequent increases in sugar intake (Avena et al. 2008), despite the changes to dopamine circuitry observed by others from repeated sugar intake (Bello et al. 2002; Kelley et al. 2003). To date, no PET imaging study has tested whether obese humans show greater dopamine release in response to food intake relative to lean humans. Accordingly, it will be important to investigate dopamine release in response to food intake in obese versus lean individuals. A second gap in the literature is that research has not tested whether abnormal responsivity of reward circuitry increases risk for future weight gain. Therefore, it is important for future studies to characterize dopaminergic synaptic function in humans and to examine its effects on future weight gain. Finally, striatal dopamine is thought to encode a reward prediction error signal (the difference between actual and expected reward), as demonstrated by numerous animal and human studies (Bayer and Glimcher 2005). Therefore, reduced BOLD signal upon food receipt may result from increased reward anticipation and hence a reduced reward prediction error signal (Hare et al. 2008).

3 Genetic Variation in Dopaminergic Reward in Humans

Given that dopamine plays a key role in reward circuitry and is involved in feeding behavior (Small et al. 2003; Yamamoto 2006), it follows that genetic polymorphisms that affect the availability and release of dopamine and the expression or functioning of dopamine receptors could influence the risk for weight gain. Research has identified several genes that influence dopamine functioning, including those that affect dopamine receptors, transport, and breakdown.

3.1 *The* Taq1A *Polymorphism of the DRD2 Gene*

To date, the most empirical support has emerged for the *Taq1A* polymorphism of the DRD2 gene. The *Taq1A* polymorphism has three allelic variants: A1/A1, A1/A2, and A2/A2. *Taq1A* was originally thought to be located in the 3′-untranslated region of DRD2, but it actually resides in the neighboring ANKK1 gene (Fossella et al. 2006). Postmortem and PET studies suggest that individuals with one or two copies of the A1 allele have 30–40% fewer D2 receptors when compared to those without an A1 allele (Ritchie and Noble 2003; Tupala et al. 2003). There is emerging evidence that the relation between abnormalities in food reinforcement and amount eaten is moderated by the A1 allele. Epstein and colleagues (Epstein et al. 2004, 2007) found an interaction between A1 allele and individual differences in food reward in adults, such that the greatest ad lib food intake occurred in those who reported more reinforcement from food and who had the A1 allele. Individuals with the A1 allele report greater food craving and work harder for food than those without this allele (Epstein et al. 2007; Comings et al. 1993). Those with the A1 allele versus those without it are also more likely to abuse drugs and to experience greater cue-induced cravings for cigarettes and heroin (Li et al. 2006; Erblich et al. 2005).

Some (Spitz et al. 2000; Thomas et al. 2001), but not all (Southon et al. 2003), studies have found positive correlations between the A1 TaqIA allele and body mass index (BMI; weight in kg/height in m^2). Stice et al. (2008a) found that the relation between fMRI activation in response to anticipated and actual eating (e.g., weaker caudate activation in response to milkshake receipt) was significantly more strongly related to current BMI and future weight gain over a 1-year follow-up in those with the A1 allele versus those without it. Using another fMRI experimental paradigm, Stice et al. (2010) found that weaker activation of the frontal operculum, lateral orbitofrontal cortex, and striatum in response to imagined eating of appetizing foods, versus imagined eating of less palatable foods or drinking water, predicted elevated weight gain for those with the A1 allele. Kirsch et al. (2006) found an increase of reward circuitry activation in response to the dopamine D2 receptor agonist bromocriptine in individuals with an A1 allele, but not A2/A2 carriers, and furthermore that the former showed improved behavioral performance in response to bromocriptine, implying that individuals with an A1 allele are more sensitive to dopamine agonists. Collectively, these data suggest that low D2 receptor density associated with the A1 *TaqIA* DRD2 allele is related to increased ad lib food intake, weight gain, and risk for obesity overeating and risk for obesity.

3.2 *Variants in the DRD4 Gene*

DRD4 is a postsynaptic dopamine receptor whose principal action is to inhibit the second messenger adenylate cyclase. D4 receptors are predominantly localized in areas that are innervated by meso-cortical projections from the ventral tegmental area (VTA), including the prefrontal cortex, cingulate gyrus, and insula (Noain et al. 2006). The 7-repeat or longer allele of this gene (DRD4-L) has been linked to reduced dopamine functioning in an in vitro study (Asghari et al. 1995), to weaker response to dopamine-stimulating drugs (Hamarman et al. 2004), and to less dopamine release in the ventral caudate and nucleus accumbens after nicotine use (Brody et al. 2006), suggesting it may affect reward sensitivity. Adults with the DRD4-L allele versus without it have shown stronger food cravings in response to food cues (Sobik et al. 2005), stronger smoking cravings and fMRI activation of the superior frontal gyrus and insula in response to smoking cues (McClernon et al. 2007), stronger alcohol cravings and fMRI activation in the OFC, anterior cingulate gyrus, and striatum in response to alcohol (Hutchison et al. 2002; Filbey et al. 2008), and greater heroin craving in response to heroin cues (Shao et al. 2006). Weaker activation of the frontal operculum in response to imagined eating of appetizing foods, versus imagined eating of less palatable foods or drinking water, predicted elevated weight gain for those with the DRD4-L allele (Stice et al. 2010). Presence of the DRD4-L allele has also been associated with higher maximum BMI in humans at risk for obesity, including individuals with seasonal affective disorder who report overeating, individuals with bulimia nervosa (Kaplan et al. 2008; Levitan et al. 2004), and African-American adolescents (Guo et al. 2007).

3.3 Variants in the Dopamine Transporter

Phasically released dopamine is normally eliminated by rapid reuptake into presynaptic terminals through the dopamine transporter (DAT), which is abundant in the striatum (Floresco et al. 2003). Lower DAT expression, which is associated with a 10 repeat allele (DAT-L), may reduce synaptic clearance and produce higher basal, or tonic, dopamine levels and blunted phasic dopamine release (Brody et al. 2006). Pecina et al. (2003) found that a DAT knockdown mouse with increased extracellular dopamine displayed increased energy intake and preference for palatable foods. In a study of normal mice, a high-fat diet significantly decreased DAT density in the dorsal and ventral parts of the caudate and putamen compared to a low-fat diet (South and Huang 2008). Lower striatal DAT levels also have been associated with elevated BMI in humans (Chen et al. 2008). DAT-L has been associated with obesity in African-American smokers, versus smokers of other ethnic groups (Epstein et al. 2002).

3.4 Variants in the Catechol-O-Methyltransferase Gene

Catechol-*O*-methyltransferase (COMT) also has been implicated in the regulation of dopamine signaling. COMT regulates extra-synaptic dopamine breakdown, which occurs predominantly in the prefrontal cortex but also plays a factor in the striatum (Matsumoto et al. 2003). A single nucleotide exchange in the COMT gene, a valine to methionine substitution (Val/Met-158), is associated with a fourfold reduction in COMT activity in humans. There is some evidence that individuals with the Met-allele (low enzyme activity) had relatively higher tonic dopamine levels in prefrontal and striatal regions, but reduced phasic dopamine release in the striatum (Bilder et al. 2004). However, both pharmacological and genetic manipulations of COMT (Gogos et al. 1998) have clarified its role in dopamine metabolism in rodents. Under basal physiological circumstances, both COMT inhibition and genetic deficiencies of COMT have little effect on DA concentrations in the striatum (Bilder et al. 2004). Indeed, individuals with the Met-allele versus without it showed less phasic release of dopamine in response to cigarette smoking (Brody et al. 2006). Although one study involving patients with eating disorders found that individuals with the Met-allele reported elevated bulimic symptoms than those without this allele (Sobik et al. 2005), another study found that individuals with bulimia nervosa were marginally less likely to have a Met-allele than healthy controls (Mikolajczyk et al. 2006). Interestingly, individuals with the Val/Val genotype also showed a preference for an immediate reward versus a larger, delayed reward as well as greater activity levels in the dorsal prefrontal cortex and posterior parietal cortex (Boettiger et al. 2007), suggesting an important interplay between reward sensitivity and impulsivity.

3.5 AKT

AKT (also known as protein kinase B) is a protein kinase effector that acts downstream of the phosphatidylinositol 3-kinase (PI3K)-dependent intracellular signaling pathway. Over-expression of AKT decreases DAT expression (Garcia et al. 2005). Furthermore, the functioning of AKT is regulated by insulin; indeed, AKT appears to represent the primary pathway through which insulin affects dopamine transmission, including the amount of dopamine released during exposure to amphetamine (Williams et al. 2007). This effect is mediated through changes in both the membrane-surface expression and the functionality of DAT. The importance of the AKT for amphetamine responsivity is demonstrated by the ability of the AKT1 inhibitor LY294002 to block amphetamine-induced dopamine release and reuptake (Williams et al. 2007). There is also evidence that phosphorylation of the AKT substrate PRAS40 is markedly reduced in rats fed a high-fat diet (Nascimento et al. 2007), suggesting a mechanism through which diet may influence DAT expression and dopamine release. Given this combination of clinical and preclinical data, AKT1 may be an important candidate for understanding genetic relations between dopamine functioning and overeating.

4 Appetite-Regulating Hormones and Peptides

Beyond the genetic polymorphisms affecting dopamine functioning, there are several hormones and peptides that may exert a direct influence on dopamine functioning and therefore moderate the relation between dopamine and overeating. For instance, the phasically released orexigenic agents ghrelin and orexin A (hypocretin A), both of which enhance appetite, have direct influences on dopamine functioning. Ghrelin binds to neurons in the VTA, increases dopamine neuronal activity, and increases dopamine turnover in the nucleus accumbens (Abizaid et al. 2006). Orexin A increases dopamine firing (Korotkova et al. 2006) and causes increased release of dopamine both in prefrontal regions (Vittoz and Berridge 2006) and in the nucleus accumbens shell (Narita et al. 2006). In addition to increasing dopamine firing and release, injection of orexin into the VTA leads to behavioral changes associated with dopamine stimulation (Narita et al. 2006). In contrast, neuropeptide Y has been found to inhibit dopamine neurons in the VTA (Korotkova et al. 2006). Leptin and insulin, which produce more tonic influences on appetite, also show relations with dopamine functioning in animal studies. Critically, leptin robustly modulates brain reward circuitry, as it has been found to alter operant response rates for rewarding brain stimulation (Fulton et al. 2000; Fulton et al. 2004), reverse the effect of food deprivation on the reinstatement of drug-seeking behavior (Shalev et al. 2001), and block conditioned place-preference for high-fat diets (Figlewicz et al. 2003). It may also provide direct or indirect effects on dopamine responsiveness to psychostimulants, although its immediate effects in normal ad lib-fed rodents are limited (Cota et al. 2006; Hao et al. 2004). Insulin exerts a direct effect on the DAT, by increasing DAT function (and dopamine uptake) (Patterson et al. 1998; Carvelli et al. 2002). Indeed, there may be a direct effect of insulin in the activation of reward circuitry by amphetamine, as insulin appears to prevent amphetamine-induced cell surface redistribution of DAT (Garcia et al. 2005; Carvelli et al. 2002). As noted, insulin's action on DAT is mediated by AKT1 (Garcia et al. 2005). Taken together, these data highlight the potential links between appetite-regulating hormones, dopamine functioning, and striatal responsivity. Yet, few studies have examined these relations in humans.

5 Conclusion

Overall data suggest that low levels of dopamine D2 receptors and attenuated responsivity of dopamine-target regions (e.g., striatum) to food intake are associated with increased eating and elevated weight. There is also growing (although partially) mixed evidence that genotypes that appear to lead to reduced dopamine signaling (e.g., DRD2, DRD4, and DAT) and certain appetite-related hormones and peptides (e.g., ghrelin, orexin A, leptin) moderate the relation between dopamine signaling, overeating, and obesity. However, there are several gaps in the literature for future studies to address. First, because existing studies in humans have used either fMRI measures of responses to food or PET measures of dopamine binding, but have never measured both in the same participants, it is unclear to what extent the fMRI findings are dependent on dopamine mechanisms and whether this explains the differential responsivity in obese versus lean individuals. It will be important for future studies to integrate measurement of dopamine functioning with fMRI measures of striatal and cortical responses to food. Second, although there is emerging evidence that reduced responsivity of brain regions implicated in food reward increases risk for future weight gain in individuals who appear to be at genetic risk for attenuated dopamine signaling by virtue of DRD2 and DRD4 genotypes, it will be important to replicate these relations in independent studies with larger sample sizes. Third, it is also important to characterize the relations between specific parameters of dopaminergic synaptic function, such as dopamine receptor density, basal extracellular dopamine levels, and phasic dopamine release, and genotypes, hormones, and peptides that influence dopamine functioning to eating and unhealthy weight gain. Finally, future research should continue exploring factors that moderate the risk conveyed by abnormalities in reward circuitry in response to food reward, as such abnormalities have been implicated in obesity as well as substance abuse and pathological gambling. Interestingly, there is some evidence that obese individuals show a decreased risk of substance use and abuse (Simon et al. 2006; Warren et al. 2005), suggesting that routinely engaging in one reinforcing behavior may reduce the probability of turning to another reinforcing behavior. It is hoped that an improved understanding of the role of dopamine-based reward circuitry as well as factors that influence the functioning of this circuitry will inform the design of more effective preventive and treatment interventions for obesity.

References

Abizaid A, Liu ZW, Andrews ZB, Shanabrough M, Borok E, Elsworth JD, Roth RH, Sleeman MW, Picciotto MR, Tschop MH, Gao XB, Horvath TL (2006) Ghrelin modulates the activity and synaptic input organization of midbrain dopamine neurons while promoting appetite. J Clin Investig 116:3229–3239

Asghari V, Sanyal S, Buchwaldt S, Paterson A, Jovanovic V, Van Tol HH (1995) Modulation of intracellular cyclic AMP levels by different human dopamine D4 receptor variants. J Neurochem 65:1157–1165

Avena NM, Rada P, Hoebel BG (2008) Underweight rats have enhanced dopamine release and blunted acetylcholine response in the nucleus accumbens while bingeing on sucrose. Neuroscience 16:865–871

Bayer HM, Glimcher PW (2005) Midbrain dopamine neurons encode a quantitative reward prediction error signal. Neuron 47(1):129–141

Bello NT, Lucas LR, Hajnal A (2002) Repeated sucrose access influences dopamine D2 receptor density in the striatum. Neuroreport 13(12):1575–1578

Berridge KC, Robinson TE (1998) What is the role of dopamine in reward: hedonic impact, reward learning, or incentive salience? Brain Res Brain Res Rev 28(3):309–369

Bilder R, Volavka J, Lachman H, Grace A (2004) Relations to the tonic-phasic dopamine hypothesis and neuropsychiatric phenotypes. Neuropsychopharmacology 29(11):1943–1961
Blackburn JR, Phillips AG, Jakubovic A, Fibiger HC (1989) Dopamine and preparatory behavior: a neurochemical analysis. Behav Neurosci 103:15–23
Boettiger CA, Mitchell JM, Tavares VC, Robertson M, Joslyn G, D'Esposito M, Fields HL (2007) Immediate reward bias in humans: fronto-parietal networks and a role for the catechol-O methyltransferase 158(Val/Val) genotype. J Neurosci 27(52):14383–14391
Bowirrat A, Oscar-Berman M (2005) Relationship between dopaminergic neurotransmission, alcoholism, and reward deficiency syndrome. Am J Med Genet B Neuropsychiatr Genet 132 (1):29–37
Brody AL, Mandelkern MA, Olmstead RE, Scheibal D, Hahn E, Shiraga S, Zamora-Paja E, Farahi J, Saxena S, London ED, McCracken JT (2006) Gene variants of brain dopamine pathways and smoking-induced dopamine release in the ventral caudate/nucleus accumbens. Arch Gen Psychiatry 63:808–816
Carvelli L, Moron JA, Kahlig KM, Ferrer V (2002) PI 3-kinase regulation of dopamine uptake. J Neurochem 81:859–869
Chen PS, Yang YK, Yeh TL, Lee IH, Yao WJ, Chiu NT, Lu RB (2008) Correlation between body mass index and striatal dopamine transporter availability in healthy volunteers: a SPECT study. Neuroimage 40:275–279
Comings DE, Flanagan SD, Dietz G, Muhleman D, Knell E, Gysin R (1993) The dopamine D2 receptor (DRD2) as a major gene in obesity and height. Biochem Med Metab Biol 50(2): 176–185
Cota D, Barrera JG, Seeley RJ (2006) Leptin in energy balance and reward: two faces of the same coin? Neuron 51:678–680
Davis JF, Tracy AL, Schurdak JD, Tschöp MH, Lipton JW, Clegg DJ, Benoit SC (2008) Exposure to elevated levels of dietary fat attenuates psychostimulant reward and mesolimbic dopamine turnover in the rat. Behav Neurosci 122(6):1257–1263
Epstein LH, Jaroni JL, Paluch RA, Leddy JJ, Vahue HE, Hawk L, Pl W, Shields PG, Lerman C (2002) Dopamine transport genotype as a risk factor for obesity in smokers. Obesity 10: 1232–1240
Epstein LH, Wright SM, Paluch RA, Leddy JJ, Hawk LW, Jaroni JL, Saad FG, Crystal-Mansour S, Shields PG, Lerman C (2004) Relation between food reinforcement and dopamine genotypes and its effect on food intake in smokers. Am J Clin Nutr 80:82–88
Epstein LJ, Leddy JJ, Temple JL, Faith MS (2007) Food reinforcment and eating: a multilevel analysis. Psychol Bull 133:884–906
Erblich J, Lerman C, Self DW, Diaz GA, Bovbjerg DH (2005) Effects of dopamine D2 receptor (DRD2) and transporter (SLC6A3) polymorphisms on smoking cue-induced cigarette craving among African American smokers. Mol Psychiatry 10:407–414
Fetissov SO, Meguid MM, Sato T, Zhang LH (2002) Expression of dopaminergic receptors in the hypothalamus of lean and obese Zucker rates and food intake. Am J Physiol Regul Integr Comp Physiol 283:905–910
Figlewicz DP, Evans SB, Murphy J, Hoen M, Baskin DG (2003) Expression of receptors for insulin and leptin in the ventral tegmental area/substantia nigra (VTA/SN) of the rat. Brain Res 964(1):107–115
Filbey FM, Ray L, Smolen A, Claus ED, Audette A, Hutchison K (2008) Differential neural response to alcohol priming and alcohol taste cues is associated with DRD4 VNTR and OPRM1 genotypes. Alcohol Clin Exp Res 32:1–11
Floresco SB, West AR, Ash B, Moore H, Grace AA (2003) Afferent modulation of dopamine neuron firing differentially regulates tonic and phasic dopamine transmission. Nat Neurosci 6:968–973
Fossella J, Green AE, Fan J (2006) Evaluation of a structural polymorphism in the ankyrin repeat and kinase domain containing 1 (ANKK1) gene and the activation of executive attention networks. Cogn Affect Behav Neurosci 6:71–78

Fulton S, Woodside B, Shizgal P (2000) Modulation of brain reward circuitry by leptin. Science 287:125–128

Fulton S, Richard D, Woodside B, Shizgal P (2004) Food restriction and leptin impact brain reward circuitry in lean and obese Zucker rats. Behav Brain Res 155:319–329

Garcia BG, Wei Y, Moron JA, Lin RZ, Javitch JA, Galli A (2005) Akt is essential for insulin modulation of amphetamine-induced human dopamine transporter cell-surface redistribution. Mol Pharmacol 68:102–109

Geiger BM, Haburcak M, Avena NM, Moeyer MC, Hoebel BG, Photos EN (2009) Deficits of mesolimbic dopamine neurotransmission in rat dietary obesity. Neuroscience 159(4): 1193–1199

Gogos JA, Morgan M, Luine V, Santha M, Ogawa S, Pfaff D et al (1998) Catechol-O-methyl-transferase-deficient mice exhibit sexually dimorphic changes in catecholamine levels and behavior. Proc Natl Acad Sci U S A 95:9991–9996

Goldstein R, Klein A, Tomasi D, Zhang L, Cottone L, Maloney T et al (2007) Is decreased prefrontal cortical sensitivity to monetary reward associated with impaired motivation and self-control in cocaine addiction? Am J Psychiatry 164:43–51

Guo G, North KE, Gorden-Larsen P, Bulik CM, Choi S (2007) Body mass, DRD4, physical activity, sedentary behavior, and family socioeconomic status. Obesity 15:1199–1206

Hamarman S, Fossella J, Ulger C, Brimacombe M, Dermody J (2004) Dopamine receptor 4 (DRD4) 7-repeat allele predicts methylphenidate dose response in children with attention deficit hyperactivity disorders: a pharmacogenetic study. J Child Adolesc Psychopharmacol 14:564–574

Hao J, Cabeza DV, Carr KD (2004) Effects of chronic ICV leptin infusion on motor-activating effects of D-amphetamine in food-restricted and ad libitum fed rats. Physiol Behav 83:377–381

Hare TA, O'Doherty J, Camerer CF, Schultz W, Rangel A (2008) Dissociating the role of the orbitofrontal cortex and the striatum in the computation of goal values and prediction errors. J Neurosci 28(22):5623–5630

Huang XF, Zavitsanou K, Huang X, Yu Y, Wang H, Chen F, Lawrence AJ, Deng C (2006) Dopamine transporter and D2 receptor binding densities in mice prone or resistant to chronic high fat diet-incuded obesity. Behav Brain Res 175:415–419

Hutchison KE, McGeary J, Smolen A, Bryan A, Swift RM (2002) The DRD4 VNTR polymorphism moderates craving after alcohol consumption. Health Psychol 21:139–146

Kaplan AS, Levitan RD, Yilmaz Z, Davis C, Tharmalingam S, Kennedy JL (2008) A DRD4/BDNF gene-gene interaction associated with maximum BMI in women with bulimia nervosa. Int J Eat Disord 41:22–28

Kelley AE, Will MJ, Steininger TL, Zhang M, Haber SN (2003) Restricted daily consumption of a highly palatable food (chocolate Ensure(R)) alters striatal enkephalin gene expression. Eur J Neurosci 18(9):2592–2598

Kirsch P, Reuter M, Mier D, Lonsdorf T, Stark R, Gallhofer B et al (2006) Imaging gene-substance interactions: the effect of the DRD2 TaqIA polymorphism and the dopamine agonists bromocriptine on the brain activation during the anticipation of reward. Neurosci Lett 405:196–201

Kiyatkin EA, Gratton A (1994) Electrochemical monitoring of extracellular dopamine in nucleus accumbens of rats lever-pressing for food. Brain Res 652:225–234

Knutson B, Gibbs SE (2007) Linking nucleus accumbens dopamine and blood oxygenation. Psychopharmacology (Berl) 191(3):813–822

Korotkova TM, Brown RE, Sergeeva OA, Ponomarenko AA, Haas HL (2006) Effects of arousal and feeding-related neuropeptides on dopaminergic and GABAergic neurons in the ventral tegmental area of the rat. Eur J Neurosci 23:2677–2685

Levin BE, Meynell AA (2002) Defense of body weight depends on dietary composition and palatability in rats with diet-induced obesity. Am J Physiol Regul Integr Comp Physiol 282: R46–54

Levitan RD, Masellis M, Lam RW, Muglia P, Basile VS, Jain U, Kaplan AS, Tharmakingam S, Kennedy SH, Kennedy JL (2004) Childhood in-attention and dysphoria and adult obesity associated with the dopamine D4 receptor gene in overeating women with seasonal affective disorder. Neuropsychopharmacology 29:179–186

Li Y, Shao C, Zhang D, Zhao M, Lin L, Yan P, Xie Y, Jiang K, Jin L (2006) The effect of dopamine D2, D5 receptor and transporter (SLC6A3) polymorphisms on the cue-elicited heroin craving in Chinese. Am J Med Genet B Neuropsychiatr Genet 141B:269–273

Matsumoto M, Weickert C, Beltaifa S, Kolachana B, Chen J, Hyde TM, Herman MM, Weinberger DR, Kleinman JE (2003) Catechol 0-methyltransferase (COMT) mRNA expression in the dorsolateral prefrontal cortex of patients with schizophrenia. Neuropsychopharmacology 28(8):1521–30

McClernon FJ, Hutchison KE, Rose JE, Kozink RV (2007) DRD4 VNTR polymorphism is associated with transient fMRI-BOLD responses to smoking cues. Psychopharmacology (Berl) 194:433–441

Mikolajczyk E, Smiarowska M, Grzywacz A, Samochowiec J (2006) Association of eating disorders with catechol-O-methyltransferase gene functional polymorphism. Neuropsychobiology 54(1):82–86

Narita M, Nagumo Y, Hashimoto S, Narita M, Khotib J, Miyatake M et al (2006) Direct involvement of orexinergic systems in the activation of the mesolimbic dopamine pathway and related behaviors induced by morphine. J Neurosci 26:398–405

Nascimento EBM, Fodor M, van der Zon GCM, Jazet IM, Meinders AE, Voshol PJ, Vlasblom R, Baan B, Eckel J, Maassen JA, Diamant M, Ouwens DM (2007) Insulin-mediated phosphorylation of the proline-rich AKT substrate PRAS40 is impaired in insulin target tissues of high-fat diet-fed rats. Diabetes Nutr Metab 55(12):3221–3228

Noain D, Avale ME, Wedemeyer C, Calvo D, Peper M, Rubinstein M (2006) Identification of brain neurons expressin the dopamine D4 receptor gene using BAC transgenic mice. Eur J Neurosci 24:2429–2438

Orosco M, Rouch C, Nicolaidis S (1996) Rostromedial hypothalamic monoamine changes in response to intravenous infusions of insulin and glucose in freely feeding obese Zucker rats: a microdialysis study. Appetite 26:1–20

Patterson TA, Brot MD, Zavosh A, Schenk JO, Szot P, Figlewicz DP (1998) Food deprivation decreases mRNA and activity of the rat dopamine transporter. Neuroendocrinology 68:11–20

Pecina S, Cagniard B, Berridge KC, Aldrigde JW, Zhuang X (2003) Hyperdopaminergic mutant mice have higher "wanting" but not "liking" for sweet rewards. J Neurosci 23:9395–9402

Ritchie T, Noble EP (2003) Association of seven polymorphisms of the D2 dopamine receptor gene with brain receptor-binding characteristics. Neurochem Res 28:73–82

Robinson TE, Berridge KC (2000) Intra-accumbens amphetamine increases the conditioned incentive salience of sucrose reward: enhancement of reward "wanting" without enhanced "liking" or response reinforcement. J Neurosci 20:s91–s117

Schultz W, Apicella P, Ljungberg T (1993) Responses of monkey dopamine neurons to reward and conditioned stimuli during successive steps of learning a delayed response task. J Neurosci 13(3):900–913

Shalev U, Yap J, Shaham Y (2001) Leptin attenuates acute food deprivation-induced relapse to heroin seeking. J Neurosci 21:RC129

Shao C, Li Y, Jiang K, Zhang D, Xu Y, Lin L, Wang Q, Zhao M, Jin L (2006) Dopamine D4 receptor polymorphism modulates cue-elicited heroin craving in Chinese. Psychopharmacology (Berl) 186:185–190

Simon G, von Korff M, Saunders K, Miglioretti D, Crane P et al (2006) Association between obesity and psychiatric disorders in the US population. Arch Gen Psychiatry 63:824–830

Small DM, Jones-Gotman M, Dagher A (2003) Feeding-induced dopamine release in dorsal striatum correlates with meal pleasantness ratings in healthy human volunteers. Neuroimage 19:1709–15

Sobik L, Hutchison K, Craighead L (2005) Cue-elicited craving for food: a fresh approach to the study of binge eating. Appetite 44:253–61

South T, Huang XF (2008) High-fat diet exposure increases dopamine D2 receptor and decreases dopamine transporter receptor binding density in the nucleus accumbens and caudate putamen of mice. Neurochem Res 33:598–605

Southon AWK, Sanigorski AM, Zimmet P, Nicholson GC, Kotowicz MA, Collier G (2003) The Taq IA and Ser311 Cys polymorphisms in the dopamine D2 receptor gene and obesity. Diabetes Nutr Metab 16:72–76

Spitz MR, Detry MA, Pillow P, Hu YH, Amos CI, Hong WK, Wu X (2000) Variant alleles of the D2 dopamine receptor gene and obesity. Nutr Res 20:371–380

Stice E, Spoor S, Bohon C, Small DM (2008a) Relation between obesity and blunted striatal response to food is moderated by TaqIA A1 allele. Science 322:449–452

Stice E, Spoor S, Bohon C, Veldhuizen M, Small DM (2008b) Relation of reward from food intake and anticipated intake to obesity: a functional magnretic resonance imaging study. J Abnorm Psychol 117:924–935

Stice E, Yokum S, Bohon C, Marti N, Smolen A (2010) Reward circuitry responsivity to food predicts future increases in body mass: moderating effects of DRD2 and DRD4. Neuroimage 50(4):1618–1625

Thanos PK, Volkow ND, Freimuth P, Umegaki H, Ikari H, Roth G, Ingram DK, Hitzemann R (2001) Overexpression of dopamine receptors reduces alcohol self-administration. J Neurochem 78(5):1094–1103

Thomas GN, Critchley JA, Tomlinson B, Cockram CS, Chan JC (2001) Relationships between the TaqI polymorphism of the dopamine D2 receptor and blood pressure in hyperglycemic and normoglycemic Chinese subjects. Clin Endocrinol 55:605–611

Tupala E, Hall H, Bergström K, Mantere T, Räsänen P, Särkioja T, Tiihonen J (2003) Dopamine D2 receptors and transporters in type 1 and 2 alcoholics measured with human whole hemisphere autoradiography. Hum Brain Mapp 20(2):91–102

Vittoz NM, Berridge CW (2006) Hypocretin/orexin selectively increases dopamine efflux within the prefrontal cortex: involvement of the ventral tegmental area. Neuropsychopharmacology 31:384–395

Volkow ND, Fowler JS, Wang GJ (2002) Role of dopamine in drug reinforcement and addiction in humans: results from imaging studies. Behav Pharmacol 13:355–366

Volkow ND, Wang GJ, Telang F, Fowler JS, Thanos PK, Logan J, Alexoff D, Ding YS, Wong C, Ma Y, Pradhan K (2008) Low dopamine striatal D2 receptors are associated with prefrontal metabolism in obese subjects: possible contributing factors. Neuroimage 42(4):1537–1543

Wang GJ, Volkow ND, Logan J, Pappas NR, Wong CT, Zhu W, Netusil N, Fowler JS (2001) Brain dopamine and obesity. Lancet 357(9253):354–357

Warren M, Frost-Pineda K, Gold M (2005) Body mass index and marijuana use. J Addict Dis 24:95–100

Wellman PJ, Nation JR, Davis KW (2007) Impairment of cocaine self-administration in rats maintained on a high fat diet. Pharmacol Biochem Behav 88:89–93

Williams JM, Owens WA, Turner GH, Saunders C, Dipace C, Blakely RD, France CP, Gore JC, Daws LC, Avison MJ, Galli A (2007) Hypoinsulinemia regulates amphetamine-induced reverse transport of dopamine. PLoS Biol 5(10):e274

Yamamoto T (2006) Neural substrates for the processing of cognitive and affective aspects of taste in the brain. Arch Histol Cytol 69:243–255

Reward and Neurocomputational Processes

Guido K.W. Frank

Contents

Abstract The neurobiology of eating disorders (EDs) is largely unknown. However, brain imaging studies over the past decade have identified neurotransmitter alterations that could be part of dysfunctional behavior characteristics of EDs. In this chapter we focus on a specific behavioral construct, the brain reward system, and demonstrate a functional brain imaging approach toward identifying dopamine

G.K.W. Frank
Department of Psychiatry, Developmental Brain Research Program, University of Colorado Denver, The Children's Hospital, Gary Pavilion A036/B-130, 13123 East 16th Avenue, Aurora, CO 80045, USA
Department of Neuroscience, University of Colorado Denver, The Children's Hospital, Gary Pavilion A036/B-130, 13123 East 16th Avenue, Aurora, CO 80045, USA
e-mail: Guido.Frank@ucdenver.edu

R.A.H. Adan and W.H. Kaye (eds.), *Behavioral Neurobiology of Eating Disorders*, Current Topics in Behavioral Neurosciences 6, DOI 10.1007/7854_2010_81

function in anorexia nervosa (AN). We demonstrate how human brain reward activation can be used in a translational approach to test whether computer models, based on basic science research, can predict expected in vivo reward system activation, and how such an approach can identify specific biologic alterations in a psychiatric population.

Keyword Anorexia nervosa · Brain Imaging · Dopamine · Food reward · Learning signal · Neurocomputational Modeling · Neurotransmitter · Reward processing · Reward system · Reinforcement · Substance use · Taste · Temporal Difference Model

1 Introduction

Anorexia nervosa (AN) is a severe psychiatric disorder associated with food avoidance, severe emaciation, and the highest mortality rate among the psychiatric disorders (Sullivan 1995; APA 2000). Clinical evidence suggests that individuals with AN have disturbances in the processing of naturally rewarding stimuli, such as food. The processing of food reward is complex and modulated by cognitive, emotional, and biologic factors that involve learned behaviors and genetic predisposition. In the brain, cortical and subcortical networks integrate external food stimuli with the internal drive and motivation to eat, resulting in behavior activation and approach or avoidance of food. Brain regions implicated in this process include the insula, the basal ganglia (anteroventral and dorsal striatum, nucleus accumbens), the ventral tegmental area (VTA), the anterior cingulate, orbitofrontal, and mesial temporal cortex, as well as the hypothalamus. This network, called the reward system, processes food and other stimuli in order to "achieve an optimal stimulus response" (Cooper et al. 2003). The function of the reward system in patients with AN has been inadequately studied. Clinically, AN subjects are able to refrain from eating over prolonged periods of time and do not respond to mechanisms that would, in most people, stimulate the drive to eat and approach food. Recent studies in AN subjects suggest abnormal brain activity in striatal parts of the basal ganglia that are related to dopamine receptor function (Frank et al. 2005; Wagner et al. 2005). These brain regions have been associated with the processing of immediate and delayed rewards in control subjects, how much reward someone may expect from a stimulus in the future, as well as the amount of motivation a person has to approach a possible reward. In this chapter we describe the use of a model founded on neuroscience that is primarily based on brain dopamine neurotransmission and brain imaging, and we discuss how such an approach can be used to identify pathways directly linked to underlying neurobiologic abnormalities in AN. Specifically, computer models based on basic science research can identify specific neuronal brain function, and results can be used to validate fMRI results and identify disorder-specific abnormalities.

2 Possible Mechanisms Underlying Reward System Abnormalities in AN

The motivation to eat and approach food is an important part of the reward pathway and is particularly disturbed in AN. AN women like sweet stimuli and do not deny having an appetite (Schweiger and Fichter 1997). Clinically, AN individuals frequently express some desire to eat and gain weight but usually continue to be drawn to the goal of weight loss and food restriction. Moreover, acting on the desire to eat is even more difficult, and thus AN individuals seem unable to direct their behavior and essential motor function toward necessary goals of weight gain. While the AN patient still has the ability to eat, move, experience taste, and even likes sweet stimuli more than controls (Drewnowski et al. 1987), she cannot activate herself sufficiently to eat adequately. This pattern is reminiscent of dopamine-deficient mice (Cannon and Palmiter 2003). These rodents like sweet sucrose or noncaloric artificial sweetener solution more than water, but show more infrequent initiation of drinking compared to controls. They are hypophagic (Zhou and Palmiter 1995) and will die of starvation even with readily available food in their cage, without motor impairment being the cause for the hypophagia. It was hypothesized that these mice have a reduced capacity to respond to rewarding stimuli ["lower basal and maximum response capacity" (Cannon and Bseikri 2004)]. Thus, dopamine-deficient mice may have a reduced motivation or capacity to obtain rewards and may not be able to sufficiently direct their behaviors toward the appropriate goals (Cannon and Bseikri 2004), such as food intake and maintenance of a normal body weight. AN, in fact, has been associated with reduced brain dopamine and abnormal dopamine receptor function in the anteroventral striatum (Kaye et al. 1999; Frank et al. 2005). Many cognitive prefrontal processes, such as fear of weight gain, cannot be addressed in the dopamine-deficient mouse model. However, low anteroventral striatal dopamine-related activity may be associated with less-afferent prefrontal input (Montague et al. 2004) and decreased motivation to approach food rewards. While speculative, it is possible that AN individuals possess a reduced drive to approach food stimuli in conjunction with overvalued ideas and desires for thinness and fear of weight gain. Together, such a constellation may lead to an imbalance of cortical and subcortical drives resulting in reduced food appetence and subsequent weight loss. Another function of the dopamine system is alerting the individual for novel stimuli and initiating learning from those experiences (Schultz 1998, 2002). It appears that the initial phase of food reduction and weight loss is associated with a sense of excitement (Bergh and Sodersten 1996); however, long-term or chronic anorexic subjects do not experience such happiness or excitement but are merely trying to avoid or ameliorate perceived aversive states, such as weight gain or feelings of being fat (APA 2000). This behavior seems to have aspects in common with substance use withdrawal. Koob and Le Moal (Koob and Le Moal 2005) describe a compulsive state that follows the intoxication of a drug addict characterized by withdrawal and negative affect, emotional pain, alexithymia, and dysphoria. Instead of the positive rewards experienced during the initial phase of drug use, plastic changes in the reward pathways create an antireward system that "drives aversive states" (Koob and Le Moal 2005). This negative antireward system has been associated with altered dopaminergic states. AN and drug users may lie on the opposite ends of the spectrum with increased dopamine receptor binding in AN but reduced dopamine receptors in the substance users (Volkow et al. 2004). AN subjects are chronically dysphoric, have difficulties expressing their emotions, and cannot enjoy things as much as others (Anderluh et al. 2003), and we believe that AN subjects might be stuck in an antireward system with reduced dopamine function, whereas such a state is only temporary in drug users before they self-medicate with another hit or drink. Interestingly, alcoholism (Kampov-Polevoy et al. 2003) and benzodiazepine use (Yasoshima and Yamamoto 2005) have also been associated with altered sweet taste response. In summary, there may be a combination of dopamine abnormalities, either trait or state related, and psychologic factors that control eating behavior and influence the motivational system.

3 Taste and Reward System Activation Research in AN

Taste by itself is a reward system activator, at least in part through dopamine activation in the brain. Taste receptors on the tongue project to the brain stem nucleus of the solitary tract (NST) and from there to the parabrachial nucleus (PBN). Then the taste pathway bifurcates to a thalamic/cortical tract and the hypothalamic/limbic pathway with projections to the amygdala and nucleus accumbens (Norgren et al. 2006; Hajnal et al. 2009). Thus, taste directly stimulates the dopamine pathway. Taste paradigms in AN can be applied as long as disorder-specific mechanisms are taken into account. Sweet taste perception is overall preserved in AN, but the hedonic experience of sweet taste is particularly biased by the fear of weight gain. Several studies have investigated taste perception in AN in the past. Some found in AN subjects reduced (Toth et al. 2004) or altered (Drewnowski et al. 1987) taste sensitivity while ill and after refeeding. Others found no disturbance in the ability to rate sweet perception (Di Costanzo et al. 1998) but found altered hedonic ratings compared to controls, where AN preferred sweeter stimuli (Sunday and Halmi 1990). Another group did not find such abnormalities (Simon et al. 1993). Exercising is a possible confound of hedonic taste perception studies as well, since more than 3 h of activity per week decreased preference ratings for high-sucrose and high-fat stimuli in controls (Crystal et al. 1995). This is important since women with AN frequently exercise. Hedonic taste response is heavily biased by the perception of calorie intake and the possibility of weight gain as a deterrent to enjoy food. In fact, one study found that sweet taste pleasantness ratings were preserved in a paradigm where AN did not swallow but spit out the sweet (sugar) taste stimulus (Eiber et al. 2002). In the same paradigm, hedonic ratings were reduced when subjects had to swallow the sugar stimulus, suggesting strong cognitive bias toward the high-caloric fluid. Taken together, there is limited evidence for taste perception alterations per se in AN, although the studies performed are few with small sample sizes. Potentially more important, the hedonic rating results tend to be biased by an overriding fear of weight gain.

In summary, taste response can be used in AN to test reward system abnormalities. However, the cognitive and emotional activation specific to AN has to be taken into account in the model.

4 Reward Pathway Function and Neurobiology

The ideal function of the brain reward system is to ensure that we make the best choices in response to environmental stimuli (Cooper et al. 2003) in order to support our long- and short-term goals. Natural rewarding food-related stimuli have been studied extensively (Saper et al. 2002). A large number of studies have also examined recreational drugs, alcohol, monetary reward, and other potentially rewarding stimuli (Martin-Soelch et al. 2001; Grigson 2002). The mechanisms of reward processing in the brain, though, are shared across stimuli (Grigson 2002). The main psychological components of this system are motivation, learning, and emotion or affect (Berridge and Robinson 2003). One line of research distinguished conscious "liking," relating to an emotional response to stimuli, from "wanting" which relates to intrinsic motivation reflective of how much an individual is willing to work for a reward. For the concept of wanting, the term "incentive salience" of stimuli has been proposed (Berridge 1996; Berridge and Robinson 1998). An underlying principle of reward processing seems to be "the anticipation of outcomes of behavior in situations with varying degrees of uncertainty" (Schultz 2006). That means we frequently have to make choices in an environment with various types of stimuli with respect to which of those stimuli might fit best our desires and needs. A stimulus can have several rewarding effects for an organism (Wise 2002). The brain tries to assess and anticipate the outcome obtained from a stimulus, weighing between available stimuli in order to determine which should be approached based on the highest reward value. While there are genetically determined contingencies of reward, for instance the innate liking of sweet taste (Bartoshuk and Beauchamp 1994), learning, and conditioning, reward-predicting experiences are important for the development of personal preference and reward system activation (Schultz 2006). From our exposure to stimuli and positive or negative reward experience, we learn what to approach and what to avoid. While we develop representations in the brain of what we like and what we do not, new experiences update these representations continuously based on their reward value.

Various neurotransmitters and regions in the brain have been found to process rewarding stimuli. Corticostriatal–hypothalamic pathways have been implicated in the receipt of sensory stimuli and subsequent processing of those stimuli's reward value, followed by behavioral activation in order to approach or avoid stimuli (Kelley et al. 2005). Dopamine pathways, mainly in the VTA, striatum, and nucleus accumbens, extending to the orbitofrontal cortex, may be involved in the approach and motivational aspects of related behavior (Kelley et al. 2005; Schultz 2006). Thus, dopamine activation may be associated with the "wanting" concept described by Berridge (Berridge and Robinson 2003). Opioid neurotransmission is localized closely to dopaminergic neurons in the ventral and dorsal striatum. Opioid activation in these areas, regulated by cholinergic interneurons, may mediate the pleasure that is associated with stimuli, and GABA neurotransmission into the anteroventral striatum may prevent excessive response to rewarding stimuli (Kelley et al. 2005). In addition, there appears to be a fine balance of opioid receptor-driven activation of dopamine and GABA neuronal activation of the reward system via glutamatergic neurons (Mansvelder 2005; O'Reilly et al. 2007).

The prefrontal cortex exerts voluntary control over behavior (Price 2005) and is an important input for the reward system (Wise 2005) in terms of deliberate control and bias toward behavior. It appears that afferent phasic dopamine input to the prefrontal cortex may be needed in order to modulate goals based on rewarding experiences (Montague et al. 2004). These dopamine pathways are influenced by various factors. Food restriction may increase dopamine receptor binding in the nucleus accumbens and caudate/putamen (Carr et al. 2003), and dopamine D2 receptors may regulate dopamine release (Cooper et al. 2003). Alterations in this mechanism may be particularly involved in substance use but also food-related disorders (Blum et al. 1995). In addition, nonpredictability is an important determinant for the activation of the anteroventral striatum dopamine system since unexpected stimuli elicit a greater response in functional dopamine transmission (Berns et al. 2001). Furthermore, serotonin pathways have been hypothesized to alter the mediation of brain reward response (Higgins and Fletcher 2003) with, for instance, serotonin 2C receptor activation attenuating drug-seeking behavior.

5 Reward Brain Imaging Studies and Anorexia Nervosa

Elements of the brain reward system have been identified through anatomic and electrophysiologic studies in nonhuman primates and neuroimaging studies in humans. Single-cell firing rates within the basal ganglia, VTA, nucleus accumbens – a part of the ventral striatum, amygdala, orbital–frontal and prefrontal cortex – are modulated by rewarding stimuli (Apicella et al. 1991; Hikosaka and Watanabe 2000; Schultz et al. 2000; Wise 2002). Neuroimaging studies using a variety of reward stimuli have corroborated the electrophysiology data. In particular, studies using monetary rewards have revealed patterns of reward-related activity in the amygdala, orbital–frontal and mesial prefrontal cortex, cingulate cortex, and portions of the striatum to which these limbic regions project (Delgado et al. 2000, 2003; Knutson et al. 2000; Breiter et al. 2001; May et al. 2004). The ventral striatum is involved in reward processing per se, while the dorsal striatum is involved in linking action to outcome (O'Doherty et al. 2004; Tricomi et al. 2004). Like the ventral striatum, more dorsal sectors of the striatum are also modulated by dopamine. However, these regions are also implicated in motor and cognitive control, specifically the learning of stimulus response associations. The caudate nucleus is activated by those tasks in which there exist both a perceived connection between action and outcome and some uncertainty about whether the action will lead to the desired outcome (Tricomi et al. 2004). O'Doherty and colleagues (O'Doherty et al. 2004) postulated that projections to the ventral striatum might be involved in reward prediction ("critic") and instrumental learning, whereas projections to dorsal striatum ("actor") might be involved in the modulation of stimulus response in order to choose the best behavioral response based on previous experience. Most recently, immediately rewarding processes have been related to ventral striatal activation, whereas delayed gratification may activate dorsal striatal brain response (Tanaka et al. 2004). This is important since, phenotypically, AN have a very high degree of self-control and are able to delay gratification in various aspects of life. Aside from delaying eating, they are extremely goal oriented and perfectionistic, they tend to work harder than peers, and they seem to prefer delaying reward over not meeting their immediate goals and standards. Abnormalities in the anteroventral and dorsal striatal reward system in AN could contribute to food restraint and the ability to refrain from commonly desired pleasurable experiences for long periods of time. Taken together, brain imaging in human controls show that testing reward pathway activation in vivo is possible, and that results obtained are consistent with the animal literature and single neuron-recording studies. It is debated whether brain imaging such as fMRI can identify pathways directly related to neurotransmitter function. Recent data provide evidence suggesting that this is in fact the case (see limitations).

Very few studies exist that specifically assessed reward system brain activation in AN. One study suggested dopamine abnormalities in AN after recovery (Frank et al. 2005), while functional magnetic resonance imaging studies in recovered AN also showed abnormalities in response to a monetary gain task (Wagner et al. 2007). Ill AN were recently shown to have increased ventral striatal response to pictures of underweight bodies. Taken together, the few studies available suggest reward system abnormalities in AN, but the neurobiologic underpinnings remain largely unknown.

6 Neurocomputational Modeling and Brain Imaging

The integration of animal models with human in vivo studies remains a crucial step in translational research. Computer models should be able to predict human behavior (based on the basic research findings) using specific tasks that apply the animal model in the human. That is, the computer model is used to test whether the brain imaging results gathered during the task application indeed reflect the brain activation obtained in the previously validated basic research. If this is the case, then one can make actual inferences on human neurobiologic brain processes related to disease.

6.1 Principals of the Temporal Difference Model

Behavioral models have been developed that mathematically describe aspects of reward under conditions that can be related to brain imaging results. This sets the stage for "Hypothesis-driven research" that "has been shown to be an excellent model for pursuing investigations in neuroscience" (Koslow 2005). Such neuro-computational models link psychological theories with underlying neural substrates. In this chapter, we describe the application of the temporal difference (TD) model (Schultz 2006) and will provide preliminary data.

Various models describe concepts of reinforcement learning (Sutton and Barto 1981), which both concerns the anticipatory control of actions and prediction of potential reward values (Worgotter and Porr 2005). The TD model has been intensively studied and applied to the neural reward system and is a valid model for studying reinforcement learning and dopamine neuronal pathway activity in the basal ganglia. It has been studied in vivo in relation to dopamine system activation in rodents using single-neuron recordings (Schultz 1998, 2002; Schultz et al. 2000) as well as in humans using brain imaging techniques (O'Doherty et al. 2003, 2004; Tanaka et al. 2004). This model (Sutton and Barto 1981; Worgotter and Porr 2005) integrates elements of the individual "value" of a reward and the prediction of its rewarding property in the future. Since rewards do not always occur and are not entirely predictable, a temporal difference error is introduced in the model that functions as a "learning signal." This error is dependent on whether the predicted reward is consistent with the received reward or if there is a discrepancy between prediction and reality (error). Another element in the model is the discount factor, which takes into account how many times a reward has been received and how many times the subject has had the opportunity to learn from experience to derive predictions.

6.2 Temporal Difference Model Study Design

The study design that we use was originally described by O'Doherty et al. (2003). Each trial consists of the presentation of one of three arbitrary visual stimuli (see Fig. 1) followed 2 s later by 1 ml of a pleasant sweet taste (1 M sucrose) (CS+), a neutral tasteless solution [CSneut: 25 mM KCl, 2 mM $NaHCO_3$, (Francis et al. 1999; Frank et al. 2003)], or no taste (CS−). The visual stimuli are presented on a gray background and are removed from the screen after 2 s to coincide with taste delivery. A fixation cross is then presented for the remainder of the trial (4 s). The allocation of each stimulus to a given trial type is counterbalanced across subjects. There is a total of 280 trials in the experiment, 100 each of CS+ and CS− and 80 of CSneut. The whole experiment lasts a total of ~28.5 min. The first ten CS+ stimuli presented are paired on each occasion with reward (sweet taste), in 20 out of 90 subsequent CS+ presentations, the reward is omitted (CS+ omit). Further, in 20

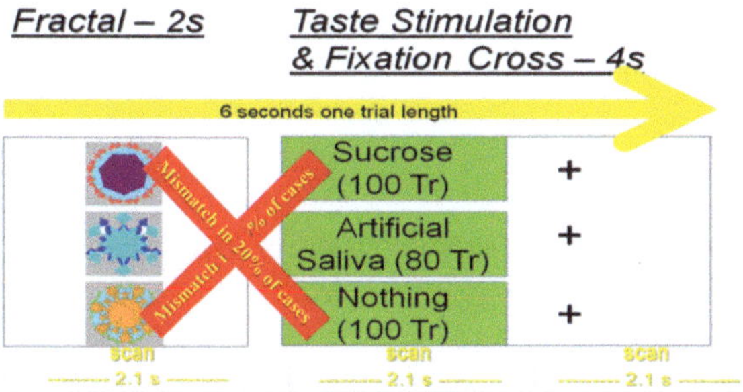

Fig. 1 Schematic tasks application, modeled after O'Doherty et al. (2003). Subjects learned to associate a specific fractal with a particular taste. In 20% of cases, there was a mismatch of fractal and expected taste ("Sucrose fractal" but no taste delivery; "Nothing fractal" but Sucrose delivery). A total of 280 trials are applied; every 2.1 s a brain scan is acquired

presentations of the CS−, a reward is unexpectedly delivered (CS-unexpreward). The CSneut condition is primarily included to provide a low valence rinse for the glucose taste during the experiment. The presentation order is randomized and delivered to subjects using E-Prime (Psychology Software Tools, Pittsburgh, PA).

6.3 Data Analysis

Image processing is done with the SPM 5 software package (http://www.fil.ion.ucl.ac.uk/spm/doc/spmbib.html). We performed whole brain analyses and also used region of interest-based analyses that include the VTA, ventral striatum (VS), amygdala, orbitofrontal cortex, and cingulate cortex for percent signal change extraction. There is a primary and a secondary analysis pathway in the paradigm. *The primary analysis pathway* identifies brain region activation in response to the conditioned fractal or unconditioned sweet taste application as well as brain response when taste is expected by not received. Main effect results are obtained for condition and subject group, as well as time activity curve data that illustrate brain activation over time in response to the individual stimulus.

The *secondary analysis* pathway involves the mathematical models that try to bridge brain function and actual behavior. As described earlier, computational modeling tries to explain a real-world problem in a simplified model that then gets translated into a mathematical equation that can be used to correlate with brain imaging results. This serves the purpose of predicting brain response following a stimulus, and depending on factors such as learning rate and received reward, different brain regions are implicated. That is, the subject performs a specific task or Reward Paradigm and learns under which conditions reward occurs. The time course of reward experience is known by the experimenter. In the taste experiment, reward delivery or omission is part of the paradigm and is recorded by the stimulus-applying computer. Those reward receipt or loss experiences are assigned a reward value in the computational model, and those reward values are then correlated with the functional magnetic resonance imaging brain response.

7 Use of Neurocomputational Models and Testing of Alternate Models (Fig. 2)

On each trial, the predicted value (V) at any time (t) within a trial is calculated as a linear product of the weights (w_i) and the presence or absence of a CS stimulus at time t, coded in the stimulus representation vector $x_i(t)$:

$$\hat{V}(t) = \sum_i w_i x_i(t).$$

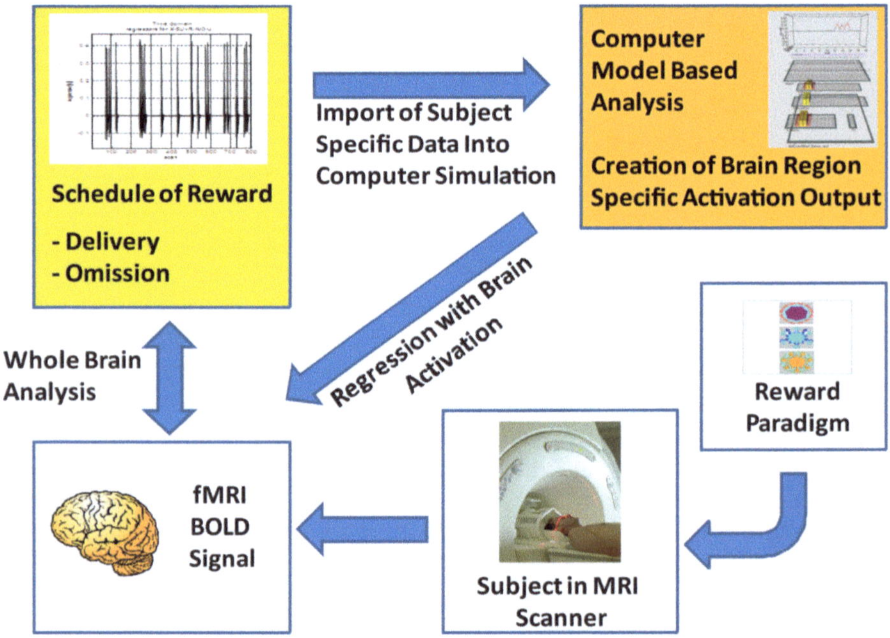

Fig. 2 Principals of the Neurocomputational Modeling Approach. A study participant in the functional magnetic resonance brain imaging (fMRI) scanner performs the taste task. The blood oxygen level-dependent (BOLD) brain response is measured. The fist brain activation analysis tests in the whole brain activations that relate to the schedule of reward application and omission, reward expectation, etc. In the second analysis pathway, we use the individual reward task schedule data for each participant and import those data into a computer algorithm that computes the expected brain response based on the animal literature. Those outcomes are then regressed with the human in vivo data in order to (1) test whether significant regression results indeed occur in the expected brain regions and (2) make group comparisons whether controls have a stronger or weaker activation pattern to a particular learning rate than the anorexia nervosa group

Learning occurs by updating the predicted value of each time point t in the trial by comparing the value at time $t + 1$ to that at time t, leading to a prediction error or $\delta(t)$:

$$\delta(t) = r(t) + \gamma\hat{V}(t+1) - \hat{V}(t),$$

where $r(t)$ = reward at time t. The parameter γ is a discount factor, which determines the extent to which rewards arriving earlier are more important than rewards that arrive later on. Similar to O'Doherty's study (O'Doherty et al. 2003), we set $\gamma = 0.99$. The weights w_i are then updated on a trial-by-trial basis according to the correlation between prediction error and the stimulus representation:

$$\Delta w_i - \alpha \sum_t x_i(t)\delta(t),$$

where α = learning rate. We assign six time points to each trial and use each subject's individual event history as input. On each trial, the CS (visual fractal) is delivered at time point 1, and the UCS reward (sweet taste stimulus) is delivered at time point 3. Like O'Doherty (O'Doherty et al. 2003), we use as learning rate of the model (α) two values for α: a lower learning rate ($\alpha = 0.2$) and a higher learning rate ($\alpha = 0.7$). In the original study, the lower learning rate best modeled brain activation in the ventral striatum and orbitofrontal cortex. The temporal difference error $\delta(t)$ is created for the CS and the positive and negative UCS for each time point, and an ideal activation curve is created and convolved with the hemodynamic response curve. That curve is then modeled to the functional magnetic resonance imaging data in order to generate brain regions that respond according to the model. We cannot predict whether restricting-type anorexia nervosa women have simply reduced brain response to the stimuli or if a different learning rate explains altered learning response. This is an empirical task during data analysis. *Reward prediction error*: The TD error is calculated from the difference between the actual reward $r(t)$ and the temporal difference of the estimated value function $V(t)$. We vary the discount factor γ as 0, 0.3, 0.6, 0.8, 0.9, and 0.99: small γ for immediate reward prediction and large γ for long future reward prediction. We then use either $V(t)$ or $\delta(t)$ as the explanatory variable in a regression analysis in the SPM brain imaging analysis program for correlation with brain activity.

8 Preliminary Results

We recruited 15 ill AN and 17 CW for this study. Both groups were similar in age, but AN individuals had higher scores on the Eating Disorders Inventory 3, Beck Depression Inventory, and lower body mass indices (BMI; data not shown). Between-group pleasantness and sweetness ratings were similar for both the Sucrose and Artificial Saliva Solutions.

Within groups, the unexpected receipt or omission of the sucrose taste stimulus activated dopamine-related brain regions such as the striatum, VTA, insula, and cingulate cortex. The whole brain group comparison indicated that unexpected receipt of the sweet taste stimulus activated brain dopamine regions more in AN than in controls, and unexpected omission of that stimulus caused more of a negative response in AN compared to the CW. These results suggest that brain taste-reward pathways linked to central dopamine transmission are functional but are overactive in ill AN compared to CW when receiving the stimulus unexpectedly, or during the omission of the taste. This response may thus suggest an increased temporal difference response in AN compared to CW. Such an over-responsiveness is consistent with clinical observations in EDs of heightened fear to novel or unexpected food stimuli, which results in stimulus avoidance and behavioral inflexibility.

In support of our biologic finding, a recent report showed that ill AN are hypersensitive to both rewarding as well as punishing stimuli (Jappe et al. 2010).

Furthermore, we imported the subject-specific reward schedule data (reward schedule during the brain imaging procedures) into a computer simulation and generated preliminary data for the VTA and AVS. As expected, the regression of the outcome activation data regressed in both CW and AN individuals significantly not only with VTA and AVS but also the insula and parts of the orbitofrontal cortex. This suggests that this dopamine-proposed mechanism involves other higher order function brain regions.

9 Limitations of the Model

The use of model-based fMRI is still in its infancy in psychiatry and there is still a lot to be learned. Several questions are still debated. The exact relationship between blood oxygen level dependent (BOLD) and brain function is not clear, such as the relationship of BOLD to action or response for instance. Evidence suggests that BOLD may reflect local processing of incoming stimuli rather than neuronal outflow to other brain areas (Logothetis 2002). Another key issue here is whether the BOLD fMRI does, in fact, reflect the proposed dopamine brain activation. The TD model used in our studies is dopamine based (Schultz 2002) and O'Doherty adapted the model to the fMRI environment (O'Doherty et al. 2003). Another carefully designed study showed that the VTA also can reflect dopamine activation when using the appropriate tasks (D'Ardenne et al. 2008). While results suggest dopamine is responsible for the BOLD response, the mechanism for such a relationship is not clear. Recent studies, however, indicate that BOLD fMRI in fact does reflect dopamine activation. Knutson for instance proposed that dopamine nucleus accumbens activation would stimulate dopamine D1 receptors and activate the BOLD signal via membrane potential changes (Knutson and Gibbs 2007). In summary, while there is still uncertainty about the exact mechanisms that determine BOLD response, increasing evidence indicates that DA function directly relates to BOLD fMRI response.

Another issue with our underlying dopamine concept is the number of brain regions involved in the model. The classic TD model focuses on the VTA and AVS and does not take other areas into account. This is a rather reductionistic model considering that emotional and cognitive processes, like learning and associations, impact the reward system activation. In response, O'Reilly created an extension to that model that takes into account the amygdala, hypothalamus, and orbitofrontal cortex (O'Reilly et al. 2007; Hazy et al. 2010). We are currently adapting computer models such as the Pavlovian Primary Reward/learned Reward model (O'Reilly et al. 2007; Hazy et al. 2010) to our imaging studies.

10 Future Outlook

We are still at the beginning of this line of research. However, results are encouraging and support the notion that we can actually study neurotransmitter-related brain pathways and identify patterns that are disorder related. This is an important step forward in the use of brain imaging, linking specific neurobiology in a psychiatric population. Such neurobiologic alterations could be used to improve treatment. For instance, making treatment very predictable may help ameliorate the dopamine hyper-responsiveness in AN, and we might be able to target specific problem behaviors with medication moderating dopamine activity.

References

Anderluh MB, Tchanturia K et al (2003) Childhood obsessive–compulsive personality traits in adult women with eating disorders: defining a broader eating disorder phenotype. Am J Psychiatry 160(2):242–247

APA (2000) Diagnostic & statistical manual of mental disorders: DSM-IV-TR. American Psychiatric Association

Apicella P, Ljungberg T et al (1991) Responses to reward in, monkey dorsal and ventral striatum. Exp Brain Res 85(3):491–500

Bartoshuk LM, Beauchamp GK (1994) Chemical senses. Annu Rev Psychol 45:419–449

Bergh C, Sodersten P (1996) Anorexia nervosa, self-starvation and the reward of stress. Nat Med 2 (1):21–22

Berns G, McClure S et al (2001) Predictability modulates human brain response to reward. J Neurosci 21(8):2793–2798

Berridge KC (1996) Food reward: brain substrates of wanting and liking. Neurosci Biobehav Rev 20(1):1–25

Berridge KC, Robinson TE (1998) What is the role of dopamine in reward: hedonic impact, reward learning, or incentive salience? Brain Res Brain Res Rev 28(3):309–369

Berridge KC, Robinson T (2003) Parsing reward. Trends Neurosci 26(9):507–513

Blum K, Sheridan PJ et al (1995) Dopamine D2 receptor gene variants: association and linkage studies in impulsive–addictive–compulsive behaviour. Pharmacogenetics 5(3):121–141
Breiter HC, Aharon I et al (2001) Functional imaging of neural responses to expectancy and experience of monetary gains and losses. Neuron 30(2):619–639
Cannon C, Bseikri M (2004) Is dopamine required for natural reward? Physiol Behav 81(5): 741–748
Cannon CM, Palmiter RD (2003) Reward without dopamine. J Neurosci 23(34):10827–10831
Carr K, Tsimberg Y et al (2003) Evidence of increased dopamine receptor signaling in food-restricted rats. Neuroscience 119:1157–1167
Cooper J, Bloom F et al (2003) The biochemical basis of neuropharmacology. Oxford University Press, Oxford
Crystal S, Frye CA et al (1995) Taste preferences and sensory perceptions in female varsity swimmers. Appetite 24(1):25–36
D'Ardenne K, McClure SM et al (2008) BOLD responses reflecting dopaminergic signals in the human ventral tegmental area. Science 319(5867):1264–1267
Delgado MR, Nystrom LE et al (2000) Tracking the hemodynamic responses to reward and punishment in the striatum. J Neurophysiol 84:3072–3077
Delgado MR, Locke HM et al (2003) Dorsal striatum responses to reward and punishment: effects of valence and magnitude manipulations. Cogn Affect Behav Neurosci 3(1):27–38
Di Costanzo V, Rodde G et al (1998) Food preferences in anorectic girls at the beginning of therapy. Diabetes Metab 24(3):262–271
Drewnowski A, Halmi KA et al (1987) Taste and eating disorders. Am J Clin Nutr 46(3):442–450
Eiber R, Berlin I et al (2002) Hedonic response to sucrose solutions and the fear of weight gain in patients with eating disorders. Psychiatry Res 113:173–180
Francis S, Rolls ET et al (1999) The representation of pleasant touch in the brain and its relationship with taste and olfactory areas. Neuroreport 10(3):453–459
Frank G, Kaye W et al (2003) The evaluation of brain activity in response to taste stimuli – a pilot study and method for central taste activation as assessed by event related fMRI. J Neurosci Methods 131(1–2):99–105
Frank G, Bailer UF et al (2005) Increased dopamine D2/D3 receptor binding after recovery from anorexia nervosa measured by positron emission tomography and [^{11}C]raclopride. Biol Psychiatry 58(11):908–912
Grigson PS (2002) Like drugs for chocolate: separate rewards modulated by common mechanisms? Physiol Behav 76:389–395
Hajnal A, Norgren R et al (2009) Parabrachial coding of sapid sucrose: relevance to reward and obesity. Ann N Y Acad Sci 1170:347–364
Hazy TE, Frank MJ et al (2010) Neural mechanisms of acquired phasic dopamine responses in learning. Neurosci Biobehav Rev 34:701–720
Higgins GA, Fletcher PJ (2003) Serotonin and drug reward: focus on 5-HT2C receptors. Eur J Pharmacol 480(1–3):151–162
Hikosaka K, Watanabe M (2000) Delay activity of orbital and lateral prefrontal neurons of the monkey varying with different rewards. Cereb Cortex 10(3):263–271
Jappe LM, Frank GKW et al (2010) Heightened sensitivity to reward and punishment in anorexia nervosa. International Journal of Eating Disorders (Article first published online: 28 JUN)
Kampov-Polevoy AB, Ziedonis D et al (2003) Association between sweet preference and paternal history of alcoholism in psychiatric and substance abuse patients. Alcohol Clin Exp Res 27(12):1929–1936
Kaye WH, Frank GK et al (1999) Altered dopamine activity after recovery from restricting-type anorexia nervosa. Neuropsychopharmacology 21(4):503–506
Kelley AE, Baldo BA et al (2005) Corticostriatal-hypothalamic circuitry and food motivation: integration of energy, action and reward. Physiol Behav 86(5):773–795
Knutson B, Gibbs SE (2007) Linking nucleus accumbens dopamine and blood oxygenation. Psychopharmacology (Berl) 191(3):813–822

Knutson B, Westdorp A et al (2000) FMRI visualisation of brain activity during a monetary incentive dealy task. Neuroimage 12:20–27
Koob GF, Le Moal M (2005) Plasticity of reward neurocircuitry and the 'dark side' of drug addiction. Nat Neurosci 8(11):1442–1444
Koslow SH (2005) Discovery and integrative neuroscience. Clin EEG Neurosci 36(2):55–63
Logothetis NK (2002) The neural basis of the blood-oxygen-level-dependent functional magnetic resonance imaging signal. Philos Trans R Soc Lond B Biol Sci 357(1424):1003–1037
Mansvelder HD (2005) Yin and yang of VTA opioid signaling. Focus on "both kappa and mu opioid agonists inhibit glutamatergic input to ventral tegmental area neurons". J Neurophysiol 93(6):3046–3047
Martin-Soelch C, Leenders KL et al (2001) Reward mechanisms in the brain and their role in dependence: evidence from neurophysiological and neuroimaging studies. Brain Res Brain Res Rev 36(2–3):139–149
May JC, Delgado MR et al (2004) Event-related functional magnetic resonance imaging of reward-related brain circuity in children and adolescents. Biol Psychiatry 55:359–366
Montague R, Hyman S et al (2004) Computational roles for dopamine in behavioural control. Nature 431:760–767
Norgren R, Hajnal A et al (2006) Gustatory reward and the nucleus accumbens. Physiol Behav 89(4):531–535
O'Doherty JP, Dayan P et al (2003) Temporal difference models and reward-related learning in the human brain. Neuron 38(2):329–337
O'Doherty J, Dayan P et al (2004) Dissocaible roles of ventral and dorsal striatum in instrumental conditioning. Science 304:452–454
O'Reilly RC, Frank MJ et al (2007) PVLV: the primary value and learned value Pavlovian learning algorithm. Behav Neurosci 121(1):31–49
Price JL (2005) Free will versus survival: brain systems that underlie intrinsic constraints on behavior. J Comp Neurol 493(1):132–139
Saper CB, Chou TC et al (2002) The need to feed: homeostatic and hedonic control of eating. Neuron 36(2):199–211
Schultz W (1998) Predictive reward signal of dopamine neurons. J Neurophysiol 80(1):1–27
Schultz W (2002) Getting formal with dopamine and reward. Neuron 36(2):241–263
Schultz W (2006) Behavioral theories and the neurophysiology of reward. Annu Rev Psychol 57:87–115
Schultz W, Tremblay L et al (2000) Reward processing in primate orbitofrontal cortex and basal ganglia. Cerebral Cortex 10(3):272–284
Schweiger U, Fichter M (1997) Eating disorders: clinical presentation, classification and etiologic models. In: Jimerson DC, Kaye WH (eds) Balliere's clinical psychiatry. Balliere's Tindall, London, pp 199–216
Simon Y, Bellisle F et al (1993) Taste responsiveness in anorexia nervosa. Br J Psychiatry 162:244–246
Sullivan PF (1995) Mortality in anorexia nervosa. Am J Psychiatry 152(7):1073–1074
Sunday SR, Halmi KA (1990) Taste perceptions and hedonics in eating disorders. Physiol Behav 48(5):587–594
Sutton RS, Barto AG (1981) Toward a modern theory of adaptive networks: expectation and prediction. Psychol Rev 88(2):135–170
Tanaka SC, Doya K et al (2004) Prediction of immediate and future rewards differentially recruits cortico-basal ganglia loops. Nat Neurosci 7(8):887–893
Toth E, Kondakor I et al (2004) Nonlinear and linear EEG complexity changes caused by gustatory stimuli in anorexia nervosa. Int J Psychophysiol 51(3):253–260
Tricomi EM, Delgado MR et al (2004) Modulation of caudate activity by action contingency. Neuron 41:281–292
Volkow ND, Fowler JS et al (2004) Dopamine in drug abuse and addiction: results from imaging studies and treatment implications. Mol Psychiatry 9(6):557–569

Wagner A, May C et al (2005) Reward-related neural responses in anorexia and bulimia nervosa after recovery using functional magnetic resonance imaging. Biol Psychiatry 57(S7):709

Wagner A, Aizenstein H et al (2007) Altered reward processing in women recovered from anorexia nervosa. Am J Psychiatry 164(12):1842–1849

Wise RA (2002) Brain reward circuitry: insights from unsensed incentives. Neuron 36(2):229–240

Wise RA (2005) Forebrain substrates of reward and motivation. J Comp Neurol 493(1):115–121

Worgotter F, Porr B (2005) Temporal sequence learning, prediction, and control: a review of different models and their relation to biological mechanisms. Neural Comput 17(2):245–319

Yasoshima Y, Yamamoto T (2005) Effects of midazolam on the expression of conditioned taste aversion in rats. Brain Res 1043(1–2):115–123

Zhou QY, Palmiter RD (1995) Dopamine-deficient mice are severely hypoactive, adipsic, and aphagic. Cell 83(7):1197–1209

Cognitive-Behavioral Flexibility in Anorexia Nervosa

Hans-Christoph Friederich and Wolfgang Herzog

Contents

Abstract Anorexia nervosa (AN) patients are characterized by perfectionism and obsessional personality traits. This anorectic personality type is associated with an exaggerated cognitive control and impaired cognitive-behavioral flexibility. Neuropsychological studies addressing flexibility have supported an impaired cognitive set-shifting (i.e., concrete and rigid behaviors to changing rules) as well as an impaired behavioral response shifting (i.e., stereotyped or perseverative behaviors) in AN patients independent of nutritional status and body weight. Furthermore, impaired set-shifting was found in healthy sisters of AN patients suggesting that cognitive inflexibility is a trait marker in AN patients. Brain imaging studies have provided new insights in striatocortical circuit dysfunctions that may underlie both the clinical symptoms of obsessive-compulsive personality traits and the neuropsychological observations of impaired cognitive-behavioral flexibility. The conceptualization of AN as a neurodevelopmental striatocortical disorder may help to develop new promising treatment approaches for this severe disorder.

H.-C. Friederich (✉) and W. Herzog
Department of General Internal Medicine and Psychosomatics Im Neuenheimer Felds 410, Medical Hospital, University of Heidelberg, 69120, Heidelberg, Germany
e-mail: hans-christoph.friederich@med.uni-heidelberg.de, wolfgang.herzog@med.uni-heidelberg.de

R.A.H. Adan and W.H. Kaye (eds.), *Behavioral Neurobiology of Eating Disorders*, Current Topics in Behavioral Neurosciences 6, DOI 10.1007/7854_2010_83

Keywords Anorexia nervosa · Cognitive flexibility · Neuroimaging · Neuropsychology

1 Introduction

Anorexia nervosa (AN) patients are characterized by obsessive-compulsive temperament traits that manifest predominantly in maladaptive preoccupation with food, weight, and body shape. In accordance with these preoccupations, patients show rigid and stereotyped behaviors controlling eating and weight. Besides compulsions related to food, patients with AN show regularly symptoms of symmetry obsessions and order/arrangement compulsions. These clinical observations indicate general alterations in cognitive control processes in AN patients. Given the very effective control over food intake in AN patients, an exaggerated cognitive-behavioral control is considered relevant to this disorder. One of the manifestations of exaggerated cognitive control is an impaired cognitive-behavioral flexibility. AN patients stick to behavioral patterns, routines, and attitudes they are familiar with and show low novelty-seeking and exploratory behavior. Thus, AN patients show phenomenological similarities with obsessive-compulsive disorder (OCD) patients. However, in contrast to an exaggerated cognitive control in AN patients, patients with OCD show an impaired cognitive control associated with a difficulty in controlling ego-dystonic thoughts and behaviors. The striatocortical circuits play a pivotal role in the capacity of cognitive control in both health and illness (Alexander et al. 1990). In OCD, functional neuroimaging studies have consistently shown a dysfunctional involvement of striatocortical circuits in cognitive control tasks (Friedlander and Desrocher 2006). This has encouraged the investigation of neural correlates of impaired cognitive-behavioral flexibility in AN patients.

This chapter reviews and integrates findings from personality research as well as from neuropsychological, neuroimaging, and neurochemical studies that contribute to the understanding of cognitive-behavioral flexibility in AN.

A specificity of psychobiological research in AN patients is that starvation and weight loss have profound effects on the brain that complicates the differentiation between causes and consequences of the disorder.

1.1 Cognitive-Behavioral Inflexibility: A State or Trait Marker?

Due to the underweight factor, chronic AN is associated with severe structural brain alterations, which make it difficult to differentiate between state and trait markers. Structural neuroimaging studies in underweight AN patients show a volume reduction in gray matter with enlarged ventricular volumes and dilated sulci (Treasure and Friederich 2007). The functional relevance of these structural brain alterations in starvation is largely unknown. However, starvation and irregular menstrual function in AN are associated with neurocognitive impairments covering several cognitive domains including cognitive flexibility (Katzman et al. 2001; Roberts et al. 2007). Whether structural brain abnormalities are completely reversible with weight gain is a controversial topic in the literature (Katzman et al. 2001; Muhlau et al. 2007; Wagner et al. 2006). Therefore, it is important to differentiate between (1) neurobiological trait alterations that precede the onset of the disorder and that contribute to an increased risk of developing the disorder and (2) state alterations that are secondary to starvation and that may play a role in sustaining the disorder by reinforcing neurobiological mechanisms.

There is evidence from the Minnesota Starvation Experiment (Keys et al. 1950) that food restriction leads to food obsessions and increased rigidity of thinking and behavior. One of the participants of this classic starvation experiment remembered the frustration of preoccupation with food:

> "...food became the one and central and only thing really in one's life. And life is really dull if that's the only thing. I mean if you went to a movie, you were not particularly interested in the love scenes, but you noticed every time they ate and what they ate" (Keys et al. 1950).

This finding in young healthy men corroborates starvation-induced impairments in flexibility of cognitions and behavior independent of a premorbid vulnerability. However, AN patients show impaired cognitive-behavioral flexibility irrespective of confounding factors of weight loss and starvation. Due to the low prevalence and the long duration of the disorder, there are no prospective longitudinal studies of cognitive flexibility in AN patients that have reached complete weight restoration. Cross-sectional investigation of long-term recovered AN patients is a suitable and feasible method for investigating trait alterations and controlling for confounding factors of starvation and emaciation. However, findings in long-term recovered AN patients might also represent a "scar" effect caused by starvation in the past.

2 Temperament and Cognitive-Behavioral Flexibility

Theory and research on personality in AN suggest an association between impaired cognitive-behavioral flexibility and the anorectic personality type. This personality type is characterized by obsessionality, perfectionism, and avoidance of novel situations.

Clinical characteristics of intense interests and preoccupation of AN patients with food, cooking, and recipes as well as routines and ritualistic behaviors of eating, weight checking, and exercising have inspired the comparison between AN and obsessive-compulsive personality disorder (OCDP) and OCD, respectively. Based on clinical interviews, the estimated prevalence of a comorbid OCDP in AN patients ranges between 20% and 30% (Halmi et al. 2005; Matsunaga et al. 1999). About the same prevalence is reported for OCD. AN patients predominantly show obsessions and compulsions with a need for symmetry and order, whereas

OCD patients frequently show aggressive obsessions and checking compulsions. In total, about half of the patients with AN fulfill the criteria either for an OCDP, an OCD or for both. According to the retrospective reports, childhood obsessive-compulsive traits are considered an important risk factor for the development of an eating disorder (Anderluh et al. 2003). Obsessive-compulsive personality traits seem to be involved in the onset, symptomatic expression, and maintenance of anorectic behaviors and they persist after recovery (Cassin and von Ranson 2005; Srinivasagam et al. 1995).

The level of perfectionism is closely associated with the magnitude of obsessions and compulsions, although a closer relationship has been observed for obsessive-compulsive personality traits compared to OCD. Pathological perfectionism in AN is regularly found in currently ill AN patients (Bulik et al. 2003), in recovered AN patients (Srinivasagam et al. 1995), and in twin sisters of AN patients without an eating disorder (Wade et al. 2008). Premorbid perfectionism is also considered as a risk factor for the development of AN (Anderluh et al. 2003; Fairburn et al. 1999). The anorectic temperament is further characterized by avoidance of novel situations and of behavioral modifications (Holliday et al. 2006; Klump et al. 2000). In summary, research on temperament traits consistently showed that the anorectic personality type was detectable irrespective of confounding influences of starvation and weight loss on behavior. The described trait characteristics have motivated the investigation of cognitive-behavioral flexibility using neuropsychological paradigms.

3 Neuropsychology

Cognitive flexibility is a fundamental component of cognitive ("executive") control together with inhibition, planning, attentional control, and working memory. Cognitive flexibility allows us to adapt our behavior making us flexible to ever-changing environmental demands. It is understood as the ability to override automatic or "prepotent" behaviors in favor of planned and intended behaviors (Miller and Cohen 2001). Cognitive inflexibility instead is associated with perseverative errors, stereotyped reactions, and rigid behaviors.

Cognitive flexibility is a broad and underspecified psychological construct that has been addressed with different neuropsychological paradigms. Paradigms used in AN can be differentiated in cognitive set-shifting (i.e., concrete and rigid behaviors to changing rules) and behavioral response shifting (i.e., stereotyped or perseverative behaviors) paradigms.

Set-shifting has most frequently been assessed with the Wisconsin Card Sorting Test (WCST: Berg (1948)) and the Trial Making Test A, B (TMT A, B: Reitan (1958)) in AN patients. A recent meta-analysis in AN patients showed an impaired task performance with a small-to-medium effect size in these tasks (Roberts et al. 2007). TMT B contains letters (A–L) and figures (1–13) that are spread over one page. The participant is required to draw as quickly as possible a line connecting letters and figures in an ascending order. This involves shifting cognitive "sets" between numbers and letters (i.e., 1-A-2-B-3-C etc.). The time needed to complete trail B is used as a measure of set-shifting ability. TMT A contains exclusively ascending figures and serves to control for visuospatial and visuomotor processing. Similarly, in the WCST, subjects are instructed to sort cards according to the shape, color, or number of symbols appearing on them. The sorting rule varies periodically and unpredictably, and rules have to be learned implicitly during the course of the task. The number of perseverative errors is used as a measure of set-shifting ability. Various alternative set-shifting tasks have been used that mostly support an impaired set-shifting in AN (Roberts et al. 2007).

A general problem in neuropsychological research of more complex tasks such as set-shifting is that they involve various cognitive functions. The above-mentioned tasks can also be described as tapping the cognitive function of working memory, and the WCST also requires functions of implicit learning. This is of relevance to AN research, as underweight is associated with cognitive impairments covering several cognitive domains.

Recovery in AN patients seems to be associated with an improvement of set-shifting ability, although their performance is still lower compared to that of noneating disordered controls (Jones et al. 1991; Roberts et al. 2007; Tchanturia et al. 2004). Further evidence for impaired set-shifting as a trait marker comes from observations of impaired set-shifting in healthy sisters of AN patients (Holliday et al. 2005). Longitudinal studies are difficult given the rarity of the disorder and the requirement of long-term follow-up . There are longitudinal studies that assessed AN patients before and after weight gain, but findings are not representative as most patients were still underweight after weight gain (Kingston et al. 1996; Tchanturia et al. 2004).

In addition to set-shitting, our research group has studied behavioral response shifting. Behavioral response shifting requires one to inhibit a prepotent impulse in favor of an alternative, less automatic behavior (Zastrow et al. 2009). In this study, AN patients were less accurate and showed more perseverative errors in behavioral response shifting compared to healthy controls. There was no support for greater impulsivity in AN patients, as reaction times were not different between patients and controls (Zastrow et al. 2009). Long-term recovered AN patients showed an improved performance that was significantly different from neither healthy controls nor underweight AN patients, but had accuracy scores intermediate to these groups (Friederich et al., unpublished results). To respond more accurately than currently ill AN patients, recovered AN patients needed significantly more time for response shifting compared to currently ill AN patients and healthy controls (Friederich et al., unpublished results). The finding of impaired behavioral response shifting in AN patients is supported by impairments in neuromotor function tests, such as diadochokinesis and two-finger tapping (Green et al. 1996). A follow-up study after 18 years of adolescent-onset AN patients indicates that this deficit of neuromotor function seems to persist after weight restoration (Gillberg et al. 2009). In summary, these findings suggest that clinical observations of impaired cognitive-behavioral flexibility in AN patients underlie impaired set-shifting and impaired behavioral response shifting. Starvation and malnutrition seem to exaggerate these neuropsychological traits. While these paradigms each have their distinctive features, they share some processes in common, related, for example, to response shifting and response inhibition in the context of behavioral change and cognitive plasticity. There is a need for future studies that assess the links and differences between cognitive and behavioral aspects of impaired flexibility in the same sample of AN patients. These studies should include a systematic hierarchical neurocognitive testing to delineate the different cognitive and behavioral mechanisms involved in flexibility tasks.

4 Neuroimaging

The functional neuroanatomy of cognitive-behavioral flexibility has extensively been studied in nonclinical samples. These studies show that striatocortical pathways play a pivotal role in shifts of cognitive-set and behavioral response. The corticosubcortical reentrant circuits are differentiated in ventral striatocortical pathways including the amygdala, ant. insula, ventral ACC and OFC, and dorsal striatocortical pathways including the dorsolateral prefrontal cortex, the parietal cortex, and the dorsal insular region (Alexander et al. 1990; Heimer 2003). The ventral circuit sends afferents to the ventral striatum and is involved in motivational processing and reinforcement learning, whereas the dorsal circuit is more closely connected to dorsolateral parts of the striatum and reflects effortful cognitive functions of goal-directed behaviors. In functional neuroimaging studies, the lateral prefrontal cortex and the anterior cingulate/ventromedial prefrontal cortex have been regularly found to be related to cognitive set-shifting and inhibitory control (Aron et al. 2003; Cools et al. 2004; Shafritz et al. 2005). Subcortical regions such as the thalamus, the subthalamic nucleus, and the striatum facilitate shifts in behavioral response mediated by the prefrontal cortex (Block et al. 2007; Monchi et al. 2006). The more cognitive demanding the flexibility task is, the greater the top-down control of the frontoparietal network (Miller and Cohen 2001). The dorsal parietal cortex plays a pivotal role in attention shifting and in detection of specific or salient targets (Gurd et al. 2002; Hampshire et al. 2007; Snyder et al. 2000).

Our research group has investigated neural correlates of behavioral response shifting in AN patients using functional magnetic resonance imaging (MRI). The paradigm required patients to inhibit a prepotent impulse in favor of an alternative, less automatic behavior. Impaired performance in behavioral response shifting in patients was associated with task-related hypoactivations at different levels (thalamus, ventral striatum, rostral anterior cingulate cortex) of the fronto-striato-thalamic network. These alterations were specific to shifts in behavioral response (except for thalamus activation), as they were not observed in response to nontarget trials that required recognition of the stimulus but no alteration in behavioral response (Fig. 1, Panel B).

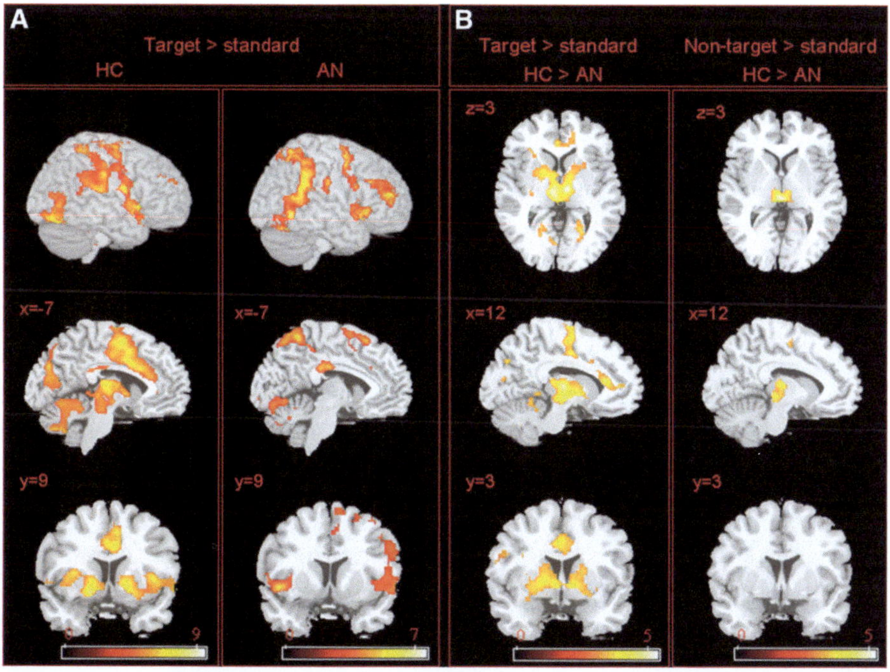

Fig. 1 Brain activation in response to target and nontarget relative to standard trials in healthy comparison subjects ($N = 15$) and anorexia nervosa patients ($N = 15$). *Panel A* shows representative brain slices and surface renderings (*right hemisphere*) of brain activation in response to target relative to standard trials in both groups. *Panel B* shows brain regions that showed greater activations in comparison subjects relative to patients in response to target and nontarget trials relative to standard trials. Brain slices are shown in neurological convention (*right hemisphere* is on *the right-hand side* of the axial and coronal image). The color of the cluster represents the intensity of activation (T value) according to the gradation of the *color bar*. Reprinted with permission from the American Journal of Psychiatry, (Copyright 2009). American Psychiatry Association

AN patients showed preserved neural metabolism in the right ventral frontoparietal network (the right middle frontal gyrus and the temporoparietal junction bilaterally) that was not observed in the comparison group (Fig. 1, Panel A). The ventral frontoparietal network system is involved in allocating and directing attention, when infrequent behaviorally relevant stimuli occur (Corbetta and Shulman 2002). Similar findings of a dominance of higher-order association cortices (the frontoparietal network) over impaired ascending motivational processes have been reported in a study of recovered AN patients using a reward paradigm (Wagner et al. 2007), suggesting a task-independent pattern of predominant effortful and supervisory control in AN patients.

Compared to healthy controls, AN patients showed an impaired activation of the ventral striatum during behavioral response shifting. As mentioned above, the ventral striatocortical circuit is involved in motivational and reward-related processes. This is in accord with studies of AN that have reported reduced ventral striatal activation to wins and losses, altered dopamine D2/D3 receptor binding of the ventral striatum, or a generally reduced sensitivity of the appetitive motivational system (Frank et al. 2005; Friederich et al. 2006; Wagner et al. 2007). However, in this study we have not found an increased metabolism in the dorsal striatum of AN patients as a central structure of the dorsal striatocortical network. This would have further supported the hypothesis of an imbalance between the ventral and dorsal striatocortical system in AN patients (Kaye et al. 2009). Further evidence for an altered striatal activity (especially caudate activity) comes from several positron imaging studies in the resting state (Delvenne et al. 1997; Herholz et al. 1987).

AN patients also showed impaired activation of the rostral anterior cingulate cortex. The anterior cingulate cortex plays an important role in performance monitoring and conflict detection. It has been implicated in emotional processing to control and initiate behavior. An altered anterior cingulate cortex function is further supported by previous morphometric and functional brain imaging studies in AN patients (Muhlau et al. 2007; Naruo et al. 2001; Uher et al. 2003).

In summary, recent neuroimaging studies indicate that impaired cognitive-behavioral flexibility in AN patients is associated with functional disturbances within striatocortical circuits.

AN typically occurs during puberty – a critical time span for the maturation of striatocortical pathways (Sowell et al. 1999). Neurodevelopmental studies on cognitive control in healthy subjects show that performance continues to improve over childhood into adolescence and adulthood (Marsh et al. 2006; Rubia et al. 2000; Sowell et al. 1999). Children have a more immature cognitive control network than adults that is more susceptible to interference regardless of salience. Adult AN patients show a decreased cognitive control performance in the presence of distracters. Findings were irrespective of whether the stimuli were disorder related (i.e., food), emotionally salient or neutral (Dickson et al. 2008). Female subjects with a vulnerability for AN may show disturbances in the maturation of striatocortical pathways complicated by pubertal and starvation-induced neurochemical changes.

5 Neuromodulation

Neurochemically, the functioning of the striatocortical loops is influenced by a number of ascending neurotransmitter systems, including catecholamines (dopamine and noradrenaline) and 5-hydroxytryptamine (serotonin) (Robbins 2000). These neurochemical systems are modulated by stress, mood, and motivational processes. Selective neuropsychopharmacological manipulations have been used to explore the distinct functions of the monoamine systems. The modulatory role for dopamine in the ventral striatum for set-shifting tasks (i.e., reversal learning) is well established. The ventral striatum is considered a primary locus for the coding of the reward predication error, which contributes to the processes of reward selection and anticipation. In contrast, serotonin seems to modulate the processing of punishment-related information or other aversive signals, and thus may serve as an opponent to dopamine. However, the possible role of serotonin in cognitive control seems complex and is not very well understood (Robbins and Arnsten 2009). Norepinephrine has been mainly implicated in response inhibition and sustained attention (Chamberlain et al. 2009). The presented findings support the hypothesis that functions of cognitive flexibility are susceptible to modulation by ascending neurotransmitter systems.

To our knowledge, there are no neuropsychopharmacological studies in AN investigating cognitive control processes. However, neurotransmitter functions and pathways, as well as their relation to pathological behavior have been investigated with brain imaging studies using neurotransmitter ligands. There is a substantial body of evidence that alterations in serotonin neurotransmission may be involved in the etiology and pathogenesis of AN (Kaye et al. 2009). Disturbed serotonin (5-HT) neurotransmission has been reported in different brain regions including the anterior cingulate cortex (Bailer et al. 2005) and ventral striatum (Bailer et al. 2007) of recovered AN patients. Noteworthy is that studies of 5-HT_{1A} and 5HT_{2A} receptor binding have shown consistently close correlations with harm avoidance (Bailer et al. 2004, 2005). Harm avoidance is a temperament trait associated with anxiety and cognitive inflexibility. For more detailed information on serotonin neurotransmission in AN, see chapter "Serotonin –Imaging Findings" in this book.

There is also preliminary evidence for disturbed dopamine neurotransmission in the ventral striatum suggesting a disturbance of reward mechanisms. Frank et al. (2005) demonstrated an increased D2/D3 receptor binding in recovered AN subjects in the ventral striatum. Furthermore, dopamine binding capacity in the dorsal striatum was positively associated with harm avoidance.

6 Conclusion and Future Directions

Both temperament and neuropsychological studies support an impaired cognitive and behavioral flexibility in both AN subtypes. Impaired flexibility has been found independent of the disease stage in currently ill and recovered AN patients. Furthermore, it was also found in healthy sisters of AN patients, suggesting that cognitive-behavioral inflexibility is a trait marker in AN patients. Despite these consistencies, the direct association between personality measures of the anorectic phenotype (e.g., obsessional personality traits) and neuropsychological measures of cognitive flexibility in AN patients seem to be variable.

Neuropsychological studies have predominantly used set-shifting tasks to assess flexibility to changing and alternating rules. Besides a concrete and rigid approach in set-shifting tasks, there is evidence for impaired behavioral response shifting (i.e., stereotyped and perseverate behaviors) in AN patients. Thus, further research is needed to delineate the distinct subcomponents of flexibility that are impaired in AN patients.

Functional brain imaging in currently ill AN patients has provided novel insights into neural mechanisms of impaired behavioral flexibility that may represent a functional disturbance within the ventral striatocortical network. This network has been implicated in motivational and incentive processes to control cognitive and behavioral operations. Further research is necessary to evaluate whether cognitive inflexibility, like behavioral inflexibility is also secondary to impaired signaling of incentive salience and motivational processes or whether it involves different neural pathways. One hypothesis that needs further evidence is whether AN patients show an imbalance within striatocortical systems with hypoactivation of ventral striatal systems involved in motivational processing and hyperactivation of dorsal striatal systems involved in cognitive control and reflective processing. There are no studies that have assessed neural correlates of cognitive-behavioral flexibility in recovered AN patients. Therefore, it is unknown whether these functional alterations are reversible with weight restoration and long-term recovery. However, findings from a gambling task in recovered AN patients support an impaired ventral striatocortical circuit independent of confounding factors of starvation. These neuropsychological and neuroimaging findings suggest that impaired flexibility is exacerbated in the acute phase of the illness but is also present in recovery stages of the disease.

It is to be assumed that trait- and state-related disturbances of cognitive-behavioral flexibility are related to genetic vulnerabilities of altered monoamine neuromodulation. In AN patients, disturbances in the serotonin and dopamine system have been described most consistently. These neurochemical alterations may strengthen and attain significance during hormonal changes and brain maturation in puberty. Puberty is a vulnerable phase for the maturation of cognitive control systems. The characteristic onset of AN around puberty, associated with noticeable problems of cognitive control, supports the conceptualization of AN as a neurodevelopmental disorder of the striatocortical system (Connan et al., 2003; Marsh et al., 2009). The neurodevelopmental disorders of the striatocortical system are highly comorbid, which may explain the overlap of diagnoses within the spectrum of OCDs.

The conceptualization of AN as a neurodevelopmental disorder of striatocortical neurocircuitries may be a promising working model for further research to improve our understanding of this complex and paradox disorder.

References

Alexander GE, Crutcher MD, DeLong MR (1990) Basal ganglia-thalamocortical circuits: parallel substrates for motor, oculomotor, "prefrontal" and "limbic" functions. Prog Brain Res 85:119–146

Anderluh MB, Tchanturia K, Rabe-Hesketh S, Treasure J (2003) Childhood obsessive-compulsive personality traits in adult women with eating disorders: defining a broader eating disorder phenotype. Am J Psychiatry 160:242–247

Aron AR, Fletcher PC, Bullmore ET, Sahakian BJ, Robbins TW (2003) Stop-signal inhibition disrupted by damage to right inferior frontal gyrus in humans. Nat Neurosci 6:115–116

Bailer UF, Price JC, Meltzer CC, Mathis CA, Frank GK, Weissfeld L, McConaha CW, Henry SE, Brooks-Achenbach S, Barbarich NC, Kaye WH (2004) Altered 5-HT(2A) receptor binding after recovery from bulimia-type anorexia nervosa: relationships to harm avoidance and drive for thinness. Neuropsychopharmacology 29:1143–1155

Bailer UF, Frank GK, Henry SE, Price JC, Meltzer CC, Weissfeld L, Mathis CA, Drevets WC, Wagner A, Hoge J, Ziolko SK, McConaha CW, Kaye WH (2005) Altered brain serotonin 5-HT1A receptor binding after recovery from anorexia nervosa measured by positron emission tomography and [carbonyl11C]WAY-100635. Arch Gen Psychiatry 62:1032–1041

Bailer UF, Frank GK, Henry SE, Price JC, Meltzer CC, Becker C, Ziolko SK, Mathis CA, Wagner A, Barbarich-Marsteller NC, Putnam K, Kaye WH (2007) Serotonin transporter binding after recovery from eating disorders. Psychopharmacology 195:315–324

Berg EA (1948) A simple objective technique for measuring flexibility in thinking. J Gen Psychol 39:15–22

Block AE, Dhanji H, Thompson-Tardif SF, Floresco SB (2007) Thalamic-prefrontal cortical-ventral striatal circuitry mediates dissociable components of strategy set shifting. Cereb Cortex 17:1625–1636

Bulik CM, Tozzi F, Anderson C, Mazzeo SE, Aggen S, Sullivan PF (2003) The relation between eating disorders and components of perfectionism. Am J Psychiatry 160:366–368

Cassin SE, von Ranson KM (2005) Personality and eating disorders: a decade in review. Clin Psychol Rev 25:895–916

Chamberlain SR, Hampshire A, Muller U, Rubia K, Del CN, Craig K, Regenthal R, Suckling J, Roiser JP, Grant JE, Bullmore ET, Robbins TW, Sahakian BJ (2009) Atomoxetine modulates right inferior frontal activation during inhibitory control: a pharmacological functional magnetic resonance imaging study. Biol Psychiatry 65:550–555

Connan F, Campbell IC, Katzman M, Lightman SL, Treasure J (2003) A neurodevelopmental model for anorexia nervosa. Physiol Behav 79:13–24

Cools R, Clark L, Robbins TW (2004) Differential responses in human striatum and prefrontal cortex to changes in object and rule relevance. J Neurosci 24:1129–1135

Corbetta M, Shulman GL (2002) Control of goal-directed and stimulus-driven attention in the brain. Nat Rev Neurosci 3:201–215

Delvenne V, Goldman S, De MV, Wikler D, Damhaut P, Lotstra F (1997) Brain glucose metabolism in anorexia nervosa and affective disorders: influence of weight loss or depressive symptomatology. Psychiatry Res 74:83–92

Dickson H, Brooks S, Uher R, Tchanturia K, Treasure J, Campbell IC (2008) The inability to ignore: distractibility in women with restricting anorexia nervosa. Psychol Med 38:1741–1748

Fairburn CG, Cooper Z, Doll HA, Welch SL (1999) Risk factors for anorexia nervosa: three integrated case-control comparisons. Arch Gen Psychiatry 56:468–476

Frank GK, Bailer UF, Henry SE, Drevets W, Meltzer CC, Price JC, Mathis CA, Wagner A, Hoge J, Ziolko S, Barbarich-Marsteller N, Weissfeld L, Kaye WH (2005) Increased dopamine D2/D3 receptor binding after recovery from anorexia nervosa measured by positron emission tomography and [11c]raclopride. Biol Psychiatry 58:908–912

Friederich HC, Kumari V, Uher R, Riga M, Schmidt U, Campbell IC, Herzog W, Treasure J (2006) Differential motivational responses to food and pleasurable cues in anorexia and bulimia nervosa: a startle reflex paradigm. Psychol Med 36:1327–1335

Friedlander L, Desrocher M (2006) Neuroimaging studies of obsessive-compulsive disorder in adults and children. Clin Psychol Rev 26:32–49

Gillberg IC, Billstedt E, Wentz E, Anckarsater H, Rastam M, Gillberg C (2009) Attention, executive functions, and mentalizing in anorexia nervosa eighteen years after onset of eating disorder. J Clin Exp Neuropsychol 1–8

Green MW, Elliman NA, Wakeling A, Rogers PJ (1996) Cognitive functioning, weight change and therapy in anorexia nervosa. J Psychiatr Res 30:401–410

Gurd JM, Amunts K, Weiss PH, Zafiris O, Zilles K, Marshall JC, Fink GR (2002) Posterior parietal cortex is implicated in continuous switching between verbal fluency tasks: an fMRI study with clinical implications. Brain 125:1024–1038

Halmi KA, Tozzi F, Thornton LM, Crow S, Fichter MM, Kaplan AS, Keel P, Klump KL, Lilenfeld LR, Mitchell JE, Plotnicov KH, Pollice C, Rotondo A, Strober M, Woodside DB, Berrettini WH, Kaye WH, Bulik CM (2005) The relation among perfectionism, obsessive-compulsive personality disorder and obsessive-compulsive disorder in individuals with eating disorders. Int J Eat Disord 38:371–374

Hampshire A, Duncan J, Owen AM (2007) Selective tuning of the blood oxygenation level-dependent response during simple target detection dissociates human frontoparietal subregions. J Neurosci 27:6219–6223

Heimer L (2003) A new anatomical framework for neuropsychiatric disorders and drug abuse. Am J Psychiatry 160:1726–1739

Herholz K, Krieg JC, Emrich HM, Pawlik G, Beil C, Pirke KM, Pahl JJ, Wagner R, Wienhard K, Ploog D (1987) Regional cerebral glucose metabolism in anorexia nervosa measured by positron emission tomography. Biol Psychiatry 22:43–51

Holliday J, Tchanturia K, Landau S, Collier D, Treasure J (2005) Is impaired set-shifting an endophenotype of anorexia nervosa? Am J Psychiatry 162:2269–2275

Holliday J, Uher R, Landau S, Collier D, Treasure J (2006) Personality pathology among individuals with a lifetime history of anorexia nervosa. J Pers Disord 20:417–430

Jones BP, Duncan CC, Brouwers P, Mirsky AF (1991) Cognition in eating disorders. J Clin Exp Neuropsychol 13:711–728

Katzman DK, Christensen B, Young AR, Zipursky RB (2001) Starving the brain: structural abnormalities and cognitive impairment in adolescents with anorexia nervosa. Semin Clin Neuropsychiatry 6:146–152

Kaye WH, Fudge JL, Paulus M (2009) New insights into symptoms and neurocircuit function of anorexia nervosa. Nat Rev Neurosci 10:573–584

Keys A, Brozek J, Henschel A, Mickelsen O, Taylor HL (1950) The biology of human starvation, Vols. I–II. University of Minnesota Press, Minneapolis, MN

Kingston K, Szmukler G, Andrewes D, Tress B, Desmond P (1996) Neuropsychological and structural brain changes in anorexia nervosa before and after refeeding. Psychol Med 26:15–28

Klump KL, Bulik CM, Pollice C, Halmi KA, Fichter MM, Berrettini WH, Devlin B, Strober M, Kaplan A, Woodside DB, Treasure J, Shabbout M, Lilenfeld LR, Plotnicov KH, Kaye WH (2000) Temperament and character in women with anorexia nervosa. J Nerv Ment Dis 188:559–567

Marsh R, Zhu H, Schultz RT, Quackenbush G, Royal J, Skudlarski P, Peterson BS (2006) A developmental fMRI study of self-regulatory control. Hum Brain Mapp 27:848–863

Marsh R, Maia TV, Peterson BS (2009) Functional disturbances within frontostriatal circuits across multiple childhood psychopathologies. Am J Psychiatry 166:664–674

Matsunaga H, Kiriike N, Iwasaki Y, Miyata A, Yamagami S, Kaye WH (1999) Clinical characteristics in patients with anorexia nervosa and obsessive-compulsive disorder. Psychol Med 29:407–414

Miller EK, Cohen JD (2001) An integrative theory of prefrontal cortex function. Annu Rev Neurosci 24:167–202

Monchi O, Petrides M, Strafella AP, Worsley KJ, Doyon J (2006) Functional role of the basal ganglia in the planning and execution of actions. Ann Neurol 59:257–264

Muhlau M, Gaser C, Ilg R, Conrad B, Leibl C, Cebulla MH, Backmund H, Gerlinghoff M, Lommer P, Schnebel A, Wohlschlager AM, Zimmer C, Nunnemann S (2007) Gray matter decrease of the anterior cingulate cortex in anorexia nervosa. Am J Psychiatry 164:1850–1857

Naruo T, Nakabeppu Y, Deguchi D, Nagai N, Tsutsui J, Nakajo M, Nozoe S (2001) Decreases in blood perfusion of the anterior cingulate gyri in Anorexia Nervosa Restricters assessed by SPECT image analysis. BMC Psychiatry 1:2

Reitan RM (1958) Validity of the trail making test as indicator of organic brain damage. Perceptual and Motor skills 8:271–276

Robbins TW (2000) Chemical neuromodulation of frontal-executive functions in humans and other animals. Exp Brain Res 133:130–138

Robbins TW, Arnsten AF (2009) The neuropsychopharmacology of fronto-executive function: monoaminergic modulation. Annu Rev Neurosci 32:267–287

Roberts ME, Tchanturia K, Stahl D, Southgate L, Treasure J (2007) A systematic review and meta-analysis of set-shifting ability in eating disorders. Psychol Med 37:1075–1084

Rubia K, Overmeyer S, Taylor E, Brammer M, Williams SC, Simmons A, Andrew C, Bullmore ET (2000) Functional frontalisation with age: mapping neurodevelopmental trajectories with fMRI. Neurosci Biobehav Rev 24:13–19

Shafritz KM, Kartheiser P, Belger A (2005) Dissociation of neural systems mediating shifts in behavioral response and cognitive set. Neuroimage 25:600–606

Snyder LH, Batista AP, Andersen RA (2000) Intention-related activity in the posterior parietal cortex: a review. Vis Res 40:1433–1441

Sowell ER, Thompson PM, Holmes CJ, Jernigan TL, Toga AW (1999) In vivo evidence for post-adolescent brain maturation in frontal and striatal regions. Nat Neurosci 2:859–861

Srinivasagam NM, Kaye WH, Plotnicov KH, Greeno C, Weltzin TE, Rao R (1995) Persistent perfectionism, symmetry, and exactness after long-term recovery from anorexia nervosa. Am J Psychiatry 152:1630–1634

Tchanturia K, Morris RG, Anderluh MB, Collier DA, Nikolaou V, Treasure J (2004) Set shifting in anorexia nervosa: an examination before and after weight gain, in full recovery and relationship to childhood and adult OCPD traits. J Psychiatr Res 38:545–552

Treasure J, Friederich HC (2007) Neuroimaging. In: Jaffa T, McDermott B (eds) Eating disorders in children and adolescents. Cambridge, Cambridge University Press, p 95–107

Uher R, Brammer MJ, Murphy T, Campbell IC, Ng VW, Williams SC, Treasure J (2003) Recovery and chronicity in anorexia nervosa: brain activity associated with differential outcomes. Biol Psychiatry 54:934–942

Wade TD, Tiggemann M, Bulik CM, Fairburn CG, Wray NR, Martin NG (2008) Shared temperament risk factors for anorexia nervosa: a twin study. Psychosom Med 70:239–244

Wagner A, Greer P, Bailer UF, Frank GK, Henry SE, Putnam K, Meltzer CC, Ziolko SK, Hoge J, McConaha C, Kaye WH (2006) Normal brain tissue volumes after long-term recovery in anorexia and bulimia nervosa. Biol Psychiatry 59:291–293

Wagner A, Aizenstein H, Venkatraman VK, Fudge J, May JC, Mazurkewicz L, Frank GK, Bailer UF, Fischer L, Nguyen V, Carter C, Putnam K, Kaye WH (2007) Altered reward processing in women recovered from anorexia nervosa. Am J Psychiatry 164:1842–1849

Zastrow A, Kaiser S, Stippich C, Walther S, Herzog W, Tchanturia K, Belger A, Weisbrod M, Treasure J, Friederich HC (2009) Neural correlates of impaired cognitive-behavioral flexibility in anorexia nervosa. Am J Psychiatry 166:608–616

Neural Circuits, Neurotransmitters, and Behavior

Serotonin and Temperament in Bulimic Syndromes

Howard Steiger, Kenneth R. Bruce, and Patricia Groleau

Contents

Abstract In bulimia nervosa (BN), and in related binge–purge syndromes, factors affecting central serotonin (5-hydroxytryptamine, 5-HT) function appear to contribute not only to appetitive dysregulation but also to temperamental and personality manifestations. Drawing upon findings from neurobiological, molecular-genetic, and brain-imaging studies, we present an integrative model of the role of 5-HT function in bulimic syndromes. At the core of our model is a consideration of the ways in which diverse hereditary and environmental influences impact the action of the 5-HT system. We believe that our model helps account for heterogeneous traits seen in

H. Steiger (✉), K.R. Bruce, and P. Groleau
Eating Disorders Program, Douglas University Institute, 6875 LaSalle Blvd, Montreal (Verdun), QC, Canada H4H 1R3
e-mail: stehow@douglas.mcgill.ca

Department of Psychiatry, McGill University, Montreal, QC, Canada

Department of Psychology, McGill University, Montreal, QC, Canada

R.A.H. Adan and W.H. Kaye (eds.), *Behavioral Neurobiology of Eating Disorders*, Current Topics in Behavioral Neurosciences 6, DOI 10.1007/7854_2010_88

the bulimic population, for disproportionate representation of individuals displaying pathological personality traits and exposure to severe environmental stressors, and for interindividual variations as to treatment response.

Keywords Bulimia · Genes · Gene–environment interactions · Personality traits · Serotonin · Temperament

1 Heterogeneous Expressions

Bulimia nervosa (BN), and related syndromes characterized by binge-eating and purging, coaggregate with diverse temperamental traits, including perfectionism, harm avoidance, affective instability, negative urgency, novelty seeking, and impulsivity (Cassin and von Ranson 2005; Lilenfeld et al. 2006; Steiger and Bruce 2007). With respect to such features, people with bulimic syndromes are, however, quite heterogeneous. Studies characterize about one-third of those affected as being frankly "dysregulated" (displaying mood lability, cognitive impulsivity, and behavioral disinhibition), another third as being "over-regulated" (displaying over attention to detail, difficulty switching perceptual or problem-solving approaches and compulsivity), and yet another third as being quite free of observable psychopathology (Steiger et al. 2009; Wonderlich et al. 2005). Such differences have led to the proposal that temperamental variations correspond to distinct patterns of exposure to developmental and constitutional risks, and define clinically and etiologically distinct bulimic variants (Lilenfeld et al. 2006; Steiger and Bruce 2007). In this chapter, we discuss the possible involvement of the serotonin (5-hydroxytryptamine, 5-HT) system in the different temperamental-trait manifestations seen in bulimic individuals.

2 General Behavioral Effects of 5-HT

Serotonin regulates cell proliferation, migration, and differentiation in developing humans, and cognition, emotion, sleep, appetite, pain, circadian rhythms, and endocrine activity in mature ones (Lesch and Mossner 1998). Synthesized from tryptophan under rate-limiting control of tryptophan hydroxylase (TPH), reuptake of extracellular 5-HT (which controls the magnitude and duration of serotonergic neurotransmission) is carried out by the 5-HT transporter. Through its action in the medial and lateral hypothalamus, 5-HT reduces carbohydrate and protein ingestion and promotes feelings of satiety (Blundell 1986). Serotonin is thus likely to be an etiopathologic factor in bulimic syndromes, characterized as they are by appetitive dyscontrol (Brewerton 1995; Kaye 2008; Steiger 2004).

Of importance in the present context, the 5-HT system is also thought to influence expressions of "personality" or "temperament" (Reif and Lesch 2003;

Van Gestel and Van Broeckhoven 2003), creating an impetus for the hypothesis that alterations in 5-HT activity also moderate some of the behavioral-trait expressions seen in people with bulimic syndromes. Serotonin mediates anxiety reactions, having an especially well-established role in harm assessment and avoidance (Cloninger et al. 1993). Decreasing 5-HT transmission impedes individuals' ability to adopt cautious waiting attitudes or to inhibit responding (Soubrie 1986), leading to problems with response inhibition. Now-classic findings, involving peripheral and central 5-HT indices, indicate a fundamental association between reduced 5-HT tone or transmission, on one hand, and impulsive aggression, suicidality, and self-injuriousness, on the other hand (Stein et al. 1993; Frankle et al. 2005). Inconsistent findings have, conversely, associated elevated 5-HT tone with tendencies toward inhibition or compulsivity, leading to the (partially supported) over-generalization that higher levels of 5-HT activity correspond to heightened vigilance, inflexibility, and preference for order and sameness, whereas lower levels coincide with greater impulsivity, sensation seeking, and hostility (Cloninger et al. 1993; Hollander 1998). Serotonin also moderates such "social" aspects of personality as propensities toward interpersonal dominance, submissiveness, or aggression (Carrillo et al. 2009; Young and Leyton 2002).

3 Sources of Variation in 5-HT Activity

3.1 Nutritional Status

Nutritional status influences 5-HT function in important ways: In humans, dieting has been associated with lower plasma tryptophan levels, a decreased ratio of tryptophan to other competing amino acids (Anderson et al. 1990), downregulation of the density of the 5-HT transporter (Huether et al. 1997), hypersensitivity of postsynaptic receptors (Goodwin et al. 1987), and increased prolactin secretion following intravenous tryptophan (Anderson et al. 1990). Suggesting that effects are relevant to BN, and confirming serotonergic mediation, several studies demonstrate that dietary manipulations that transiently deplete tryptophan levels (and in turn, 5-HT synthesis in brain) produce increased bulimic urges or depression in currently (Bruce et al. 2009; Weltzin et al. 1995) and formerly (Smith et al. 1999) bulimic individuals. Furthermore, given disproportionate occurrence of bulimic syndromes in females, it is noteworthy that interventions aimed at lowering 5-HT activity (e.g., a 3-week calorie-reducing diet or acute tryptophan depletion) will produce greater alterations of 5-HT function or lowering of mood in women than in men (Goodwin et al. 1987; Young and Leyton 2002). In other words, dietary effects upon 5-HT activity seem to preferentially affect females.

3.2 *Genetic Factors Influencing 5-HT Activity: General Population*

Since an initial report on an association between the 5-HT transporter promoter polymorphism, 5HTTLPR, and traits of neuroticism, harm avoidance, and impulsivity (Lesch et al. 1996), 5HTTLPR has been a popular candidate gene for personality researchers. Individual studies yield mixed results, but a recent meta-analysis of 26 trials ($N = 7657$) by Schinka et al. (2004) supports an association between low-function 5HTTLPR genotypes and neuroticism. Likewise, a quantitative review by Munafò et al. (2008) concludes that the low-function 5HTTLPR allele is associated with greater amygdala activation in response to emotional stimuli and hence, greater emotionality. There is, furthermore, inconsistent evidence to suggest that 5-HTTLPR low-function alleles coincide with traits of impulsivity (Lesch et al. 1996), novelty seeking (Sander et al. 1998), affective instability, and suicidality (Anguelova et al. 2003) – all arguably characterized by "dysregulation." In contrast, high-function alleles reportedly correspond to obsessive–compulsive disorder (OCD) (Hu et al. 2006) and "hyperfrontality" (Heinz et al. 2005) – arguably associated with "over-regulation." In a related vein, other 5-HT system genes (i.e., influencing TPH2, 5HT2a, and 5HT2c activity) have been thought to modulate personality traits such as impulsivity, anxiety, obsessive–compulsiveness, and reward dependence (Reif and Lesch 2003; Van Gestel and Van Broeckhoven 2003).

We note that available findings are characterized by inconsistent replication and lack of trait specificity, suggesting that we are probably (at best) seeing generalized (and often relatively small) influences upon personality traits of individual 5-HT system genes – not highly trait-specific effects. Nonetheless, molecular-genetic findings are such as to corroborate a general role of the 5-HT system in the expression of personality traits – with links to neuroticism and negative emotionality being quite well confirmed.

4 Serotonin and the EDs

A sizable literature now exists to indicate that individuals who are ill with a bulimic syndrome, or who have recovered from one, display disorder-relevant alterations of 5-HT activity – as evinced by measures of 5-HT metabolism (Brewerton 1995; Kaye 2008; Steiger 2004), receptor sensitivity (Kaye 2008; Bailer et al. 2004), and transporter activity (Steiger and Bruce 2007). Studies conducted using platelet measures have suggested that peripheral 5-HT reuptake is altered, not only in active bulimics (Ramacciotti et al. 2003; Steiger et al. 2005), but also in binge–purge free former bulimics (Steiger et al. 2005), and even in ED patients' unaffected relatives (Steiger et al. 2006). Likewise, brain-imaging techniques reveal anomalous 5-HT function in actively ill and recovered patients, with available data showing:

(a) reduced central 5-HT transporter availability in women with BN (Tauscher et al. 2001) or who have recovered from bulimic anorexia nervosa (AN) (Bailer et al. 2007); (b) reduced 5HT2a receptor binding (in subgenual cingulate, mesial temporal, and parietal cortical regions) in women recovered from BN (Bailer et al. 2004); (c) increased presynaptic 5-HT1A autoreceptor activity (in the dorsal raphe) of people recovered from bulimic-type AN, and in postsynaptic 5HT1A receptors in various brain regions in people with active BN (Tiihonen et al. 2004). We refer interested readers to available, comprehensive reviews on 5-HT function in BN (Brewerton 1995; Kaye 2008; Steiger 2004) and to relevant chapters of this volume (Kaye et al. 2010; Bailer and Kaye 2010). For the moment, we assert that findings implicate altered 5-HT kinetics in the pathogenesis of bulimic syndromes.

5 Serotonin and Temperament in the EDs

5.1 Neurobiological Indices

An accumulating literature indicates that, in bulimic individuals, 5-HT indices covary with personality-trait manifestations, much as they seem to do in other populations. Waller et al. (1996) reported that in women with BN, self-reported hostility was inversely associated with the size of neuroendocrine responses following buspirone (presumed to be a 5-HT1A agonist). In bulimic women, our group has found reduced paroxetine binding and prolactin response following administration of the postsynaptic 5-HT agonist, meta-chlorophenylpiperazine (mCCP) to be associated with self-reported impulsivity (Steiger et al. 2001a) and self-destructiveness (Steiger et al. 2001b). We also found bulimic individuals with comorbid avoidant personality disorder to have especially blunted prolactin responses following m-CPP, associating reduced serotonergic activation in the expected direction with submissiveness (Bruce et al. 2004). Ramacciotti et al. (2003) found no association between impulsive acts and platelet paroxetine binding in a mixed anorexic/bulimic sample – but after excluding individuals with comorbid personality disorders and, possibly, the variability needed to detect covariations on indices of impulsivity and 5-HT.

In apparent corroboration of the notion (raised earlier) that impulsivity and compulsivity may (in some respects) occur at opposite poles of a continuum spanning serotonergic under- versus over-activity, a study by our group compared bulimic patients classified along dimensions reflecting trait compulsivity and impulsivity, and showed the "high compulsive/low impulsive" individuals to have unusually high prolactin responses following m-CPP (Steiger et al. 2003) – suggesting abnormally high postsynaptic 5-HT sensitivity. In another study, increased perfectionism and compulsivity were similarly associated with elevated paroxetine-binding density (Steiger et al. 2004).

5.2 *Brain Imaging*

Recent brain-imaging findings are also compatible with the idea that in BN, variations in "trait" features coincide with variations in 5-HT function. Findings in BN, as in other populations, localize behavioral impulsivity (i.e., problems with response inhibition) to hypoactive frontostriatal brain circuits, in which 5-HT is, along with dopamine, an important neurotransmitter (Marsh 2009). Likewise, a positron emission tomography (PET) study in individuals recovered from AN, binge–purge subtype has shown a theoretically expected, inverse relationship between 5HT2a receptor binding (in the cingulate and temporal regions) and novelty seeking, and a positive association (in the left subgenual cingulate, left lateral temporal, and mesial temporal cortices) between 5HT2A binding potential and harm avoidance (Bailer et al. 2004). In other words, as in other contexts, decreased 5-HT neurotransmission has been associated with greater novelty seeking, and increased 5-HT activity with harm avoidance.

5.3 *Genetic Indices*

Various association studies have addressed the hypothesis that candidate 5-HT system genes influence personality-trait expressions in BN, much as they appear to do in other populations. As seems often to be the case with single-gene studies, these findings are characterized by inconsistent replication. We suspect that one basis for such inconsistency may be that candidate genes, rather than conveying direct risk for categorical syndromes (like BN), influence the expression of personality traits that indirectly impact susceptibility to an ED. Indeed, our reading of the available literature (review to follow) is that the "fit" of genetic information to personality-trait variations is somewhat better than the fit to categorical diagnostic distinctions (such as "bulimic" versus "not bulimic"). We review findings for individual 5-HT relevant candidate genes in turn.

5.3.1 5HTTLPR

Some candidate-gene studies associate BN with low-function alleles of 5HTTLPR (Di Bella et al. 2000); other do not (cf. Monteleone et al. 2006). However, resembling findings suggesting association of 5HTTLPR with temperamental traits in other populations, recent studies have reported that bulimic individuals who do carry low-function 5HTTLPR alleles display more harm avoidance (Monteleone et al. 2006), anxiety (Ribases et al. 2008), affective instability, impulsivity (Steiger et al. 2005b), sensation seeking (Steiger et al. 2007), and dissocial behavior – implying a combination of hostile and reckless tendencies Steiger et al. (2008a, b). Similar trends seem to apply to people showing binge-eating disorder in a community sample (Akkermann et al. 2010). In apparent contrast to the preceding, we note that one study (implicating an unusually high proportion of low-function 5HTTLPR allele carriers, and conducted using a biallelic rather than triallelic 5HTTLPR formulation) found no expected association (in 178 bulimic women) between 5HTTLPR variations and latent profile analysis-derived personality clusters characterized as "low psychopathology," "affective perfectionistic," and "impulsive" (Wonderlich et al. 2005). However, a subsequent replication using a triallelic 5HTTLPR assay in 185 women with heterogeneous EDs documented a more familiar pattern: increased rates of high-function alleles in individuals classified as being "inhibited-compulsive;" more low-function alleles in individuals classified as being "dissocial-impulsive" (Steiger et al. 2009). Taken together, we read the available evidence as indicating that 5HTTLPR variations do predict personality-trait variations in bulimic individuals.

5.3.2 Tryptophan Hydroxylase

Too scarce to support firm generalizations, studies in BN have suggested that genes influencing TPH activity may vary systematically with personality. Monteleone et al. (2007) reported that bulimic individuals carrying the AA genotype of the TPH gene A218C polymorphism exhibited more disturbed binging behaviors and higher harm avoidance scores than did bulimics with the CC genotype. Significance of the finding is difficult to ascertain, as A218C is expressed in peripheral (rather than brain) tissues. However, a study from our group, examining the influence on eating symptoms and pathological personality traits in BN-spectrum disorders of the rs4570625 polymorphism of the TPH-2 gene (which is expressed in brain), found individuals with the GG genotype to display more perfectionism than did individuals with other genotypes (Groleau et al. 2009).

5.3.3 5HT Receptors (5-HT$_{2A}$ and 5-HT$_{2c}$)

A couple of studies involving bulimic samples has supported an association between the G allele of the −1428G/A promoter polymorphism of the 5HT2A receptor gene and heightened impulsivity, in much the same way as this polymorphism has been linked to impulsivity in noneating-disordered populations. In a study of 182 Japanese patients with heterogeneous eating disorders, Nishiguchi et al. (2001) found the G allele to be associated with more dietary disinhibition (binge-eating and/or purging), more impulsivity (higher Barrat Impulsivity Scale scores), and greater likelihood of borderline personality disorder. Consistent with this, a study from our group found bulimic GG homozygotes to be more impulsive and less sensitive to postsynaptic 5-HT activation (Bruce et al. 2005a, b). Finally, we are aware of one study in BN reporting an association between the −995A/−759T/−697C/Cys23 haplotype of the 5HT2C gene and traits of hostility, obsessive–compulsiveness, somatization, depression, anxiety, phobic anxiety, and paranoid ideation (Ribases et al. 2008).

5.4 Gene–Environment Interactions and Personality-Trait Expressions

In nonhuman primates, peer-rearing commonly serves as a proxy for developmental stress: available findings indicate peer-reared rhesus macaque carriers of the low-function allele of the 5-HT transporter (rh-5HTTLPR) to display altered 5-HT metabolism and stress-responsiveness (Barr et al. 2003). An implication is that developmental stress, in genetically susceptible individuals, can precipitate alterations in 5-HT activity and adaptation. Primate-research paradigms are noteworthy as, being prospective, they allow for controls on the directionality of effects – i.e., they ascertain that developmental stress activates genetic propensities, and not that genetic propensities cause individuals to seek out stressful life situations, or otherwise to "come to harm." Similarly, these paradigms allow for differentiation of serotonergic sequelae of stress and the serotonergic causes (through such associated traits as risk-taking) of stress exposure.

The preceding lends credence to the idea that when gene–environment effects implicating 5-HT polymorphisms and stress are observed in humans, they also implicate stress-induced activation of genetically primed, serotonergic susceptibilities. Although contradictions and controversies exist (Risch et al. 2009), several reports have indicated that childhood maltreatment (and other life stressors) confers disproportionate vulnerability to depression (Caspi et al. 2003), behavioral disinhibition (Paaver et al. 2008), and impulsivity (Wagner et al. 2009) in human carriers of low-function alleles of 5HTTLPR. Consistent with such observations, some of our recent findings have linked severe stressors to heightened expression of psychopathological traits in eating-disordered women who carry low-function 5HTTLPR alleles. We observed that women with bulimic syndromes, when they both carried 5HTTLPR low-function alleles and reported severe childhood abuse, tended to display more pronounced novelty seeking (Steiger et al. 2007) and dissocial behavior (Steiger et al. 2008a, b). Likewise, we observed that bulimic individuals displaying comorbid anxiety, substance-abuse, and conduct disorders, when compared to bulimic individuals with lesser comorbidity, were more likely to be carriers of 5HTTLPR low-function alleles and survivors of childhood abuse (Richardson et al. 2008).

6 Theoretical and Clinical Implications

Findings on associations among 5-HT alterations and temperamental traits in bulimic individuals are likely to have important implications for etiological modeling in the area. To begin, we emphasize that in the bulimic population, anomalies in 5-HT activity are observed to predict severity of concurrent psychopathological-trait manifestations far more consistently than they do for severity of bulimic symptoms (see Steiger 2004 and Steiger and Bruce 2007 for full treatments of this question). An implication is that 5-HT alterations may not directly cause bulimic eating problems, but rather may indirectly influence susceptibility to bulimic syndromes, by heightening traits (such as affective instability or impulsivity) that play indirectly into risk. We also assume that abnormal 5-HT status in bulimic individuals reveals an "end state" associated with diverse causal paths – sometimes reflecting the cumulative effects of chronic dieting, sometimes a constitutional tendency based on heredity, sometimes the consequence of exposure to intense developmental or current stress, and sometimes a combination of all these things. The preceding means that, with respect to the 5-HT system, bulimic syndromes likely have heterogeneous causes – sometimes a circumscribed erosion of appetitive controls (following prolonged dieting) in relatively intact individuals, and at other times, a more fundamental, primary, and more pervasive disturbance in serotonergic controls over mood, impulses, and appetite, related to constitutional defects, the sequelae of severe stressors, or both.

Observed gene–environment interaction effects in bulimic individuals seem to have similar implications. Documented effects imply environmental activation (via childhood abuse and other stressors) of genetically determined (5-HT mediated) vulnerabilities in some bulimia-prone individuals. However, at the same time, findings are such as to suggest that the vulnerabilities in question are more pertinent to the expression of comorbid clinical traits than they are to bulimia-specific symptoms (Steiger and Bruce 2007). In this respect, genetic factors (e.g., the 5HTTLPR low-function alleles) and developmental experiences (e.g., childhood abuse) may be of greater value in explaining personality-trait expressions (e.g., impulsive vs. nonimpulsive tendencies) occurring in bulimic individuals than they are of main phenotypic status (i.e., bulimic or not). An implication of the preceding is that attention to gene × environment interactions may reveal heterogeneous etiopathologies that help explain heterogeneous clinical manifestations within the bulimic population.

The notion that BN affects individuals who often display 5-HT disturbances, but in whom such alterations may have heterogeneous causes, leads to an obvious clinical inference. Bulimia sufferers whose 5-HT abnormalities are mainly secondary consequences of dietary factors may have relatively focal treatment needs (i.e., nutritionally oriented, eating-symptom-focused therapies), whereas individuals showing more marked 5-HT dysregulation, especially when it is partly determined by trait-linked hereditary factors mediated by 5-HT functioning (or by severe developmental "damage" that activates 5-HT mediated trait expressions), might require more elaborate or more intense forms of intervention. Our conjecture is indirectly supported by recent findings in the literature. Fairburn et al. (2009) reported that among patients with heterogeneous EDs, those with low psychiatric comorbidity respond best to a traditional (eating, weight, and body-image focused) version of cognitive-behavioral therapy (CBT), whereas those with more pronounced comorbidity respond better to an "enhanced" CBT, with added components addressing problems of self-esteem, clinical perfectionism, affective instability, and interpersonal functioning. Could such findings, in part, be attributable to variations in 5-HT function (i.e., more-pronounced 5-HT disturbances in individuals with greater comorbid psychopathology)? We believe that this may be so, as follows: elevated psychopathological manifestations (especially those of a borderline or impulsive type) have been quite consistently linked to unfavorable bulimia-treatment response (Bruce and Steiger 2006). The same psychopathological symptoms have, similarly, been linked to pronounced dysregulation of the serotonin (5-HT) system – and, in bulimic patients, to low-function alleles of 5-HT linked polymorphisms, such as 5-HTTLPR or –1438G/A (Nishiguchi et al. 2001; Steiger and Bruce 2007). An obvious prediction follows: "Individuals who are genetically susceptible to 5-HT dysregulation will be less-responsive to conventional treatments for BN." Two available studies support this point. A first has linked low-function 5-HTTLPR variants to unfavorable response of BN to pharmacotherapy (Monteleone et al. 2005). A second, by our group, tracked responses of 98 bulimic women through 4- and 8-month spans of specialized, multimodal treatment (Steiger et al. 2008a, b). 5-HTTLPR low-function allele carriers showed smaller treatment reductions in binge-eating, anxiety, and depression, whereas carriers of low-function –1438G/A variants showed smaller reductions in binge-eating and impulsivity. In other words, the data corroborate an association between poorer bulimia-treatment response and low-function variants of genes influencing 5-HT system function. Such findings suggest that a better understanding of hereditary, 5-HT-mediated factors acting in bulimic syndromes may further the understanding of processes that facilitate or impede bulimic individuals' progress during therapy.

7 Discussion and Conclusions

Available findings suggest that personality traits may inform efforts to understand abnormalities and patterns of variation observed on 5-HT indices within the bulimic population. The evidence favors the idea that the momentary 5-HT status in anyone suffering a bulimic syndrome is likely to be multiply determined – and to reflect a principled convergence of state, trait, and developmental influences. Not surprisingly, given that brain 5-HT is derived exclusively from dietary tryptophan, one strong source of variation in 5-HT status is the nutritional status. However, accumulating neurobiological and genetic data point to other sources of variation, and suggest that, in bulimic individuals, some personality-trait variations coincide systematically with constitutionally determined variations in 5-HT system function. Furthermore, available evidence indicates that developmental stressors and traumatic experiences can affect both 5-HT functioning and personality-trait manifestations. Viewing these indices together, we assume that "personality" in bulimic individuals, especially those aspects of it that can be confirmed to be stable or "trait-like," provides clues to the existence of constitutional serotonergic susceptibilities, developmental "damage" to the 5-HT system, or both. Thus, bulimic psychopathology, especially in individuals displaying marked temperamental disturbances, can sometimes be understood to represent the amplification, by effects on the 5-HT system of developmental stressors, intensive dieting, or both, of underlying serotonergic susceptibilities. We believe that the preceding convergences account for the fact that the bulimic population includes a disproportionate number of individuals displaying "dysregulatory" traits (such as impulsivity or affective instability) and an unusually high proportion of people who report past exposure to traumatic stress – and above all, the fact that bulimic syndromes flourish so well in social contexts that encourage extreme dieting. Indeed, we have previously proposed that an understanding of the interplay among genetic factors coding for 5HT activity, environmental stresses that may amplify genetic susceptibilities, and environmental inducements toward dieting (which can be understood to indirectly reduce 5-HT activity at a population level) not only provides a heuristic toward the development of an integrative etiological model for BN, but also helps explain heterogeneous clinical manifestations and treatment outcomes seen in the bulimic population (Steiger and Bruce 2007).

References

Akkermann K, Nordquist N, Oreland L, Harro J (2010) Serotonin transporter gene promoter polymorphism affects the severity of binge eating in general population. Prog Neuropsychopharmacol Biol Psychiatry 34:111–114

Anderson IM, Parry-Billings M, Newsholme EA, Fairburn CG, Cowen PJ (1990) Dieting reduces plasma tryptophan and alters brain 5-HT function in women. Psychol Med 20:785–791

Anguelova M, Benkelfat C, Turecki G (2003) A systematic review of association studies investigating genes coding for serotonin receptors and the serotonin transporter: II suicidal behaviour. Mol Psychiatry 8:646–653

Bailer UF, Price JC, Meltzer CC, Mathis CA, Frank GK, Weissfeld L, McConaha CW, Henry SE, Brooks-Achenbach S, Barbarich NC, Kaye WH (2004) Altered 5-HT2A receptor binding after recovery from bulimia type anorexia nervosa: relationships to harm avoidance and drive for thinness. Neuropsychopharmacology 29:1143–1155

Bailer UF, Frank G, Henry S, Price J, Meltzer C, Mathis C et al (2007) Exaggerated 5-HT1A but normal 5-HT2A receptor activity in individuals ill with anorexia nervosa. Biol Psychiatry 61(9):1090–1099

Bailer UF, Kaye WH (2010) Serotonin: imaging findings in eating disorders. Curr Topics Behav Neurosci DOI 10.1007/7854_2010_78

Barr S, Newman TK, Becker ML, Parker CC, Champoux M, Lesch KP, Goldman D, Suomi SJ, Higley JD (2003) The utility of the non-human primate model for studying gene by environment interactions in behavioral research. Genes Brain Behav 2:336–340

Blundell JE (1986) Serotonin manipulations and the structure of feeding behaviour. Appetite 7 Suppl:39–56

Brewerton TD (1995) Toward a unified theory of serotonin dysregulation in eating and related disorders. Psychoneuroendocrinology 20:561–590

Bruce K, Steiger H (2006) Prognostic implications of personality disorders in eating disorders. In: Sansone RA, Levitt JL (eds) Personality disorders and eating disorders: exploring the frontier. Routledge, New York, pp 247–262

Bruce KR, Steiger H, Koerner N, Israel M, Young SM (2004) Bulimia nervosa with co-morbid avoidant personality disorder: behavioural characteristics and serotonergic function. Psychol Med 34:113–124

Bruce KR, Steiger H, Ng Ying Kin NMK, Israel M (2005a) Reduced platelet [3H]paroxetine binding in anorexia nervosa: relationship to eating symptoms and personality pathology. Psychiatry Res 142:225–232

Bruce K, Steiger H, Joober R, Ng Ying Kin NM, Israel M, Young SN (2005b) Association of the promoter polymorphism –1438G/A of the 5-HT2A receptor gene with behavioural impulsiveness and serotonin function in women with bulimia nervosa. Am J Med Genet B Neuropsychiatr Genet 13:40–44

Bruce KR, Steiger H, Young SN, Ng Ying Kin NMK, Israël M, Lévesque M (2009) Impact of acute tryptophan depletion on mood and eating-related urges in bulimic and nonbulimic women. J Psychiatry Neurosci 34:376–382

Carrillo M, Ricci LA, Coppersmith GA, Melloni RH Jr (2009) The effect of increased serotonergic neurotransmission on aggression: a critical meta-analytical review of preclinical studies. Psychopharmacology 205:349–368

Caspi A, Sugden K, Moffit T, Taylor A, Craig IW, Harrington HL, McClay J, Mill J, Martin J, Braithwaite A, Poulton R (2003) Influence of life stress on depression: moderation by a polymorphism in the 5-HTT gene. Science 301:386–389

Cassin SE, Von Ranson KM (2005) Personality and eating disorders: a decade in review. Clin Psychol Rev 25:895–916

Di Bella DD, Catalano M, Cavallini HC, Riboldi C, Bellodi L (2000) Serotonin transporter linked polymorphic region in anorexia nervosa and bulimia nervosa. Molec Psychiatry 5:233–241

Cloninger CR, Svrakic DM, Przybeck TR (1993) A psychobiological model of temperament and character. Arch Gen Psychiatry 50:975–990

Fairburn CG, Cooper Z, Doll HA, O'Connor ME, Bohn K, Hawker DM, Wales JA, Palmer RL (2009) Transdiagnostic cognitive-behavioral therapy for patients with eating disorders: a two-site trial with 60-week follow-up. Am J Psychiatry 166:311–319

Frankle WG, Lombardo I, New AS, Goodman M, Talbot PS, Huang Y, Hwang D-R, Slifstein M, Curry S, Abi-Dargham A, Laruelle M, Siever LJ (2005) Brain serotonin transporter distribution in subjects with impulsive aggressivity: a positron emission study with [11C]McN 5652. Am J Psychiatry 162:915–923

Goodwin GM, Fairburn CG, Cowen PJ (1987) The effects of dieting and weight loss on neuroendocrine tesponses to tryptophan, clonidine, and apomorphine in volunteers: important implications for neuroendocrine investigations in depression. Arch Gen Psychiatry 44:952–957

Groleau P, Richardson J, Steiger H, Israel M, Bruce K, Joober R, Howard H, De Guzman R. Evidence for the implication of the tryptophan hydroxylase-2 gene in eating and psychopathological symptom severity in bulimia eating syndromes. Poster presentation at the 15th annual meeting of the eating disorders research society, New York, September 2009

Heinz A, Braus DF, Smolka MN. Woaser J, Puls I, Hermann D et al (2005) Amygdala-prefrontal coupling depends on a genetic variation of the serotonin transporter. Nature Neuroscience 81:20–21

Hollander E (1998) Treatment of obsessive-compulsive spectrum disorders with SSRIs. Br J Psychiatry Suppl 35:7–12

Hu XZ, Lipsky RH, Zhu G (2006) Serotonin transporter promoter gain of function genotypes are linked to obsessive–compulsive disorder. Am J Hum Genet 78:815–826

Huether G, Zhou D, Ruther E (1997) Long-term modulation of presynaptic 5-HT-output: experimentally induced changes in cortical 5-HT-transporter density, tryptophan-hydroxylase content and 5-HT innervations density. J Neural Transm 104:993–1004

Kaye WH (2008) Neurobiology of anorexia and bulimia nervosa. Physiol Behav 94:121–135

Kaye WH, Wagner A, Fudge JL, Paulus M (2010) Neurocircuity of eating disorders. Curr Topics Behav Neurosci DOI 10.1007/7854_2010_85

Lesch KP, Mossner R (1998) Genetically driven variations in serotonin uptake: is there a link to affective spectrum, neurodevelopmental, and neurodegenerative disorders? Biol Psychiatry 44:179–192

Lesch KP, Bengel D, Heils A, Sabol SZ, Greenberg BD, Petri S (1996) Association of anxiety-related traits with a polymorphism in the serotonin transporter gene regulatory region. Science 274:1527–1530

Lilenfeld LR, Wonderlich S, Riso LP, Crosby R, Mitchell J (2006) Eating disorders and personality: a methodological and empirical review. Clin Psychol Rev 26:299–320

Marsh R (2009) Dysfunctional frontostriatal control systems in bulimia nervosa. Future Neurol 4:383–387

Monteleone P, Santonastaso P, Tortorella A, Favaro A, Fabrazzo M, Castaldo E, Caregaro L, Fuschino A, Maj M (2005) Serotonin transporter polymorphism and potential response to SSRIs in bulimia nervosa. Mol Psychiatry 10:716–718

Monteleone P, Santonastaso P, Mauri M, Bellodi L, Erzegovesi S, Fushino A, Favaro A, Rotondo A, Castaldo E, Maj M (2006) Investigation of the serotonin transporter regulatory region polymorphism in bulimia nervosa: relationships to harm avoidance, nutritional parameters, and psychiatric comorbidity. Psychosom Med 68:99–103

Monteleone P, Tortorella A, Martiadis V, Serino I, Di Filippo C, Maj M (2007) Association between A218C polymorphism of the tryptophan-hydroxylase-1 gene, harm avoidance and binge eating behavior in bulimia nervosa. Neurosci Lett 421:42–46

Munafò MR, Brown SM, Hariri AR (2008) Serotonin transporter (5-HTTLPR) genotype and amygdala activation: a meta-analysis. Biol Psychiatry 63:852–857

Nishiguchi N, Matsushita S, Suzuki K, Murayama M, Shirakawa O, Higuchi S (2001) Association between 5HT2A receptor gene promoter region polymorphism and eating disorders in Japanese patients. Biol Psychiatry 50:123–128

Paaver M, Kurrikoff T, Nordquist N, Oreland L, Havoo J (2008) The effect of 5-HTT gene promoter polymorphism on impulsivity depends on family relations in girls. Prog Neuropsychopharmacol Biol Psychiatry 32:1263–1268

Ramacciotti CE, Coli E, Paoli R, Marazziti D, Dell'Osso L (2003) Serotonergic activity measured by platelet [3H] paroxetine binding in patients with eating disorders. Psychiatry Res 118:33–38

Reif A, Lesch KP (2003) Toward a molecular architecture of personality. Behav Brain Res 139:1–20

Ribases M, Fernandez-Aranda F, Gratacos M, Mercader JM, Casasnovas C, Nunez A, Vallejo J, Estivill X (2008) Contribution of the serotoninergic system to anxious and depressive traits that may be partially responsible for the phenotypical variability of bulimia nervosa. J Psychiatr Res 42:50–57

Richardson J, Steiger H, Schmitz N, Joober N, Bruce K, Israel M et al (2008) Relevance of the 5HTTLPR polymorphism and childhood abuse to increased psychiatric comorbidity in women with bulimia-spectrum disorders. J Clin Psychiatry 69:981–990

Risch N, Herrell R, Lehner T, Liang K-Y, Eaves L, Hoh J, Griem A, Kovacs M, Ott J, Merikangas KR (2009) Interaction between the serotonin transporter gene (5-HTTLPR), stressful life events, and risk of depression: a meta-analysis. JAMA 301:2462–2471

Sander T, Harms H, Dufeu P, Kuhn S, Hoehe M, Lesch KP, Rommelspacher H, Schmidt LG (1998) Serotonin transporter gene variants in alcohol-dependent subjects with dissocial personality disorder. Biol Psychiatry 43:908–912

Schinka JA, Busch RM, Robichaux-Keene N (2004) A meta-analysis of the association between the serotonin transporter gene polymorphism (5-HTTLPR) and trait anxiety. Mol Psychiatry 9:197–202

Smith KA, Fairburn CG, Cowen PJ (1999) Symptomatic relapse in bulimia nervosa following acute tryptophan depletion. Arch Gen Psychiatry 56:171–176

Soubrie P (1986) Reconciling the role of central serotonin neurons in human and animal behavior. Behav Brain Sci 9:319–364

Steiger H (2004) Eating disorders and the serotonin connection: state, trait and developmental effects. J Psychiatry Neurosci 29:20–29

Steiger H, Bruce KR (2007) Phenotypes, endophenotypes, and genotypes in bulimia spectrum eating disorders. Can J Psychiatry 52:220–227

Steiger H, Young SM, Ng Ying Kin NMK, Koerner N, Israel M, Lageix P, Paris J (2001a) Implications of impulsive and affective symptoms for serotonin function in bulimia nervosa. Psychol Med 31:85–95

Steiger H, Koerner N, Engleberg M, Israël M, Ng Ying Kin NMK, Young SN (2001b) Self-destructiveness and serotonin function in bulimia nervosa. Psychiatry Res 103:15–26

Steiger H, Israel M, Gauvin L, Ng Ying Kin NMK, Young SM (2003) Implications of compulsive and impulsive traits for serotonin status in women with bulimia nervosa. Psychiatry Res 120:219–229
Steiger H, Gauvin L, Israel M, Ng Ying Kin NMK, Young SM, Roussin J (2004) Serotonin function, personality-trait variations, and childhood abuse in women with bulimia-spectrum eating disorders. J Clin Psychiatry 65:830–837
Steiger H, Richardson J, Israel M, Ng Ying Kin NMK, Brug K, Mansour S, Pareut AM (2005a) Reduced density of plalelet-binding sites for 3H-paroxetine in remitted, bulimic women. Neuropsychopharmacology 30:1028–1032
Steiger H, Joober R, Israel M, Young SM, Ng Ying Kin NMK, Gauvin L, Bruce KR, Joncas J, Torkaman-Zehi A (2005b) The 5HTTLPR polymorphism, psychopathological symptoms, and platelet [3H-] paroxetine binding in bulimic syndromes. Int J Eat Disord 37:1–4
Steiger H, Gauvin L, Joober R, Israel M, Ng Ying Kin NMK, Bruce K, Richardson J, Young S, Hakim J (2006) Intrafamilial correspondences on platelet [^{3}H-] paroxetine-binding indices in bulimic probands and their unaffected first-degree relatives. Neuropsychopharmacology 31:1785–1792
Steiger H, Richardson J, Joober R, Gauvin L, Israel M, Bruce KR, Ng Ying Kin NMK, Howard H, Young SN (2007) The 5HTTLPR polymorphism, prior maltreatment and dramatic–erratic personality manifestations in women with bulimic syndromes. J Psychiatry Neurosci 32:354–362
Steiger H, Richardson J, Joober R, Israel M, Bruce K, Ng Ying Kin NMK, Howard H, Anestin A, Dandurand C, Gauvin L (2008a) Dissocial behavior, the 5HTTLPR polymorphism and maltreatment in women with bulimic syndromes. Am J Med Genet B Neuropsychiatr Genet 147B:128–130
Steiger H, Joober R, Gauvin L, Bruce K, Richardson J, Israel M, Anestin A, Groleau P (2008b) Serotonin-system polymorphisms (5HTTLPR and -1438G/A) and responses of patients with bulimic syndromes to multimodal treatments. J Clin Psychiatry 69:1565–1571
Steiger H, Richardson J, Schmitz N, Joober R, Israel M, Bruce KR, Gauvin L, Dandurand C, Annestin A (2009) Association of trait-defined, eating-disorder sub-phenotypes with (biallelic and triallelic) 5HTTLPR variations. J Psychiatr Res 43:1086–1094
Stein DJ, Hollander E, Liebowitz M (1993) Neurobiology of Impulsivity and the impulse control disorders. J Neuropsychiatry Clin Neurosci 5:9–17
Tauscher J, Pirker W, Willeit M, de Zwaan M, Bailer U, Neumeister A, Asenbaum S, Lernkh C, Praschak-Rieder N, Brücke T, Kasper S (2001) Beta-CIT and single photon emission computer tomography reveal reduced brain serotonin transporter availability in bulimia nervosa. Biol Psychiatry 49:326–332
Tiihonen J, Keski-Rahkonen A, Lopponen M, Muhonen M, Kajander J, Allonen T, Någren K, Hietala J, Rissanen A (2004) Brain serotonin 1A receptor binding in bulimia nervosa. Biol Psychiatry 55:871–873
Van Gestel S, Van Broeckhoven C (2003) Genetics of personality: are we making progress? Mol Psychiatry 8:840–852
Wagner S, Baskaya O, Lieb K, Dahmen N, Tadic A (2009) The 5-HTTLPR Polymorphism modulates the association of serious life events (SLE) and impulsivity in patients with Borderline Personality Disorder. J Psychiatr Res 43:1067–1072
Waller DA, Sheinberg AL, Gullion C, Moeller FG, Cannon DS, Petty F, Hardy BW, Orsulak P, Rush AJ (1996) Impulsivity and neuroendocrine response to buspirone in bulimia nervosa. Biol Psychiatry 39:371–374
Weltzin TE, Fernstrom MH, Fernstrom JD, Neuberger SK, Kaye WH (1995) Acute tryptophan depletion and increased food intake and irritability in bulimia nervosa. Am J Psychiatry 152:1668–1671
Wonderlich SA, Crosby RD, Joiner T, Peterson CB, Bardone-Cone A, Klein M, Crow S, Mitchell JE, LeGrange D, Steiger H, Kolden G, Johnson F, Vrshek S (2005) Personality subtyping and bulimia nervosa: psychopathological and genetic correlates. Psychol Med 35:649–657
Young SN, Leyton M (2002) The role of serotonin in human mood and social interaction insight from altered tryptophan levels. Pharmacol Biochem Behav 71:857–865

Part III
Genetics, Gender and Heritability

The Heritability of Eating Disorders: Methods and Current Findings

Laura M. Thornton, Suzanne E. Mazzeo, and Cynthia M. Bulik

Contents

Abstract Family, twin, and adoption studies of anorexia nervosa (AN), bulimia nervosa (BN), binge-eating disorder (BED), and the proposed purging disorder presentation (PD) have consistently demonstrated that genetic factors contribute to the variance in liability to eating disorders. In addition, endophenotypes and component phenotypes of eating disorders have been evaluated and provide further insight regarding genetic factors influencing eating disorders and eating disorder

L.M. Thornton (✉) and C.M. Bulik
Department of Psychiatry, University of North Carolina at Chapel Hill, 101 Manning Drive, CB #7160, Chapel Hill, NC 27599-7160, USA
e-mail: laurathornton@verizon.net; cbulik@med.unc.edu

S.E. Mazzeo
Department of Psychology, Virginia Commonwealth University, PO Box 842018, Richmond, VA 23284-2018, USA
e-mail: semazzeo@vcu.edu

R.A.H. Adan and W.H. Kaye (eds.), *Behavioral Neurobiology of Eating Disorders*, Current Topics in Behavioral Neurosciences 6, DOI 10.1007/7854_2010_91

diagnostic criteria. Many of these phenotypes have demonstrated substantial heritability. This chapter reviews biometrical genetic methods and current findings from family and twin studies that investigate the role of genes and environment in the etiology of eating disorders. We review the methodology used to estimate heritability, the results of these studies, and discuss the implications of this research for the basic conceptualization of eating disorders and the future value of twin modeling in the molecular genetic era.

Keywords Anorexia nervosa · Bulimia nervosa · Eating disorders · Endophenotypes · Twin models

Abbreviations

A	Additive genetic effects
a^2	Proportion of variance due to additive genetic effects
AN	Anorexia nervosa
BED	Binge eating disorder
BMI	Body mass index
BN	Bulimia nervosa
C	Shared environmental effects
c^2	Proportion of variance due to shared environmental effects
CI	95% Confidence interval
DZ	Dizygotic
E	Unique environmental effects
e^2	Proportion of variance due to unique environmental effects
EDE	Eating Disorders Examination
EDI	Eating Disorder Inventory
IWL	Intentional weight loss
MZ	Monozygotic
PD	Purging disorder
TFEQ	Three Factor Eating Questionnaire
V	Variance in liability to a disorder

1 Overview

Knowledge of genetic influences on liability to eating disorders has grown rapidly over the past three decades. Indeed, numerous family, twin, and genetic studies have indicated that genetic effects contribute to the variance in liability to eating disorders including AN, BN, BED, and PD. The consistency of these findings and their implications for prevention and treatment programs compel professionals who treat individuals with these conditions to become familiar with genetic epidemiological research relevant to their clinical work.

This chapter reviews biometrical genetic methods and current findings from family and twin studies that investigate the role of genes and environment in the etiology of eating disorders. In addition, we discuss the implications of this research for the basic conceptualization of eating disorders and the future value of twin modeling in the molecular genetic era.

2 Genetic Epidemiology Methods: Family, Twin, and Adoption Studies

The threshold eating disorders, AN and BN, and the proposed syndromes of BED and PD are complex and often chronic illnesses. Estimates of the general population lifetime prevalence for these disorders range from approximately 0.5% for AN to 3% for BED in adult women (Hoek 2006; Hudson et al. 2007; Keski-Rahkonen et al. 2007), with lower prevalence in men (Garfinkel et al. 1995; Hudson et al. 2007; Woodside et al. 2001). Although the etiology of these conditions is generally assumed to be biopsychosocial in nature; the following sections focus on genetic factors influencing liability to these disorders.

2.1 Family Studies

Family studies assess the lifetime risk that a relative of an individual with a disorder will develop the condition himself/herself in comparison to either (1) the general risk of that disorder in the population or (2) the risk of the disorder in families of comparable individuals without the disorder. A statistically increased lifetime risk of, for example, AN in biological relatives of probands with AN, demonstrates that the trait aggregates in families. However, this increased risk is not sufficient to prove that genes influence the disorder because the resemblance among family members could be due to either genetic or environmental factors shared within families (e.g., socio-cultural factors such as an over-emphasis on appearance) or, more likely, some combination of the two.

The familial nature of AN is well established. There is a significantly greater lifetime prevalence of AN or subthreshold eating disorders in first-degree relatives of probands (3–12%) than in relatives of controls (0–4%) (Gershon et al. 1983; Lilenfeld et al. 1998; Strober et al. 1985, 1990, 2001). Specifically, relatives of probands with AN are 11.3 times more likely to have AN than relatives of controls (Strober et al. 2000).

Most research (Kassett et al. 1989; Strober et al. 2000) has found an increased incidence of BN in relatives of BN probands, with rare exception (Hudson et al. 1987). The morbid risk of BN in first-degree relatives of probands with the disorder is estimated to be between 4.4 (Strober et al. 2000) and 9.6 (Kassett et al. 1989) times greater than in controls.

BED also occurs more frequently in family members of individuals with BED than in control families (Brody et al. 1994; Fowler and Bulik 1997; Hudson et al. 2006; Javaras et al. 2008a). Family studies report odds ratios between 1.9 and 2.2 for the risk of BED in a relative of a proband with BED compared with relatives of controls (Fowler and Bulik 1997; Hudson et al. 2006; Javaras et al. 2008b; Lee et al. 1999).

PD has recently been proposed as independent, unique eating disorder presentation (Keel et al. 2005; Mehler et al. 2004). Diagnostic criteria include recurrent purging episodes for the purposes of weight control in the absence of binge eating. To date, no family studies of purging disorder have been published. However, two population based studies found that between 1.1% and 5.3% of women met criteria for a lifetime diagnosis of PD (Favaro et al. 2003; Wade et al. 2006).

In addition, family studies indicate that the prevalence of both threshold and subthreshold AN and BN are elevated in relatives of AN and of BN probands compared with that in relatives of controls (Gershon et al. 1983; Hudson et al. 1987; Lilenfeld et al. 1998; Stein et al. 1999; Strober et al. 1990, 2000; Woodside et al. 1998). This suggests that (1) eating disorders are expressed in families as a broad spectrum of eating-related pathology and (2) some liability factors are shared among the different eating disorder types.

2.2 Twin Studies

Twin studies are a useful tool to differentiate the effects of genes and the environment on behavioral characteristics and liability to psychopathology and illness (Kendler 1993). Identical twins (i.e., monozygotic or MZ twins), for most intents and purposes, have identical DNA and thus the same genetic variants at all genes, neglecting stochastic errors in DNA replication during development such as copy number variation (Bruder et al. 2008). Assuming stochastic error and gene by environment interaction are negligible, phenotypic differences observed between MZ twins can be attributed to environmental factors (Plomin et al. 1994). Similarities between twins could be either the result of genes, environmental influences coming from life experiences that twins share ("common environment"), or, in all probability, some combination of the two. Like full siblings, dizygotic or fraternal twins (DZ) share, on average, half of their genes. However, because DZ twins, like MZ twins, share an intrauterine environment and experience many events at the same age (e.g., starting school), common environmental influences are, in theory, similar in both kinds of twins. Certainly, there are many ways in which the MZ twin environment could be more similar than the DZ twin environment. One assumption of twin studies, the equal environment assumption, posits that the environment of MZ twins is no more similar than the environment of DZ twins on dimensions of etiological relevance to the trait under study (Scarr 1968). However, at this time, there is scant evidence for violations of this equal environments assumption that would increase the similarity for eating disorder liability in MZ twins relative to that for DZ twins (Bulik et al. 1998; Klump et al. 2000a). Thus, differences in concordance rates of MZ versus DZ twins inform understanding of the relative contribution of genes and environment.

Standard twin models use concordance rates of MZ and DZ twins to assess the degree to which additive genetic effects (A), shared environmental effects (C), and/or unique environmental effects (E) contribute to the liability to a particular disorder. The objective of the model is to decompose the observed variance in liability to the disorder (V) in the population into these sources of variability: $V = a^2 + c^2 + e^2$, where a^2 is the proportion of variance due to A, also called heritability (Fisher 1918), c^2 is the proportion due to C, and e^2 is the proportion due to E (Neale and Cardon 1992).

If genetic factors were to act additively, it should be evident in the contrast of MZ to DZ correlations. For example, if a trait were entirely due to additive genetic effects, then MZ twins would be perfectly correlated because they share 100% of their genes, and the DZ correlation would be 0.50 because they share, on average, 50% of their genes (Fisher 1918). The effect of common environment contributes equally to the similarity of both MZ and DZ pairs. Thus, if the MZ and DZ correlations for a trait were equal, then any positive correlation of the trait values must be due to common environment. Genetic influences are incompatible with equal correlations in MZ and DZ twins. A mix of environmental and genetic effects is inferred if the MZ correlation is greater than the DZ correlation, but less than twice as large. Unique environment refers to environmental influences to which only one member of a twin pair is exposed (e.g., participating in sports emphasizing thinness) and also incorporates measurement error. Unique environmental effects decrease the correlations of both MZ and DZ twin pairs.

The first, systematic, clinically-ascertained twin study of AN found that MZ twins had higher concordance rates than DZ twins (Holland et al. 1984, 1988; Treasure and Holland 1989). Reanalysis of these data indicates that additive genetic effects accounted for 88% of the liability to AN, unique environmental effects account for the remainder, and shared environmental effects are absent (Bulik et al. 2000).

Population-based twin studies substantiate the above results. Wade et al. (2000) reported an estimated heritability of approximately 58% for AN, with the remaining variance accounted for by unique environmental factors. However, the broad confidence intervals for heritability do not exclude the possibility of common environmental influences contributing to the liability of AN. Similarly, several other studies estimated heritability for narrow and broad definitions of AN. Estimates ranged from 28 to 58% (Bulik et al. 2006, 2010; Kortegaard et al. 2001). However, Klump et al. (2001) reported a heritability estimate of 74% (95% CI: 35%, 95%) for AN syndrome in 17-year-old female twins. The remaining variance in the latter studies was best attributed to unique rather than shared environmental factors.

Population-based studies of BN lend further support to the substantial role of additive genetic effects to the liability to eating disorders (Bulik et al. 1998, 2010; Kendler et al. 1991; Kortegaard et al. 2001; Wade et al. 1999) and have yielded heritability estimates between 54% and 83%. However, as in the studies on AN, the confidence intervals around these heritability estimates are wide. These wide confidence intervals are likely due to the low statistical power of these studies and suggest that the influence of shared environmental effects on the liability to BN cannot be completely disregarded. To address this limitation, researchers have applied a measurement model to population-based twin data to boost power. By incorporating multiple waves of measurement, this method has the added benefit of increasing diagnostic reliability. Two studies that have used this approach have yielded heritability estimates for latent liability to BN of 83% (95% CI: 49%, 100%) (Bulik et al. 1998) and 59% (95% CI: 36%, 60%) (Wade et al. 1999).

Twin studies of BED have reported heritability estimates ranging from 41 to 57% for varying definitions of this disorder (Javaras et al. 2008a; Reichborn-Kjennerud et al. 2004). Currently, there are no published twin reports on the heritability of PD.

2.3 Adoption Studies

Adoption studies, while increasingly rare, allow the contributions of genetic and environmental effects to be distinguished and have greater power than twin studies to detect shared environmental influences. Because biological relatives have only genes in common with the adopted individual and adoptive relatives have only shared environment in common with the adopted individual, the relative influence of genetic and shared environmental factors can be estimated by comparing the incidence of a disorder or the similarity of a trait in biological relatives to that in adoptive relatives.

To date, only one adoption study on eating disorder pathology has been conducted (Klump et al. 2009). Participants were biological and adopted female sibling pairs. Given the low prevalence of both AN and BN, disordered eating symptoms rather than the diagnoses themselves were explored. Heritability estimates from biometrical modeling ranged from 59 to 82%, which provide convergent evidence for the ever-increasing body of twin studies.

In aggregate, family, twin, and adoption studies provide compelling evidence that genetic factors contribute to the etiology of eating disorders and some of these factors appear to be shared across the various types of eating disorders.

3 Eating Disorder Endophenotypes and Component Phenotypes

One strategy to enhance statistical power challenged by relatively low population prevalence of eating disorders has been to examine genetic influences of putative endophenotypes or component phenotypes of eating disorders. Endophenotypes are considered to be measureable biological markers for a disease, which are associated with the illness in the general population, are observable regardless of whether the illness is active, are observed in unaffected family members of probands at a higher rate than in the general population, and are heritable (Bulik et al. 2007; Gershon and Goldin 1986). Identification of endophenotypes, which are more proximal to the genotype than the disorder (Bulik et al. 2007), can facilitate the refinement of diagnostic criteria. Specifically endophenotype identification can clarify which traits are most highly heritable, and thereby most reflective of underlying biological, neurocognitive, or psychological processes. This, in turn, might aid in identifying genes which contribute to the liability of a disorder because the endophenotypes are theoretically less complex and influenced by a smaller number of genes (Bulik et al. 2007).

Bulik et al. (2007) identified a number of potential endophenotypes and component phenotypes for eating disorders. Some are disorder specific, and others are behaviors, attitudes, and temperament characteristics associated with eating disorders. The following sections review those constructs that have been shown to be heritable.

3.1 DSM Criteria

An item factor approach has been applied to examine genetic and environmental contributions to the criteria for AN (Mazzeo et al. 2009b), BN {Mazzeo, 2009a, and BED (Mitchell et al. 2010) in twins. The advantages of this model are the following: (1) the association between the latent trait (i.e., the specific eating disorder diagnosis) and each of the symptoms is estimated via the factor scores, (2) the relative contributions of genetic, shared environmental, and unique environmental factors to liability to overall diagnosis are estimated, and (3) heritability and contribution of shared and unique environmental factors are estimated for each measured symptom (i.e., item). Thus, this model can facilitate the identification of endophenotypes by clarifying which specific criteria are most strongly related to the overall diagnosis (as indicated by the factor loadings), and are most heritable.

Mazzeo et al. (2009b) used the item-factor approach to investigate AN within a population-based twin sample. Estimates of total heritability for several of the items assessed (i.e., whether participants had ever lost a lot of weight, whether friends and relatives had said they were too thin, whether weight affected how they felt about themselves at lowest weight, and lowest body mass index (BMI)), ranged between 29% and 34%. Heritability estimates for items related to weight concern at time of low weight were lower (ranging from 18 to 23%). Amenorrhea had a heritability estimate of 16% and was most strongly influenced by unshared environment. Results from the item factor model provided additional information: factor loadings indicating how strongly the symptoms were related to AN. Surprisingly, low BMI had the weakest association with AN. This was likely influenced by the structure of the survey, in which only individuals who met certain gateway criteria were asked the remaining questions. Symptoms most strongly associated with AN were losing a lot of weight, people saying you looked too thin, and the items related to weight concern.

These results inform future research on AN endophenotypes. Specifically, on the basis of these results, it appears that a criterion such as amenorrhea, which did not load highly on the latent AN factor and also had a low heritability estimate, may not be the most promising endophenotype for this diagnosis. However, endophenotypes or liability indices worthy of further examination would be those with the higher heritability estimates that are also associated to the greatest degree with AN: including intentional and extreme weight loss (as measured by items assessing whether participants had lost a lot of weight and whether others had noted they looked much too thin). Finally, it should be noted that the use of gateway items in this study presents challenges to the identification of endophenotypes. Although this survey technique is a useful and appropriate strategy for reducing participant burden in large-scale epidemiological studies, it is important to recognize that the heritability of the AN criteria was evaluated only among the subset of the sample with a low lifetime BMI. It is possible that genetic and environmental factors operate differently within individuals who are already at a low BMI compared to the general population.

Similarly, an item factor model was applied to twin data to estimate the heritability of BN criteria and to determine how strongly each symptom was associated with BN (Mazzeo et al. 2009a). Items assessing inappropriate compensatory behaviors had heritability estimates ranging from 35 to 53%. Of these items, vomiting and laxative use were most highly associated with BN, although all items had reasonably high loadings on the overall latent construct (i.e., the BN diagnosis). For the binge eating items examined, both loss of control and frequency had the greatest association with BN and had the highest heritability estimates (39% and 41%, respectively). The influence of weight and shape on self evaluation had the lowest factor loading and lowest heritability of the 11 symptoms examined. These results indicate that the symptoms are differentially heritable and confirm the centrality of binge eating and purging to the syndrome of BN.

Finally, Mitchell et al. (2010) applied the item-factor model to BED in a population-based twin sample. Factor loadings for each criterion on the latent construct were high, suggesting that the individual items are relevant to an underlying, unidimensional diagnosis. Further, heritability estimates for the individual items were relatively similar, ranging from 29 to 43%. These results vary somewhat from those obtained in the item-factor analyses of AN and BN described previously Mazzeo, 2009b; (Mazzeo et al. 2009a), which identified greater variability in estimates of A, C, and E across component diagnostic criteria.

It should be noted that, in all three of these item-factor studies (Mazzeo et al. 2009a,b; Mitchell et al. 2010), unique environmental factors accounted for the greatest proportion of variance at the item/criterion level. As noted previously, this component of variance includes both measurement error and unique environmental experiences. Thus, refinement of eating disorder assessment (including improvements in the questions asked, based on diagnostic revisions) should help differentiate whether specific symptoms are truly influenced more strongly by unique life events (e.g., appearance-oriented activities to which only one member of a twin pair is exposed) or whether an item's psychometric properties are attenuating heritability estimates.

3.2 *Disordered Eating Behaviors*

In addition to examining the genetic and environmental influences on threshold and subthreshold eating disorders, numerous studies have investigated eating disorder attitudes and related behaviors, including intentional weight loss (IWL), eating restraint, binge eating, and purging. This approach is appropriate as there is considerable interest in these behaviors as factors influencing the development and maintenance of eating disorders.

Restrained eating has gained a great deal of attention within the eating disorders field as this behavior is hypothesized to trigger the development of eating pathology (Polivy and Herman 1985), a theory supported by longitudinal data (Killen et al. 1996; Stice 1998). However, the measurement of restraint has proven challenging, and there is extensive and long-standing controversy about this construct and its assessment (e.g., Heatherton et al. 1988; Stice et al. 2004, 2007; Westenhoefer 1991; Westenhoefer et al. 1999; Williamson et al. 2007). Nonetheless, restraint remains a focus of research and could represent an important endophenotype given its temporal relation to the development of disordered eating.

Results regarding the heritability of restraint are mixed. One study conducted with twins from the University of Washington Twin Registry found that the heritability of restrained eating (measured by the Revised Restraint Scale) was 43%, adjusting for both sex and BMI (Schur et al. 2009). These results differ from those obtained in an earlier study (Neale et al. 2003) in which an alternate measure of restraint, the Three Factor Eating Questionnaire (TFEQ), was used. Specifically, in this investigation restraint was not significantly influenced by additive genetic factors, but rather, were associated with shared ($c^2 = 0.31$, CI: 0.04, 0.42) and specific ($e^2 = 0.69$, CI: 0.58, 0.80), environmental factors. Similar results were obtained for the TFEQ Susceptibility to Hunger subscale. Scores on the TFEQ Disinhibition subscale, in contrast, (which assesses the tendency to eat or overeat in response to contextual cues) were significantly influenced by additive genetic factors ($a^2 = 0.45$, CI = 0.32–0.57); common environment did not contribute to variance in scores on this subscale. The disparate results of these two studies highlight the potential influence of psychometric issues in the derivation of heritability estimates.

IWL is a related eating symptom that has been the focus of behavioral genetic research. Results of these studies have been much more consistent than of those investigating restraint. For example, Keski-Rahkonen et al. (2005b) obtained a heritability estimate of 38% for men and 66% for women. Similarly, Wade et al. (2009) obtained a heritability estimate of 0.61 for IWL in their bivariate analysis of genetic and environmental factors influencing this construct and overeating. Wade et al. also reported a heritability estimate of 45% for overeating, and the genetic correlation between these two behaviors was 0.61, indicating that they share a number of genetic factors and are not completely independent behaviors.

A larger body of research has examined the heritability of binge eating (Bulik et al. 2003; Root et al. 2010; Sullivan et al. 1998; Wade et al. 2008). In one of the earliest of these studies (Sullivan et al. 1998), a bivariate analysis of objective binge eating and self-induced vomiting yielded substantial heritability estimates for both behaviors (46% and 72%, respectively). The genetic correlation between the two traits was 0.74; this high correlation suggests that these behaviors are significantly influenced by shared genetic factors. Similar results were obtained in a subsequent study by Bulik et al. (2003), in which the heritability of binge eating was estimated at 49%. Further, the genetic correlation between this behavior and obesity was 0.34 indicating only a modest overlap of genetic factors.

The relation between binge eating and self-induced vomiting was evaluated in a bivariate model by Wade et al. (2008). Although these authors also found a substantial genetic correlation between the behaviors, the estimates for heritability were considerably lower than those obtained by Sullivan et al. (1998) for both binge eating and vomiting: 17% and 8% respectively. These differences in heritability estimates across studies could be due to the assessment of different populations or to the use of frequency criteria in one study but not in the other.

Night eating has more recently become a focus of scientific inquiry. Root et al. (2010) estimated heritability for binge eating and night eating separately for men and women. For men, the estimates were 74% and 44%, respectively. For women, the estimates were 70% and 35%, respectively, with a genetic correlation of 0.66.

In sum, although some investigations have reported relatively high heritability estimates for specific eating disorder behaviors (e.g., self-induced vomiting, binge eating), there is considerable variability in these estimates across studies. These differences are likely attributable to the definition of the constructs used and the format used to assess them. For example, as noted previously, many large-scale epidemiological surveys have used gateway items to assess eating disorders as part of a study of multiple psychiatric disorders (Mazzeo et al. 2009b). In contrast, other studies have used measures such as the Eating Disorders Examination (EDE; e.g., Wade et al. 2008), which is administered to all participants, without the use of gateway criteria. Further, specific symptoms, such as binge eating, have been defined as occurring with loss of control (Root et al. 2010) and without this component (Bulik et al. 2003; Sullivan et al. 1998). These differences underscore the importance of clearly defining the behavior of interest and using the most reliable and valid forms of measurement available for their assessment.

3.3 *Eating Disorder Attitudes and Temperament*

Similarities in temperament and attitudes among women with eating disorders have long been noted among clinicians working with this population (Bruch 1974). These similarities include a relentless drive for thinness, body dissatisfaction, perfectionism, obsessionality, and sensitivity to reward and punishment. Given that these constructs have been found to both exist before eating disorder onset (e.g., Tyrka et al. 2002) and persist following recovery in many patients (e.g., Klump et al. 2004), they represent potential endophenotypes of interest.

Drive for thinness and body dissatisfaction are perhaps the most proximal risk factors for eating disorders and among the most frequently studied in the area of behavior genetics. One of the earliest of these studies (Holland et al. 1988) examined the Eating Disorder Inventory (EDI) (Garner et al. 1984) and reported a heritability estimate of "near 1.0" for the Drive-for-Thinness scale, although the standard errors were large. Two other studies also investigated the constructs measured by the EDI in twins and reported heritability estimates ranging from 28 to 52% for the subscales, with the remainder of the variance being attributed to unique environmental effects (Rutherford et al. 1993; Wilksch and Wade 2009). Baker et al. (2009) and Keski-Rahkonen et al. (2005a) examined heritability of EDI subscales in men and women separately. Both studies found that genetic factors contributed more to the variance of both Drive-for-Thinness and Body Dissatisfaction in women (51–61% and 57–59%, respectively) as compared with men (heritability estimates of 20–40% (Baker et al. 2009) and 0% (Keski-Rahkonen et al. 2005a)). Klump et al. (2000b) examined EDI subscales in two samples of adolescent twins, aged 11 and 17. For the younger cohort, both additive genetic effects and common environmental effects were found to significantly contribute to most subscales. However, the contribution of additive genetic effects outweighs the contribution of common environmental effects in the older cohort. Overall, these results, suggest that heritability estimates are gender and age specific.

Studies from the Australian Twin Registry examined measures of dietary restraint and concern about eating, weight, and shape from the EDE (Wade et al. 1998). Heritability estimates for the total EDE score were 62% (95% CI: 21%, 71%). Individual variation of three of the EDE subscale measures was also best explained by a model which included only additive genetic and unique environmental effects (no shared environmental effects), with heritability estimates ranging from 32 to 62%. The exception was the Weight Concern measure, best explained by a model containing shared and unique environmental effects only (no additive genetic effects). Further, in a study by Wilksch and Wade (2009), the heritability of importance of weight and shape was assessed. The heritability estimate of the importance of weight and shape was low: 15% of the variance observed in this measure was accounted for by genetic factors. Yet, Wade and Bulik (2007) obtained a heritability estimate of 25% for weight and shape concerns.

Temperament, specifically perfectionism, sensitivity to reward, sensitivity to punishment, and obsessionality, have been studied in relation to shape and weight concerns (Wade and Bulik 2007; Wilksch and Wade 2009). The temperament measures had heritability estimates ranging from 27 to 71%. The perfectionism measures and weight and shape concerns shared about 10% of their genetic factors, indicating that different genes influence these constructs. In contrast, sensitivity to punishment and weight and shape concerns had a genetic correlation of 0.52. Although these temperament measures do not share a considerable proportion of genetic influences with the eating disorders weight and shape criterion, they have demonstrated substantial heritability and are prevalent in the eating disorders population and, thus, deserve further study.

In summary, genetic effects certainly contribute to the liability of eating disorders. Using endophenotypes to identify genes implicated in the etiology of eating disorders could open up new areas of exploration of biological pathways that lead to dysregulated eating, appetite, and weight regulation, as well as anxiety, obsessionality, and core eating disorder symptoms, such as drive for thinness. Such avenues may provide refinements in diagnosis by using objective biological measures, which could ultimately inform prevention and treatment strategies.

4 The Value of Family and Twin Studies in the Molecular Genetic Era

Family and twin studies of eating disorders are valuable methods to evaluate the magnitude of genetic effects on the liability to eating disorders. Further, these approaches have been employed to identify and refine endophenotypes and component phenotypes. This research has implications for molecular genetic studies: the focus on endophenotypes can increase the size of the population available for study and, perhaps more importantly, phenotypes identified in twin studies with consistently higher heritabilities can be used for sample selection for molecular genetic studies to potentially yield a more genetically homogenous population. This approach, theoretically, may enhance our ability to identify loci associated with the phenotypes and ultimately delineate underlying biological mechanisms that influence risk for and maintenance of disordered eating behavior. In addition, advances in molecular genetics may open new avenues for exploring environmental risk factors as well as gene x environment interactions and correlations. As our library of replicated genes expands for eating disorders, we will be able to further explore gene x environment interplay in large well-characterized twin samples.

References

Baker JH, Maes HH, Lissner L, Aggen SH, Lichtenstein P, Kendler KS (2009) Genetic risk factors for disordered eating in adolescent males and females. J Abnorm Psychol 118:576–586

Brody ML, Walsh BT, Devlin MJ (1994) Binge eating disorder: reliability and validity of a new diagnostic category. J Consult Clin Psychol 62:381–386

Bruch H (1974) Perils of behavior modification in treatment of anorexia nervosa. J Am Med Assoc 230:1419–1422

Bruder CE, Piotrowski A, Gijsbers AA, Andersson R, Erickson S, de Stahl TD, Menzel U, Sandgren J, von Tell D, Poplawski A, Crowley M, Crasto C, Partridge EC, Tiwari H, Allison DB, Komorowski J, van Ommen GJ, Boomsma DI, Pedersen NL, den Dunnen JT, Wirdefeldt K, Dumanski JP (2008) Phenotypically concordant and discordant monozygotic twins display different DNA copy-number-variation profiles. Am J Hum Genet 82:763–771

Bulik C, Sullivan P, Kendler K (1998) Heritability of binge-eating and broadly defined bulimia nervosa. Biol Psychiatry 44:1210–1218

Bulik C, Sullivan P, Wade T, Kendler K (2000) Twin studies of eating disorders: a review. Int J Eat Disord 27:1–20

Bulik CM, Sullivan PF, Kendler KS (2003) Genetic and environmental contributions to obesity and binge-eating. Int J Eat Disord 33:293–298

Bulik C, Sullivan P, Tozzi F, Furberg H, Lichtenstein P, Pedersen N (2006) Prevalence, heritability and prospective risk factors for anorexia nervosa. Arch Gen Psychiatry 63:305–312

Bulik CM, Hebebrand J, Keski-Rahkonen A, Klump KL, Reichborn-Kjennerud T, Mazzeo SE, Wade TD (2007) Genetic epidemiology, endophenotypes, and eating disorder classification. Int J Eat Disord 40(Suppl):S52–S60

Bulik CM, Thornton LM, Root TL, Pisetsky EM, Lichtenstein P, Pedersen NL (2010) Understanding the relation between anorexia nervosa and bulimia nervosa in a Swedish national twin sample. Biol Psychiatry 67:71–77

Favaro A, Ferrara S, Santonastaso P (2003) The spectrum of eating disorders in young women: a prevalence study in a general population sample. Psychosom Med 65:701–708

Fisher R (1918) The correlation between relatives on the supposition of Medelian inheritance. Trans R Soc Edinb 52:399–433

Fowler S, Bulik C (1997) Family environment and psychiatric history in women with binge eating disorder and obese controls. Behav Change 14:106–112

Garfinkel P, Lin E, Goering P, Spegg C, Goldbloom D, Kennedy S, Kaplan A, Woodside D (1995) Bulimia nervosa in a Canadian community sample: prevalence and comparison of subgroups. Am J Psychiatry 152:1052–1058

Garner D, Olmsted M, Polivy J (1984) Eating disorders inventory manual. Psychological Assessment Resources, New York

Gershon ES, Goldin LR (1986) Clinical methods in psychiatric genetics. I. Robustness of genetic marker investigative strategies. Acta Psychiatr Scand 74:113–118

Gershon E, Schreiber J, Hamovit J, Dibble E, Kaye W, Nurnberger J, Anersen A, Ebert M (1983) Anorexia nervosa and major affective disorders associated in families: a preliminary report. In: Guze SB, Earls FJ, Barrett JE (eds) Childhood psychopathology and development. Raven Press, New York, pp 279–284

Heatherton TF, Herman CP, Polivy J, King GA, McGree ST (1988) The (mis)measurement of restraint: an analysis of conceptual and psychometric issues. J Abnorm Psychol 97:19–28

Hoek HW (2006) Incidence, prevalence and mortality of anorexia nervosa and other eating disorders. Curr Opin Psychiatry 19:389–394

Holland AJ, Hall A, Murray R, Russell GFM, Crisp AH (1984) Anorexia nervosa: a study of 34 twin pairs and one set of triplets. Br J Psychiatry 145:414–419

Holland AJ, Sicotte N, Treasure J (1988) Anorexia nervosa: evidence for a genetic basis. J Psychosom Res 32:561–571

Hudson JI, Pope HG, Jonas JM, Yurgelun-Todd D, Frankenburg FR (1987) A controlled family history study of bulimia. Psychol Med 17:883–890

Hudson J, Lalonde J, Pindyck L, Bulik C, Crow S, McElroy S, Laird N, Tsuang M, Rosenthal N, Pope H (2006) Familial aggregation of binge-eating disorder. Arch Gen Psychiatry 63:313–319

Hudson JI, Hiripi E, Pope HG Jr, Kessler RC (2007) The prevalence and correlates of eating disorders in the National Comorbidity Survey Replication. Biol Psychiatry 61:348–358

Javaras KN, Laird NM, Reichborn-Kjennerud T, Bulik CM, Pope HG Jr, Hudson JI (2008a) Familiality and heritability of binge eating disorder: results of a case-control family study and a twin study. Int J Eat Disord 41:174–179

Javaras KN, Pope HG, Lalonde JK, Roberts JL, Nillni YI, Laird NM, Bulik CM, Crow SJ, McElroy SL, Walsh BT, Tsuang MT, Rosenthal NR, Hudson JI (2008b) Co-occurrence of binge eating disorder with psychiatric and medical disorders. J Clin Psychiatry 69:266–273

Kassett J, Gershon E, Maxwell M, Guroff J, Kazuba D, Smith A, Brandt H, Jimerson D (1989) Psychiatric disorders in the first-degree relatives of probands with bulimia nervosa. Am J Psychiatry 146:1468–1471

Keel P, Haedt A, Edler C (2005) Purging disorder: an ominous variant of bulimia nervosa? Int J Eat Disord 38:191–199

Kendler KS (1993) Twin studies of psychiatric illness. Arch Gen Psychiatry 50:905–915

Kendler KS, MacLean C, Neale MC, Kessler RC, Heath AC, Eaves LJ (1991) The genetic epidemiology of bulimia nervosa. Am J Psychiatry 148:1627–1637

Keski-Rahkonen A, Bulik CM, Neale BM, Rose RJ, Rissanen A, Kaprio J (2005a) Body dissatisfaction and drive for thinness in young adult twins. Int J Eat Disord 37:188–199

Keski-Rahkonen A, Neale BM, Bulik CM, Pietilainen KH, Rose RJ, Kaprio J, Rissanen A (2005b) Intentional weight loss in young adults: sex-specific genetic and environmental effects. Obes Res 13:745–753

Keski-Rahkonen A, Hoek HW, Susser ES, Linna MS, Sihvola E, Raevuori A, Bulik CM, Kaprio J, Rissanen A (2007) Epidemiology and course of anorexia nervosa in the community. Am J Psychiatry 164:1259–1265

Killen J, Taylor C, Hayward C, Haydel K, Wilson D, Hammer L, Kraemer H, Blair-Greiner A, Strachowski D (1996) Weight concerns influence the development of eating disorders: a 4-year prospective study. J Consult Clin Psychol 64:936–940

Klump KL, Holly A, Iacono WG, McGue M, Willson LE (2000a) Physical similarity and twin resemblance for eating attitudes and behaviors: a test of the equal environments assumption. Behav Genet 30:51–58

Klump KL, McGue M, Iacono WG (2000b) Age differences in genetic and environmental influences on eating attitudes and behaviors in preadolescent and adolescent female twins. J Abnorm Psychol 109:239–251

Klump KL, Miller KB, Keel PK, McGue M, Iacono WG (2001) Genetic and environmental influences on anorexia nervosa syndromes in a population-based twin sample. Psychol Med 31:737–740

Klump K, Strober M, Bulik C, Thornton L, Johnson C, Devlin B, Fichter M, Halmi K, Kaplan A, Woodside D, Crow S, Mitchell J, Rotondo A, Keel P, Berrettini W, Plotnicov K, Pollice C, Lilenfeld L, Kaye W (2004) Personality characteristics of women before and after recovery from an eating disorder. Psychol Med 34:1407–1418

Klump KL, Suisman JL, Burt SA, McGue M, Iacono WG (2009) Genetic and environmental influences on disordered eating: an adoption study. J Abnorm Psychol 118:797–805

Kortegaard LS, Hoerder K, Joergensen J, Gillberg C, Kyvik KO (2001) A preliminary population-based twin study of self-reported eating disorder. Psychol Med 31:361–365

Lee Y, Abbott D, Seim H, Crosby R, Monson N, Burgard M, Mitchell J (1999) Eating disorders and psychiatric disorders in the first-degree relatives of obese probands with binge eating disorder and obese non-binge eating disorder controls. Int J Eat Disord 26: 322–332

Lilenfeld L, Kaye W, Greeno C, Merikangas K, Plotnikov K, Pollice C, Rao R, Strober M, Bulik C, Nagy L (1998) A controlled family study of restricting anorexia and bulimia nervosa: comorbidity in probands and disorders in first-degree relatives. Arch Gen Psychiatry 55:603–610

Mazzeo SE, Mitchell KS, Bulik CM, Aggen SH, Kendler KS, Neale MC (2009a) A twin study of specific bulimia nervosa symptoms. Psychol Med 40:1–11

Mazzeo SE, Mitchell KS, Bulik CM, Reichborn-Kjennerud T, Kendler KS, Neale MC (2009b) Assessing the heritability of anorexia nervosa symptoms using a marginal maximal likelihood approach. Psychol Med 39:463–473

Mehler PS, Crews C, Weiner K (2004) Bulimia: medical complications. J Womens Health (Larchmt) 13:668–675

Mitchell KS, Neale MC, Bulik CM, Aggen SH, Kendler KS, Mazzeo SE (2010) Binge eating disorder: a symptom-level investigation of genetic and environmental influences on liability. Psychol Med published online Feb 5

Neale M, Cardon L (1992) Methodology for the study of twins and families. Kluwer Academic Publisher Group, Dordrecht, the Netherlands

Neale BM, Mazzeo SE, Bulik CM (2003) A twin study of dietary restraint, disinhibition and hunger: an examination of the eating inventory (three factor eating questionnaire). Twin Res 6:471–478

Plomin R, DeFries JC, McClearn GE, Rutter M (1994) Behavioral genetics, 3rd edn. W.H. Freeman & Co, New York

Polivy J, Herman CP (1985) Dieting and binging: a causal analysis. Am Psychol 40:193–201

Reichborn-Kjennerud T, Bulik C, Tambs K, Harris J (2004) Genetic and environmental influences on binge eating in the absence of compensatory behaviours: a population-based twin study. Int J Eat Disord 36:307–314

Root TL, Pisetsky EM, Thornton L, Lichtenstein P, Pedersen NL, Bulik CM (2010) Patterns of co-morbidity of eating disorders and substance use in Swedish females. Psychol Med 40:105–115

Rutherford J, McGuffin P, Katz R, Murray R (1993) Genetic influences on eating attitudes in a normal female twin population. Psychol Med 23:425–436

Scarr S (1968) Environmental bias in twin studies. Eugen Q 15:34–40

Schur E, Noonan C, Polivy J, Goldberg J, Buchwald D (2009) Genetic and environmental influences on restrained eating behavior. Int J Eat Disord 42:765–772

Stein D, Lilenfeld L, Plotnicov K, Pollice C, Rao R, Strober M, Kaye W (1999) Familial aggregation of eating disorders: results from a controlled family study of bulimia nervosa. Int J Eat Disord 26:211–215

Stice E (1998) Relations of restraint and negative affect to bulimic pathology: a longitudinal test of three competing models. Int J Eat Disord 23:243–260

Stice E, Fisher M, Lowe MR (2004) Are dietary restraint scales valid measures of acute dietary restriction? Unobtrusive observational data suggest not. Psychol Assess 16:51–59

Stice E, Cooper JA, Schoeller DA, Tappe K, Lowe MR (2007) Are dietary restraint scales valid measures of moderate- to long-term dietary restriction? Objective biological and behavioral data suggest not. Psychol Assess 19:449–458

Strober M, Morrell W, Burroughs J, Salkin B, Jacobs C (1985) A controlled family study of anorexia nervosa. J Psychiatr Res 19:239–246

Strober M, Lampert C, Morrell W, Burroughs J, Jacobs C (1990) A controlled family study of anorexia nervosa: evidence of familial aggregation and lack of shared transmission with affective disorders. Int J Eat Disord 9:239–253

Strober M, Freeman R, Lampert C, Diamond J, Kaye W (2000) Controlled family study of anorexia nervosa and bulimia nervosa: evidence of shared liability and transmission of partial syndromes. Am J Psychiatry 157:393–401

Strober M, Freeman R, Lampert C, Diamond J, Kaye W (2001) Males with anorexia nervosa: a controlled study of eating disorders in first-degree relatives. Int J Eat Disord 29: 263–269

Sullivan PF, Bulik CM, Kendler KS (1998) The genetic epidemiology of binging and vomiting. Br J Psychiatry 173:75–79

Treasure J, Holland A (1989) Genetic vulnerability to eating disorders: evidence from twin and family studies. In: Remschmidt H, Schmidt M (eds) Child and youth psychiatry: European persoectives. Hogrefe & Huber, New York, pp 59–68

Tyrka A, Waldron I, Graber J, Brooks-Gunn J (2002) Prospective predictors of the onset of anorexic and bulimic syndromes. Int J Eat Disord 32:282–290

Wade TD, Bulik CM (2007) Shared genetic and environmental risk factors between undue influence of body shape and weight on self-evaluation and dimensions of perfectionism. Psychol Med 37:635–644

Wade T, Martin N, Tiggemann M (1998) Genetic and environmental risk factors for the weight and shape concerns characteristic of bulimia nervosa. Psychol Med 28:761–771

Wade TD, Martin N, Neale M, Tiggemann M, Trealor S, Heath A, Bucholz K, Madden P (1999) The structure of genetic and environmental risk factors for three measures of disordered eating characteristic of bulimia nervosa. Psychol Med 29:925–934

Wade TD, Bulik CM, Neale M, Kendler KS (2000) Anorexia nervosa and major depression: shared genetic and environmental risk factors. Am J Psychiatry 157:469–471

Wade T, Bergin J, Tiggemann M, Bulik C, Fairburn C (2006) Prevalence and long-term course of lifetime eating disorders in an adult Australian twin cohort. Aust N Z J Psychiatry 40:121–128

Wade TD, Treloar S, Martin NG (2008) Shared and unique risk factors between lifetime purging and objective binge eating: a twin study. Psychol Med 38:1455–1464

Wade TD, Treloar SA, Heath AC, Martin NG (2009) An examination of the overlap between genetic and environmental risk factors for intentional weight loss and overeating. Int J Eat Disord 42:492–497

Westenhoefer J (1991) Dietary restraint and disinhibition: is restraint a homogeneous construct? Appetite 16:45–55

Westenhoefer J, Stunkard AJ, Pudel V (1999) Validation of the flexible and rigid control dimensions of dietary restraint. Int J Eat Disord 26:53–64

Wilksch SM, Wade TD (2009) An investigation of temperament endophenotype candidates for early emergence of the core cognitive component of eating disorders. Psychol Med 39:811–821

Williamson DA, Martin CK, York-Crowe E, Anton SD, Redman LM, Han H, Ravussin E (2007) Measurement of dietary restraint: validity tests of four questionnaires. Appetite 48:183–192

Woodside D, Field LL, Garfinkel P, Heinmaa M (1998) Specificity of eating disorders diagnoses in families of probands with anorexia nervosa and bulimia nervosa. Compr Psychiatry 39:261–264

Woodside DB, Garfinkel PE, Lin E, Goering P, Kaplan AS, Goldbloom DS, Kennedy SH (2001) Comparisons of men with full or partial eating disorders, men without eating disorders, and women with eating disorders in the community. Am J Psychiatry 158:570–574

The Genetics of Eating Disorders

Sietske G. Helder and David A. Collier

Contents

Abstract The eating disorders anorexia nervosa (AN), bulimia nervosa (BN), binge eating disorder and allied diagnoses such as eating disorder not otherwise specified are common, complex psychiatric disorders with a significant genetic component. Aetiology is unknown, but both phenotypic characteristics and genetic factors appear to be shared across these disorders, and indeed patients often change between diagnostic categories. Molecular studies have attempted to define genetic risk factors for these disorders, including case-control and family-based candidate gene association studies and linkage analysis of multiply affected nuclear families. These have used both clinical diagnoses and eating disorder-related intermediate phenotypes such as drive-for-thinness or body dissatisfaction. Candidate gene

---

D.A. Collier (✉) and S.G. Helder
MRC Social Genetic and Developmental Psychiatry Centre, Institute of Psychiatry, King's College London, P82, Room 1.11, De Crespigny Park, Denmark Hill, London SE5 8AF, UK
e-mail: david.collier@kcl.ac.uk; sietske.helder@kcl.ac.uk

R.A.H. Adan and W.H. Kaye (eds.), *Behavioral Neurobiology of Eating Disorders*, 157
Current Topics in Behavioral Neurosciences 6, DOI 10.1007/7854_2010_79

studies have focussed on neurotransmitter and neurodevelopmental systems [e.g. serotonergic, opioid, cannabinoid and dopaminergic receptors, and brain-derived neurotrophic factor (BDNF)], appetite regulatory peptides and their receptors [leptin, ghrelin, agouti-related protein (AgRP), melanocortin receptors, neuropeptide Y], energy balance systems (e.g. uncoupling proteins), genes implicated in obesity (e.g. FTO) and sex hormone systems (e.g. oestrogen receptors), either identified on the basis of their function alone or as positional candidates from linkage analysis. Of these studies, linkage analysis implicates 1p33-36 for AN, 1q31.3 for quantitative behavioural traits related to anorexia and 10p14 for BN, as well as other behavioural phenotypes across both disorders. Candidate gene association has implicated BDNF, delta 1 opioid receptor (OPDR1) and AgRP. More recently, with the advent of genome-wide association studies (GWAS), analysis with microsatellite markers has implicated novel candidate loci for AN at 1q41 and 11q22, and further GWAS results are expected in the near future.

Keywords Anorexia nervosa · Association · Bulimia nervosa · Linkage

1 Background

1.1 The Genetic Architecture of Eating Disorders

Eating disorders are complex genetic disorders, and they have contributions to aetiology from genetic factors, as determined by twin and family studies, as well as environmental (psychosocial) risk factors including dieting and childhood abuse. However, little is known about the genetic architecture of eating disorders (Bulik et al. 2007). The most accepted hypothesis for complex genetic disorders is the common disease-common variant hypothesis, in which common genetic risk variants (allele frequency $> 5\%$ in the general population) cause a small increase in risk (between 1.1-fold and twofold) (Risch and Merikangas 1996; Schork et al. 2009). Common genetic risk variants have been identified for many common, complex genetic disorders such as type 2 diabetes (T2D) and obesity (Prokopenko et al. 2008; O'Rahilly 2009) and schizophrenia (Wray and Visscher 2010). Typically, the genotype relative risk for a common genetic variant associated with these disorders is about 1.25-fold, i.e. a very small increase in risk for each disease-associated allele. The alternative hypothesis is the common disease-rare variant hypothesis (Schork et al. 2009; Bodmer and Bonilla 2008), in which rare genetic variants (allele frequency of $<5\%$ in the general population) have a moderate effect on risk, with a genotype relative risk of between 2 and 10. In reality, complex genetic disorders are likely to be a combination of common and rare genetic risk variants, which are likely to interact with each other through epistasis (Moore and Williams 2009).

Some clues to the genetic architecture of eating disorders come from Uher (2009), addressing the apparent paradox of high heritability and reproductive disadvantage associated with mental illness. Uher suggests an alternative to the disease-common variant framework consisting mainly of gene–environment interactions (G×E) and rare genetic variants (Uher 2009). Thus, a common mental illness with mild reproductive disadvantage is likely to have a large contribution from interactions between common genetic variants and environmental exposures [e.g. bulimia nervosa (BN)]. On the other hand, severe mental illness that confers strong reproductive disadvantage [such as restricting anorexia nervosa (AN)] might have a significant contribution from rare variants of recent origin. Although there is no formal evidence for this, it seems intuitive that genes and environment interact to increase risk of eating disorders, i.e. G×E makes a contribution to aetiology, because of their moderate heritability and known role of psychosocial factors.

Another major issue for geneticists is the stability and overlap of eating disorder diagnoses, as well as difficulty with relating common diagnostic categories such as eating disorder not otherwise specified in relation to the main diagnoses. In reality, eating disorder diagnoses are not fully separable categories, as there is conversion between diagnoses, for example, from AN to BN (Fairburn and Harrison 2003) and considerable overlap in symptoms. Thus, their aetiology is likely to consist of a mixture of risk factors which may not be specific to any one eating disorder diagnosis, as well as genetic and environmental overlap with other disorders such as depression, obesity, obsessive-compulsive disorder (OCD), autism and anxiety. While challenging for geneticists, this also provides the opportunity to redefine these disorders by biological means.

The molecular genetics of obesity and T2D is currently more advanced than eating disorders; as they have a similar heritability, they might guide us on how the molecular genetics of eating disorders might proceed. Both candidate gene and genome-wide approaches have led to the identification of T2D and obesity genes, including MC4R for obesity and PPARG and KCNJ11 for T2D from candidate gene studies, and a number of variants from genome-wide association studies (GWAS), including FTO and TCF7L2 (Table 1) (Prokopenko et al. 2008; O'Rahilly 2009).

In addition, for extreme obesity, rare, pathogenic copy number variants (CNVs, deletions and duplications of more than 1,000 base pairs) have also been identified (Bochukova et al. 2010). However, together these landmark discoveries, while of considerable importance in identifying the pathways

Table 1 Abbreviations in Sect. 1.1

Abbreviations	
MC4R	Melanocortin 4 receptor
PPARG	Peroxisome proliferator-activated receptor gamma
KCNJ11	Potassium inwardly rectifying channel, subfamily J, member 11
FTO	Fat mass and obesity associated
TCF7L2	Transcription factor 7-like 2 (T-cell specific, HMG-box)

involved in T2D and obesity, thus far account for less than 10% of the heritability of these disorders, even when tens of thousands of cases are used (Manolio et al. 2009). This contrasts with macular degeneration, another complex disorder in which just a few variants account for half of the heritability. The remaining genetic variants for these complex genetic disorders are continuously emerging from GWAS and other studies, and genome sequencing efforts are currently underway to identify rare, moderate risk single nucleotide polymorphisms (SNPs) and other sequence variants. Although all complex genetic disorders may each differ in their genetic architecture, it is most likely that eating disorders fit into the same pattern as most others; a combination of common, low risk alleles and rare, moderate risk alleles, likely to be moderated by genetic background and environment.

2 Candidate Gene Studies

The candidate gene approach examines genes which are suspected of being involved in a disease, because the function of the gene product suggests that it could be related to the pathophysiology of that disease. For example, in T2D, candidate genes from the insulin pathway have been successfully examined for association with the disease (Prokopenko et al. 2008). The main flaw of candidate gene studies is the low prior probability of association, since for most diseases, the task of selecting the correct candidate gene from approximately 27,000 human genes is very difficult, even when there is detailed knowledge of pathophysiology, since the function of most genes is poorly characterised. Because of this, but also probably because of other issues such as statistical power and poor genetic coverage, candidate gene studies have had limited success (Tabor et al. 2002). For example, despite hundreds of candidate gene studies in T2D, only two have stood the test of time, the Pro12Ala variant in the PPARG gene, involved in insulin action, and the Glu23Lys variant in KCNJ11 gene, involved in β-cell dysfunction (Prokopenko et al. 2008).

In psychiatric disorders, this is even more difficult as there is even less information on pathophysiology. However, there have been some successes, notably attention-deficit hyperactivity disorder (ADHD), which is commonly treated by dopaminergic system drugs such as methylphenidate, where genes from the dopamine system, particularly the dopamine D4 and D5 receptor genes DRD4 and DRD5, have been implicated (Thapar et al. 2007).

There are also advantages to candidate gene analysis; they have greater statistical power than either linkage analysis or genome-wide approaches, because of their reliance on linkage disequilibrium (LD) and the low level of multiple testing; they can be used to test hypotheses about disease aetiology and are easy to perform (Tabor et al. 2002). There have been a series of candidate gene studies in eating disorders over the past decade or so, most of which focus on AN, and which examined neurotransmitter or other pathways related to behaviour and feeding, such as the serotonin and dopamine systems, the neuropeptides involved in feeding, or genes implicated in obesity. These studies have been extensively reviewed by others (Pinheiro et al. 2010; Rask-Andersen et al. 2010; Scherag et al. 2010) and will not be considered in detail here.

Few, if any, of these candidate-gene association studies have replicated genetic association, although some findings remain interesting if controversial, in particular AgRP (Vink et al. 2001), the serotonin receptor type 2A gene (HTR2A) (Collier et al. 1997; Gorwood et al. 2002; Ziegler et al. 1999; Kiezebrink et al. 2010), functional variants of the serotonin receptor type 3A and B genes (HTR3A and HTR3B), the serotonin receptor type 1D (HTR1D) (Bergen et al. 2003; Brown et al. 2007; Kiezebrink et al. 2010), the serotonin receptor type 1B (HTR1B) (Kiezebrink et al. 2010), BDNF, (Ribases et al. 2003, 2004) the dopamine D2 receptor (DRD2) (Bergen et al. 2005; Dmitrzak-Weglarz et al. 2007) and OPRD1 (Bergen et al. 2003; Brown et al. 2007), a candidate gene identified by its position under a linkage peak (see Sect. 3).

Since this review, several candidate gene studies have been performed, including analysis of obesity genes in AN. Brandys et al. also examined several variants influencing body mass index (BMI), identified by GWAS studies, for association with AN, in a sample of 267 AN patients and 1,636 population controls (Brandys et al. 2010). They likewise found no significant association between individual SNPs and AN, or an influence of the combined effect of BMI-increasing alleles. This study found no evidence that genetic variants regulating BMI in the general population are significantly associated with susceptibility to AN. However, one variant not tested in this study is the BDNF Val66Met variant, a functional polymorphism in the gene which reduces functional protein levels significantly. This variant is a strong candidate for association with eating disorders, although the data are not conclusive, with the low-functionality Met66 allele associated with eating disorders (Ribases et al. 2003, 2004). In obesity, GWAS found significant association with the same polymorphism, but with the wild-type val66 allele increasing the likelihood of higher BMI (Thorleifsson et al. 2009).

A major study also performed case–control association analysis in over 1,000 cases with AN and 677 controls. Pinheiro et al. (2010) used more than 5,000 SNPs in 182 candidate genes selected on the basis of previous association data, gene expression in the brain, biological plausibility and markers from regions linked to AN in family-based linkage studies. After accounting for multiple testing by Bonferonni correction, there were no statistically significant associations for any individual SNP or haplotype. However, there were a number of potentially interesting findings, including GLP2R, coding for the receptor of a 33-amino acid proglucagon-derived gut peptide (GLP2), the phenylalanine hydroxylase (PAH) gene and KCNN3, coding for a small-conductance calcium-activated potassium channel expressed in the brain. For KCNN3, a CAG repeat polymorphism in the gene has previously been associated with AN (Koronyo-Hamaoui et al. 2002, 2004).

3 Linkage Analysis

The classical approach to the detection of genes for genetic disorders is family-based linkage analysis followed by positional cloning. This was spectacularly successful for single-gene disorders such as cystic fibrosis and was consequently attempted for complex disorders such as AN and BN. The difficulties in linkage analysis of complex disorders relate to their genetic architecture; causation is assumed to result from many common, low-risk variants, exactly the type of variant linkage analysis does not detect, because the effect size is too small to see in a practical number of pedigrees (Risch and Merikangas 1996). However, if the genetic architecture of eating disorders also includes less common, moderate risk variants, linkage analysis should be able to detect these with a realistic sample size.

Several studies have attempted to detect susceptibility loci for eating disorders using genome-wide linkage analysis with short tandem repeat (microsatellite) markers. This has focussed primarily on affected relative pairs with eating disorders, as multiply affected families with AN are very rare. Grice et al. performed genome-wide linkage analysis of 192 families with at least one affected relative pair with AN and related eating disorders, including BN (Grice et al. 2002). Analysis resulted in modest evidence for linkage at marker D4S2367 on chromosome 4, and further analysis in a subset of 37 families with more than one case of restricting AN gave a multipoint nonparametric linkage (NPL) score of 3.45 at D1S3721 on chromosome 1p, providing suggestive evidence for an AN-susceptibility locus.

Since AN is an unstable diagnosis, studies have also used quantitative traits for linkage, including those related to psychiatric, personality and temperament in eating disorders. Devlin et al. used two quantitative traits, drive-for-thinness from the Eating Disorders Inventory (EDI) and obsessionality from the Yale-Brown Obsessive Compulsive Scale (Y-BOCS), in 196 multiplex families with an AN proband from the same sample set as Grice et al. (2002) for genome-wide quantitative trait linkage (Devlin et al. 2002). Using a novel method that incorporates covariates, they found several regions of suggestive linkage: one close to genome-wide significance on chromosome 1 for a combined drive-for-thinness–obsessionality phenotype [logarithm of the odds score (LOD) = 3.46], another on chromosome 2 for obsessionality only (LOD = 2.22) and a third region on chromosome 13 for drive-for-thinness only (LOD = 2.50).

Linkage studies have also been used to examine BN (Bulik et al. 2003), using microsatellite markers in 308 multiplex families with eating disorders identified through a proband with BN. The highest nonparametric multipoint maximum LOD score (MLS) was 2.92, on chromosome 10p. A symptom marker of BN, self-induced vomiting, was also used as a linkage phenotype in a subset of families with at least two affected relatives reporting this, which produced the highest MLS (3.39) at the same locus on chromosome 10p, indicating that this region may harbour susceptibility alleles for BN.

Table 2 Linkage signals in eating disorders

Study	Phenotype	Sample size	Statistic	Chromosome
Grice et al. (2002)	Restricting AN	37 families	NPL = 3.45	1p33-p36
Devlin et al. (2002)	AN: drive-for-thinness/ obsessionality	196 families	LOD = 3.46	1q31.3
Devlin et al. (2002)	AN: obsessionality	196 families	LOD = 2.22	2p11.2
Devlin et al. (2002)	AN: drive-for-thinness	196 families	LOD = 2.50	13q13.3
Bulik et al. (2003)	BN	308 families	LOD = 2.92	10p13
Bulik et al. (2003)	BN: self-induced vomiting	308 families	LOD = 2.92	10p13
Bacanu et al. (2005)	BN: concern over mistakes	308 families	LOD = 2.97	16p13.3
Bacanu et al. (2005)	BN: minimum BMI	308 families	LOD = 2.97	4q21.1
Bacanu et al. (2005)	BN: food-related obsessions	308 families	LOD = 2.97	14q21.1

Subsequently, linkage analysis of both the AN and BN linkage cohorts described above was performed using behavioural phenotypes as quantitative traits or covariates, selected from over 100 attributes thought to be related to liability to eating disorders (Bacanu et al. 2005). These were obsessionality, age at menarche and anxiety for quantitative trait locus linkage, and lifetime minimum BMI, concern over mistakes and food-related obsessions for covariate-based linkage analysis. The BN cohort produced significant linkage, for minimum BMI at 4q21.1, for concern over mistakes at 16p13.3 and 14q21.1, and for food-related obsessions at 14q21.1, as well as some signals suggestive of linkage. No significant loci were found for AN, and overlap between the two cohorts was minimal for substantial linkage signals (Table 2).

Since these studies, advances in microarray genotyping technologies have allowed the development of several GWAS including GWAS of common SNPs, which primarily analyse common genetic variation in the genome for association (Corvin et al. 2010), pathway analysis, in which the contribution of biological pathways rather than individual genes is assessed using GWAS data (Cantor et al. 2010), epistasis, the testing for gene–gene interactions (Moore and Williams 2009), methods such as comparative genome hybridisation, which can detect CNVs (Merikangas et al. 2009), and sequencing-based discovery of rare variants, including genome sequencing of the exome (the protein coding and regulatory regions of genes) (Ng et al. 2008) and whole genome sequencing in individuals, which is becoming increasingly feasible as costs fall (Bonetta 2010).

4 Genome-Wide Association

GWAS examine genetic variation across a given genome, designed to identify genetic associations with disease (Corvin et al. 2010). GWAS typically uses genotyping microarrays, which can measure more than one million SNPs in a single individual in one experiment, with the aim of measuring the majority of common genetic variation. This is achieved not by genotyping every common SNP variant, but by the use of LD, i.e. inferring association information about ungenotyped variants using genotyped variants and information on LD from the HAPMAP project (www.hapmap.org). Typically, they capture more than 60–80% of common (minor allele frequency $> 5\%$) genetic variation (Hao et al. 2008), but are poor at detecting rare variation and polymorphisms with a high mutation rate, such as some variable number of tandem repeats and simple sequence repeats.

GWAS designs are usually cross-sectional case–control studies, using thousands of cases and controls, but can also be performed using population traits in a quantitative trait locus design, or with family trios using transmission disequilibrium tests (TDTs) (Lasky-Su et al. 2010; Neale et al. 2008). Association for each individual SNP in a case–control study is typically measured by logistic regression producing an odds ratio along with a P-value that tests whether the odds ratio differs from unity, i.e. if there is significant association (Corvin et al. 2010).

4.1 Microsatellite-Based GWAS

In addition to SNP-based GWAS, other genetic markers such as microsatellite markers can be used for genome-wide association. Microsatellite markers are simple sequence repeats which are highly polymorphic and thus show high average heterozygosity, at 70% on average, compared to SNPs, which have an average heterozygosity of around 30% (Matsuzaki et al. 2004). Their mutation rates are also higher (typically 10^{-3} to 10^{-5} per generation) (Ellegren 2004; Weber and Wong 1993), compared to SNPs (10^{-8} to 10^{-9} per generation) (Nachman and Crowell 2000). Thus, their LD lengths are in the 100 kb range, compared to about 10 kb for SNPs (The International HapMap Consortium 2007) meaning that an order of magnitude fewer microsatellites should be required for GWAS than SNPs.

Nakabayashi et al. used a genome-wide set of over 23,000 microsatellite markers to perform a GWAS of AN (Nakabayashi et al. 2009). This GWAS mapping approach has previously been used in mapping novel susceptibility genes for rheumatoid arthritis (Tamiya et al. 2005). They used the DNA pooling method, where samples from a number of individuals are pooled at equimolar concentration and run as a single sample (Breen et al. 1999). Using a two-stage screening approach and individual genotyping in 320 AN cases and 341 controls, they identified ten novel loci showing association with AN, seven of which were followed up to further narrow down candidate intervals for susceptibility alleles, which identified two loci (1q41 and 11q22) remaining significantly associated with AN. The most significant association was observed at SNP rs2048332 (odds ratio 1.4, allelic P-value $= 0.00023$) located at $3'$-downstream of the spermatogenesis-associated 17 (SPATA17) gene on the 1q41 locus, and a haplotype that comprised four SNP/microsatellite markers within the contactin 5 (CNTN5) gene was also associated with AN.

5 Future Prospects

5.1 Array-Based GWAS

The first GWAS study, of age-related macular degeneration, appeared in 2005 (Klein et al. 2005), and since then, almost 400 GWAS articles have been published in the National Human Genome Research Institute GWAS Catalog [www.genome.gov/26525384, accessed 20 September 2009 (Corvin et al. 2010)]. Technology has advanced rapidly, meaning that the majority (up to 90%) of common genetic variation (minor allele frequency $> 5\%$) can be analysed using microarrays which analyse hundreds of thousands or even millions of SNPs. The GWAS approach has identified common genetic variants that predispose to a variety of complex human diseases and traits, including obesity, T2D, height, body composition, biochemical traits related to insulin and lipid metabolism, psychiatric disorders such as schizophrenia and depression, and so forth, as is reviewed extensively elsewhere (McCarthy and Hirschhorn 2008).

Currently, two GWAS of AN are underway, one funded by the Wellcome Trust analysing 4,000 cases and 4,000 controls. The results of this are unknown at the time of writing, but are expected to emerge in 2010.

5.2 Copy Number Variants

In addition to common, low-risk genetic variants, moderate risk, de novo deletions and duplications of DNA (CNVs) have recently emerged as genomic risk factors for common brain disorders, including schizophrenia, autism and mental retardation with deletions on chromosomes 1q21.1, 15q11.2, 15q13.3 and the neurexin 1 (NRXN1) gene at 2p16.3 (Stefansson et al. 2008; Rujescu et al. 2009), and a duplication on chromosome 16p13.1 (Ingason et al. 2009). These CNVs are not specific to any particular psychiatric diagnosis and can give rise to a range of phenotypes, from autism to epilepsy (Sebat et al. 2009; Merikangas et al. 2009). They also tend to be rare in the population: deletions at 1q21.1 and 15q13.3 occur in about 1/500 patients with schizophrenia compared with 1/5,000 controls without neuropsychiatric illness (Stefansson et al. 2008). In idiopathic generalised epilepsy, however, deletions at 15q13.3 occur in as many as 1/100 of cases (Mefford et al. 2010; Helbig et al. 2009). These have a moderate odds ratio for risk, at around three- to tenfold increased risk, in contrast to common variants, which show odds ratios between 1 and 2. Thus, the penetrance of the pathogenic CNVs discovered so far is intermediate between common variants and the highly penetrant genomic variants associated with syndromes commonly used in a clinical genetic setting, at between 2% and 9%. Common CNVs have also been examined for disease risk. Unlike de novo or rare CNVs, these appeared to be well tagged by common SNP variants and tend to generally not be associated with disease (Craddock et al. 2010).

At the time of writing, there have been no studies of association between CNVs and eating disorders; these are likely to emerge as GWAS and other studies progress.

5.3 Genome Sequencing

Even when high-powered GWAS for eating disorders are completed, it is likely that only a fraction of susceptibility alleles will be discovered, and many more will remain unknown, as has been the case for many GWAS studies of other disorders, including psychiatric disorders. GWAS methods can identify common low-risk variants, but they are at present limited to studying relatively high allele frequency variants (around 5% allele frequency and above), because of the composition of arrays, making it likely that the "missing heritability" consists at least in part of rare variants. In many complex traits, resequencing of genes has identified multiple, rare variants associated with the disease or trait. For example, the proprotein convertase subtilisin/kexin type 9 (PCSK9) gene has several variants with around 0.2–1.5% allele frequency and 20–40% effect on low density lipoprotein cholesterol (LDL-C) levels (Kotowski et al. 2006), and in the Niemann–Pick disease type C1-like 1 (NPC1L1) gene (Cohen et al. 2006) there appear to be many rare variants with effects on LDL-C and lipid absorption.

Sequencing technology from high-throughput DNA sequencing platforms (such as Roche 454, Solexa/Illumina and AB SOLiD) has enabled cheap whole genome and whole-exome sequencing (a few thousand dollars per genome), along with efficient methods to analyse the data (Bonetta 2010; Bentley et al. 2008; Li and Durbin 2009). Other initiatives such as the 1,000 Genomes Project (www.1000genomes.org), launched in 2008 to sequence over 1,000 people from different populations, are finding essentially all genetic polymorphisms in these populations down to 1% allele frequency, consisting eventually of tens of millions of variants. In addition, novel statistical methods for the detection of complex trait associations with rare variants detected through large-scale resequencing data have been developed (Morris and Zeggini 2010). These are now being applied on a large scale to sequence thousands of people with complex disorders and traits such as early-onset severe obesity and schizophrenia. As a disorder with a probable rare-variant component, AN is a strong candidate for genome sequencing experiments (Uher 2009).

5.4 Gene–Environment Interaction

Interaction between genes and environmental factors is thought to increase the risk of several psychiatric disorders (Rutter et al. 2006) including major depression (Caspi et al. 2003), IQ (Caspi et al. 2007) and schizophrenia (Caspi et al. 2005; Arseneault et al. 2002). Given that eating disorders are complex psychiatric disorders with an affective component with both genetic and environmental risk factors including psychosocial stress, it seems likely that G×E should occur in eating disorders (Bulik 2005). For example, in depression, where the repeat length polymorphism 5-HTTLPR, in the promoter of the serotonin transporter (SLC6A4), alone has a negligible or weak impact on the risk of developing the disease, individuals with the short (S) allele of this polymorphism are at greater risk for depression (Caspi et al. 2003) following stressful life events. Several reports [reviewed by (Caspi et al. 2010)] have replicated the initial finding of G×E between the 5-HTTLPR and stressful life events by Caspi et al. (2003). Others, however, have not, as reviewed in Risch et al. (2009), but the overall picture is supportive of G×E between the 5-HTTLPR and depression (Caspi et al. 2010), especially those studies which use better quality measures of environmental data, such as prospective measurement or face-to-face interview.

The discovery of additional functional polymorphisms in the 5-HTTLPR region is another possible explanation for inconsistent associations (Hu et al. 2005; Kraft et al. 2005; Wendland et al. 2006, 2008). In addition to several rare functional variants, there is at least one relatively common SNP (rs25531 A > G) within the 5-HTTLPR insertion (Hu et al. 2005; Kraft et al. 2005; Wendland et al. 2006; Rasmussen and Werge 2007). Hu et al. show that only the A variant of the L allele (designated La) yields high serotonin transporter mRNA levels; thus, the Lg variant apparently behaves equivalent to the low expressing S allele (Hu et al. 2005). There is some evidence that the high expressing La allele is associated with inhibited/compulsive traits in patients with eating disorders (Steiger et al. 2009). Indeed, the high expressing La allele is also associated with increased genetic susceptibility to OCD (Wendland et al. 2008); however, this was observed only when multiple variants were analysed as haplotypes, not as single loci, which stresses the need for more detailed genotyping of the 5-HTTLPR region in future association studies.

5.5 *Epigenetics*

Epigenetics is the reversible regulation of various genomic functions, mediated principally through changes in DNA methylation and chromatin structure (histone modifications) but without changing the classical DNA sequence. DNA methylation is the best understood epigenetic process; the addition of a methyl group to CpG dinucleotides affects the binding of transcription factors as well as chromatin compaction and gene silencing (Jaenisch and Bird 2003). Epigenetic processes are essential for normal cellular development (e.g. the differentiation of cells into different types of neurons) through long-term *cis*-regulation of gene function. Like DNA sequence, the epigenetic profile of somatic cells is passed on during normal cell division (mitosis), but, unlike DNA sequence, can change as cells differentiate or respond to external factors. In addition, there are suggestions that the epigenetic profile can be passed on during meiosis thus potentially transmitting traits or the effects of environmental risk factors across generations. Aberrant epigenetic modifications are hypothesised to be involved in many human pathologies (Hatchwell and Greally 2007), including complex neuropsychiatric disorders such as psychosis (Mill et al. 2008), depression (Mill and Petronis 2007), drug addiction (Renthal and Nestler 2008), ADHD (Mill and Petronis 2008) and autism (Schanen 2006).

One of the best examples of transgeneration effects is that of nutrition, which has been shown to impact on health. These studies make use of natural events, such as the Dutch famine between 1944 and 1945 (Lumey and Stein 2009), or famine in China related to the Great Leap Forward (Xu et al. 2009). For example, the effects of maternal famine and malnutrition in pregnancy can be seen in offspring across generations (Xu et al. 2009; Susser and Lin 1992; Kaati et al. 2007; Pembrey 1996), probably in an epigenetic mechanism involving DNA methylation (Heijmans et al. 2008). The effects from maternal nutrition during a critical period in childhood, the slow growth period, can later influence her child's risk as an adult for cardiovascular disease, diabetes mellitus and hypertension (Kaati et al. 2002), and furthermore the nutrition of the grandmother during pregnancy might influence not only the mother's nutrition as a foetus but also the grandchild's birth weight (Stein et al. 2004). There also appears to be a similar effect in the male-line, linking paternal grandparental nutrition to male mortality from diabetes and cardiovascular disease in later generations (Pembrey et al. 2006). Thus, ancestors' nutrition appears to be a major influence on longevity, in relation to disease risk. This challenges the assumption that the "heritable" component to eating disorders, and other complex disorders, is entirely genetic.

Most epigenetic analyses examine DNA methylation, focussing on CpG islands which are thought to regulate genes. Methylation of DNA at these CpG islands can trigger a change in gene expression, usually thought to involve changes in chromatin structure which lead to inaccessibility of regulatory regions and a reduction of transcription. Methylation of DNA can be analysed by a variety of methods, such as the use of sodium bisulphite conversion on genomic DNA followed by DNA sequencing or array analysis, and can involve candidate gene or genome-wide approaches. There have been a few studies on epigenetic mechanisms in eating disorders, all examining candidate genes. For example, Frieling et al. examined the dopamine transporter (DAT1) and DRD2 for methylation changes in eating disorders, and found that patients showed elevated expression of DAT1 mRNA accompanied by hypermethylation of its promoter in the AN and BN group and reduced expression of DRD2 accompanied by significant hypermethylation of its promoter in the AN group (Frieling et al. 2009). This study indicates that there may be altered expression of dopaminergic genes accompanied by changes in DNA methylation which might have epigenetic effects. Although cause and effect will need careful examination, the dynamic nature of the epigenome suggests that, unlike genetic mutations, aberrant epigenetic modifications are likely to be reversible, which makes them realistic targets for the development of novel interventions for psychiatric disorders (Wong et al. 2010; Docherty and Mill 2008).

Reference

Arseneault L, Cannon M, Poulton R, Murray R, Caspi A, Moffitt TE (2002) Cannabis use in adolescence and risk for adult psychosis: longitudinal prospective study. BMJ 325:1212–1213

Bacanu SA, Bulik CM, Klump KL, Fichter MM, Halmi KA, Keel P, Kaplan AS, Mitchell JE, Rotondo A, Strober M, Treasure J, Woodside DB, Sonpar VA, Xie W, Bergen AW, Berrettini WH, Kaye WH, Devlin B (2005) Linkage analysis of anorexia and bulimia nervosa cohorts using selected behavioral phenotypes as quantitative traits or covariates. Am J Med Genet B Neuropsychiatr Genet 139B:61–68

Bentley DR, Balasubramanian S, Swerdlow HP, Smith GP, Milton J, Brown CG, Hall KP, Evers DJ, Barnes CL, Bignell HR, Boutell JM, Bryant J, Carter RJ, Keira CR, Cox AJ, Ellis DJ, Flatbush MR, Gormley NA, Humphray SJ, Irving LJ, Karbelashvili MS, Kirk SM, Li H, Liu X, Maisinger KS, Murray LJ, Obradovic B, Ost T, Parkinson ML, Pratt MR, Rasolonjatovo IM, Reed MT, Rigatti R, Rodighiero C, Ross MT, Sabot A, Sankar SV, Scally A, Schroth GP, Smith ME, Smith VP, Spiridou A, Torrance PE, Tzonev SS, Vermaas EH, Walter K, Wu X, Zhang L, Alam MD, Anastasi C, Aniebo IC, Bailey DM, Bancarz IR, Banerjee S, Barbour SG, Baybayan PA, Benoit VA, Benson KF, Bevis C, Black PJ, Boodhun A, Brennan JS, Bridgham JA, Brown RC, Brown AA, Buermann DH, Bundu AA, Burrows JC, Carter NP, Castillo N, Catenazzi CE, Chang S, Neil CR, Crake NR, Dada OO, Diakoumakos KD, Dominguez-Fernandez B, Earnshaw DJ, Egbujor UC, Elmore DW, Etchin SS, Ewan MR, Fedurco M, Fraser LJ, Fuentes Fajardo KV, Scott FW, George D, Gietzen KJ, Goddard CP, Golda GS, Granieri PA, Green DE, Gustafson DL, Hansen NF, Harnish K, Haudenschild CD, Heyer NI, Hims MM, Ho JT, Horgan AM, Hoschler K, Hurwitz S, Ivanov DV, Johnson MQ, James T, Huw Jones TA, Kang GD, Kerelska TH, Kersey AD, Khrebtukova I, Kindwall AP, Kingsbury Z, Kokko-Gonzales PI, Kumar A, Laurent MA, Lawley CT, Lee SE, Lee X, Liao AK, Loch JA, Lok M, Luo S, Mammen RM, Martin JW, McCauley PG, McNitt P, Mehta P, Moon KW, Mullens JW, Newington T, Ning Z, Ling NB, Novo SM, O'Neill MJ, Osborne MA, Osnowski A, Ostadan O, Paraschos LL, Pickering L, Pike AC, Pike AC, Chris PD, Pliskin DP, Podhasky J, Quijano VJ, Raczy C, Rae VH, Rawlings SR, Chiva RA, Roe PM, Rogers J, Rogert Bacigalupo MC, Romanov N, Romieu A, Roth RK, Rourke NJ, Ruediger ST, Rusman E, Sanches-Kuiper RM, Schenker MR, Seoane JM, Shaw RJ, Shiver MK, Short SW, Sizto NL, Sluis JP, Smith MA, Ernest Sohna SJ, Spence EJ, Stevens K, Sutton N, Szajkowski L, Tregidgo CL, Turcatti G, Vandevondele S, Verhovsky Y, Virk SM, Wakelin S, Walcott GC, Wang J, Worsley GJ, Yan J, Yau L, Zuerlein M, Rogers J, Mullikin JC, Hurles ME, McCooke NJ, West JS, Oaks FL, Lundberg PL, Klenerman D, Durbin R, Smith AJ (2008) Accurate whole human genome sequencing using reversible terminator chemistry. Nature 456:53–59

Bergen AW, van den Bree MB, Yeager M, Welch R, Ganjei JK, Haque K, Bacanu S, Berrettini WH, Grice DE, Goldman D, Bulik CM, Klump K, Fichter M, Halmi K, Kaplan A, Strober M, Treasure J, Woodside B, Kaye WH (2003) Candidate genes for anorexia nervosa in the 1p33-36 linkage region: serotonin 1D and delta opioid receptor loci exhibit significant association to anorexia nervosa. Mol Psychiatry 8:397–406

Bergen AW, Yeager M, Welch RA, Haque K, Ganjei JK, van den Bree MB, Mazzanti C, Nardi I, Fichter MM, Halmi KA, Kaplan AS, Strober M, Treasure J, Woodside DB, Bulik CM, Bacanu SA, Devlin B, Berrettini WH, Goldman D, Kaye WH (2005) Association of multiple DRD2 polymorphisms with anorexia nervosa. Neuropsychopharmacology 30:1703–1710

Bochukova EG, Huang N, Keogh J, Henning E, Purmann C, Blaszczyk K, Saeed S, Hamilton-Shield J, Clayton-Smith J, O'Rahilly S, Hurles ME, Farooqi IS (2010) Large, rare chromosomal deletions associated with severe early-onset obesity. Nature 463:666–670

Bodmer W, Bonilla C (2008) Common and rare variants in multifactorial susceptibility to common diseases. Nat Genet 40:695–701

Bonetta L (2010) Whole-genome sequencing breaks the cost barrier. Cell 141:917–919
Brandys MK, van Elburg AA, Loos RJ, Bauer F, Hendriks J, van der Schouw YT, Adan RA (2010) Are recently identified genetic variants regulating BMI in the general population associated with anorexia nervosa? Am J Med Genet B Neuropsychiatr Genet 153B:695–699
Breen G, Sham P, Li T, Shaw D, Collier DA, St CD (1999) Accuracy and sensitivity of DNA pooling with microsatellite repeats using capillary electrophoresis. Mol Cell Probes 13:359–365
Brown KM, Bujac SR, Mann ET, Campbell DA, Stubbins MJ, Blundell JE (2007) Further evidence of association of OPRD1 & HTR1D polymorphisms with susceptibility to anorexia nervosa. Biol Psychiatry 61:367–373
Bulik CM (2005) Exploring the gene-environment nexus in eating disorders. J Psychiatry Neurosci 30:335–339
Bulik CM, Devlin B, Bacanu SA, Thornton L, Klump KL, Fichter MM, Halmi KA, Kaplan AS, Strober M, Woodside DB, Bergen AW, Ganjei JK, Crow S, Mitchell J, Rotondo A, Mauri M, Cassano G, Keel P, Berrettini WH, Kaye WH (2003) Significant linkage on chromosome 10p in families with bulimia nervosa. Am J Hum Genet 72:200–207
Bulik CM, Slof-Op't Landt MC, van Furth EF, Sullivan PF (2007) The genetics of anorexia nervosa. Annu Rev Nutr 27:263–275
Cantor RM, Lange K, Sinsheimer JS (2010) Prioritizing GWAS results: a review of statistical methods and recommendations for their application. Am J Hum Genet 86:6–22
Caspi A, Hariri AR, Holmes A, Uher R, Moffitt TE (2010) Genetic sensitivity to the environment: the case of the serotonin transporter gene and its implications for studying complex diseases and traits. Am J Psychiatry 167:509–527
Caspi A, Moffitt TE, Cannon M, McClay J, Murray R, Harrington H, Taylor A, Arseneault L, Williams B, Braithwaite A, Poulton R, Craig IW (2005) Moderation of the effect of adolescent-onset cannabis use on adult psychosis by a functional polymorphism in the catechol-O-methyltransferase gene: longitudinal evidence of a gene X environment interaction. Biol Psychiatry 57:1117–1127
Caspi A, Sugden K, Moffitt TE, Taylor A, Craig IW, Harrington H, McClay J, Mill J, Martin J, Braithwaite A, Poulton R (2003) Influence of life stress on depression: moderation by a polymorphism in the 5-HTT gene. Science 301:386–389
Caspi A, Williams B, Kim-Cohen J, Craig IW, Milne BJ, Poulton R, Schalkwyk LC, Taylor A, Werts H, Moffitt TE (2007) Moderation of breastfeeding effects on the IQ by genetic variation in fatty acid metabolism. Proc Natl Acad Sci USA 104:18860–18865
Cohen JC, Pertsemlidis A, Fahmi S, Esmail S, Vega GL, Grundy SM, Hobbs HH (2006) Multiple rare variants in NPC1L1 associated with reduced sterol absorption and plasma low-density lipoprotein levels. Proc Natl Acad Sci USA 103:1810–1815
Collier DA, Arranz MJ, Li T, Mupita D, Brown N, Treasure J (1997) Association between 5-HT2A gene promoter polymorphism and anorexia nervosa. Lancet 350:412
Corvin A, Craddock N, Sullivan PF (2010) Genome-wide association studies: a primer. Psychol Med 40:1063–1077
Craddock N, Hurles ME, Cardin N, Pearson RD, Plagnol V, Robson S, Vukcevic D, Barnes C, Conrad DF, Giannoulatou E, Holmes C, Marchini JL, Stirrups K, Tobin MD, Wain LV, Yau C, Aerts J, Ahmad T, Andrews TD, Arbury H, Attwood A, Auton A, Ball SG, Balmforth AJ, Barrett JC, Barroso I, Barton A, Bennett AJ, Bhaskar S, Blaszczyk K, Bowes J, Brand OJ, Braund PS, Bredin F, Breen G, Brown MJ, Bruce IN, Bull J, Burren OS, Burton J, Byrnes J, Caesar S, Clee CM, Coffey AJ, Connell JM, Cooper JD, Dominiczak AF, Downes K, Drummond HE, Dudakia D, Dunham A, Ebbs B, Eccles D, Edkins S, Edwards C, Elliot A, Emery P, Evans DM, Evans G, Eyre S, Farmer A, Ferrier IN, Feuk L, Fitzgerald T, Flynn E, Forbes A, Forty L, Franklyn JA, Freathy RM, Gibbs P, Gilbert P, Gokumen O, Gordon-Smith K, Gray E, Green E, Groves CJ, Grozeva D, Gwilliam R, Hall A, Hammond N, Hardy M, Harrison P, Hassanali N, Hebaishi H, Hines S, Hinks A, Hitman GA, Hocking L, Howard E, Howard P, Howson JM, Hughes D, Hunt S, Isaacs JD, Jain M, Jewell DP, Johnson T, Jolley JD,

Jones IR, Jones LA, Kirov G, Langford CF, Lango-Allen H, Lathrop GM, Lee J, Lee KL, Lees C, Lewis K, Lindgren CM, Maisuria-Armer M, Maller J, Mansfield J, Martin P, Massey DC, McArdle WL, McGuffin P, McLay KE, Mentzer A, Mimmack ML, Morgan AE, Morris AP, Mowat C, Myers S, Newman W, Nimmo ER, O'Donovan MC, Onipinla A, Onyiah I, Ovington NR, Owen MJ, Palin K, Parnell K, Pernet D, Perry JR, Phillips A, Pinto D, Prescott NJ, Prokopenko I, Quail MA, Rafelt S, Rayner NW, Redon R, Reid DM, Renwick Ring SM, Robertson N, Russell E, St CD, Sambrook JG, Sanderson JD, Schuilenburg H, Scott CE, Scott R, Seal S, Shaw-Hawkins S, Shields BM, Simmonds MJ, Smyth DJ, Somaskantharajah E, Spanova K, Steer S, Stephens J, Stevens HE, Stone MA, Su Z, Symmons DP, Thompson JR, Thomson W, Travers ME, Turnbull C, Valsesia A, Walker M, Walker NM, Wallace C, Warren-Perry M, Watkins NA, Webster J, Weedon MN, Wilson AG, Woodburn M, Wordsworth BP, Young AH, Zeggini E, Carter NP, Frayling TM, Lee C, McVean G, Munroe PB, Palotie A, Sawcer SJ, Scherer SW, Strachan DP, Tyler-Smith C, Brown MA, Burton PR, Caulfield MJ, Compston A, Farrall M, Gough SC, Hall AS, Hattersley AT, Hill AV, Mathew CG, Pembrey M, Satsangi J, Stratton MR, Worthington J, Deloukas P, Duncanson A, Kwiatkowski DP, McCarthy MI, Ouwehand W, Parkes M, Rahman N, Todd JA, Samani NJ, Donnelly P (2010) Genome-wide association study of CNVs in 16,000 cases of eight common diseases and 3,000 shared controls. Nature 464:713–720

Devlin B, Bacanu SA, Klump KL, Bulik CM, Fichter MM, Halmi KA, Kaplan AS, Strober M, Treasure J, Woodside DB, Berrettini WH, Kaye WH (2002) Linkage analysis of anorexia nervosa incorporating behavioral covariates. Hum Mol Genet 11:689–696

Dmitrzak-Weglarz M, Skibinska M, Slopien A, Szczepankiewicz A, Rybakowski F, Kramer L, Hauser J, Rajewski A (2007) BDNF Met66 allele is associated with anorexia nervosa in the Polish population. Psychiatr Genet 17:245–246

Docherty S, Mill J (2008) Epigenetic mechanisms as mediators of environmental risks for psychiatric disorders. Psychiatry 7(12):500–506

Ellegren H (2004) Microsatellites: simple sequences with complex evolution. Nat Rev Genet 5:435–445

Fairburn CG, Harrison PJ (2003) Eating disorders. Lancet 361:407–416

Frieling H, Romer KD, Scholz S, Mittelbach F, Wilhelm J, de ZM, Jacoby GE, Kornhuber J, Hillemacher T, Bleich S (2009) Epigenetic dysregulation of dopaminergic genes in eating disorders. Int J Eat Disord. Sept 2 Epub ahead of print

Gorwood P, Ades J, Bellodi L, Cellini E, Collier DA, Di Bella D, Di Bernardo M, Estivill X, Fernandez-Aranda F, Gratacos M, Hebebrand J, Hinney A, Hu X, Karwautz A, Kipman A, Mouren-Simeoni MC, Nacmias B, Ribases M, Remschmidt H, Ricca V, Rotella CM, Sorbi S, Treasure J (2002) The 5-HT(2A) -1438G/A polymorphism in anorexia nervosa: a combined analysis of 316 trios from six European centres. Mol Psychiatry 7:90–94

Grice DE, Halmi KA, Fichter MM, Strober M, Woodside DB, Treasure JT, Kaplan AS, Magistretti PJ, Goldman D, Bulik CM, Kaye WH, Berrettini WH (2002) Evidence for a susceptibility gene for anorexia nervosa on chromosome 1. Am J Hum Genet 70:787–792

Hao K, Schadt EE, Storey JD (2008) Calibrating the performance of SNP arrays for whole-genome association studies. PLoS Genet 4:e1000109

Hatchwell E, Greally JM (2007) The potential role of epigenomic dysregulation in complex human disease. Trends Genet 23:588–595

Heijmans BT, Tobi EW, Stein AD, Putter H, Blauw GJ, Susser ES, Slagboom PE, Lumey LH (2008) Persistent epigenetic differences associated with prenatal exposure to famine in humans. Proc Natl Acad Sci USA 105:17046–17049

Helbig I, Mefford HC, Sharp AJ, Guipponi M, Fichera M, Franke A, Muhle H, de Kovel C, Baker C, von Spiczak S, Kron KL, Steinich I, Kleefuss-Lie AA, Leu C, Gaus V, Schmitz B, Klein KM, Reif PS, Rosenow F, Weber Y, Lerche H, Zimprich F, Urak L, Fuchs K, Feucht M, Genton P, Thomas P, Visscher F, de Haan GJ, Moller RS, Hjalgrim H, Luciano D, Wittig M, Nothnagel M, Elger CE, Nurnberg P, Romano C, Malafosse A, Koeleman BP, Lindhout D,

Stephani U, Schreiber S, Eichler EE, Sander T (2009) 15q13.3 microdeletions increase risk of idiopathic generalized epilepsy. Nat Genet 41:160–162
Hu X, Oroszi G, Chun J, Smith TL, Goldman D, Schuckit MA (2005) An expanded evaluation of the relationship of four alleles to the level of response to alcohol and the alcoholism risk. Alcohol Clin Exp Res 29:8–16
Ingason A, Rujescu D, Cichon S, Sigurdsson E, Sigmundsson T, Pietilainen OP, Buizer-Voskamp JE, Strengman E, Francks C, Muglia P, Gylfason A, Gustafsson O, Olason PI, Steinberg S, Hansen T, Jakobsen KD, Rasmussen HB, Giegling I, Moller HJ, Hartmann A, Crombie C, Fraser G, Walker N, Lonnqvist J, Suvisaari J, Tuulio-Henriksson A, Bramon E, Kiemeney LA, Franke B, Murray R, Vassos E, Toulopoulou T, Muhleisen TW, Tosato S, Ruggeri M, Djurovic S, Andreassen OA, Zhang Z, Werge T, Ophoff RA, Rietschel M, Nothen MM, Petursson H, Stefansson H, Peltonen L, Collier D, Stefansson K, and Clair DM (2009) Copy number variations of chromosome 16p13.1 region associated with schizophrenia. Mol Psychiatry Sept 2009 Epub ahead of print
Jaenisch R, Bird A (2003) Epigenetic regulation of gene expression: how the genome integrates intrinsic and environmental signals. Nat Genet 33(Suppl):245–254
Kaati G, Bygren LO, Edvinsson S (2002) Cardiovascular and diabetes mortality determined by nutrition during parents' and grandparents' slow growth period. Eur J Hum Genet 10:682–688
Kaati G, Bygren LO, Pembrey M, Sjostrom M (2007) Transgenerational response to nutrition, early life circumstances and longevity. Eur J Hum Genet 15:784–790
Kiezebrink K, Mann ET, Bujac SR, Stubbins MJ, Campbell DA, Blundell JE (2010) Evidence of complex involvement of serotonergic genes with restrictive and binge purge subtypes of anorexia nervosa. World J Biol Psychiatry 11:824–833
Klein RI, Zeiss C, Chew EW, Tsail Y, Sackler RS, Haynes C, Henning AK, Sanjovanni JP, Mane SM, Mayne ST et al. (2005)Complement factor H polymorphism in age-related macular degeneration. Science 308:385–389.
Koronyo-Hamaoui M, Danziger Y, Frisch A, Stein D, Leor S, Laufer N, Carel C, Fennig S, Minoumi M, Apter A, Goldman B, Barkai G, Weizman A, Gak E (2002) Association between anorexia nervosa and the hsKCa3 gene: a family-based and case control study. Mol Psychiatry 7:82–85
Koronyo-Hamaoui M, Gak E, Stein D, Frisch A, Danziger Y, Leor S, Michaelovsky E, Laufer N, Carel C, Fennig S, Mimouni M, Apter A, Goldman B, Barkai G, Weizman A (2004) CAG repeat polymorphism within the KCNN3 gene is a significant contributor to susceptibility to anorexia nervosa: a case-control study of female patients and several ethnic groups in the Israeli Jewish population. Am J Med Genet B Neuropsychiatr Genet 131B:76–80
Kotowski IK, Pertsemlidis A, Luke A, Cooper RS, Vega GL, Cohen JC, Hobbs HH (2006) A spectrum of PCSK9 alleles contributes to plasma levels of low-density lipoprotein cholesterol. Am J Hum Genet 78:410–422
Kraft JB, Slager SL, McGrath PJ, Hamilton SP (2005) Sequence analysis of the serotonin transporter and associations with antidepressant response. Biol Psychiatry 58:374–381
Lasky-Su J, Won S, Mick E, Anney RJ, Franke B, Neale B, Biederman J, Smalley SL, Loo SK, Todorov A, Faraone SV, Weiss ST, Lange C (2010) On genome-wide association studies for family-based designs: an integrative analysis approach combining ascertained family samples with unselected controls. Am J Hum Genet 86:573–580
Li H, Durbin R (2009) Fast and accurate short read alignment with Burrows-Wheeler transform. Bioinformatics 25:1754–1760
Lumey LH, Stein AD (2009) Transgenerational effects of prenatal exposure to the Dutch famine. BJOG 116:868
Manolio TA, Collins FS, Cox NJ, Goldstein DB, Hindorff LA, Hunter DJ, McCarthy MI, Ramos EM, Cardon LR, Chakravarti A, Cho JH, Guttmacher AE, Kong A, Kruglyak L, Mardis E, Rotimi CN, Slatkin M, Valle D, Whittemore AS, Boehnke M, Clark AG, Eichler EE, Gibson G, Haines JL, Mackay TF, McCarroll SA, Visscher PM (2009) Finding the missing heritability of complex diseases. Nature 461:747–753

Matsuzaki H, Dong S, Loi H, Di X, Liu G, Hubbell E, Law J, Berntsen T, Chadha M, Hui H, Yang G, Kennedy GC, Webster TA, Cawley S, Walsh PS, Jones KW, Fodor SP, Mei R (2004) Genotyping over 100, 000 SNPs on a pair of oligonucleotide arrays. Nat Methods 1:109–111

McCarthy MI, Hirschhorn JN (2008) Genome-wide association studies: past, present and future. Hum Mol Genet 17:R100–R101

Mefford HC, Muhle H, Ostertag P, von Spiczak S, Buysse K, Baker C, Franke A, Malafosse A, Genton P, Thomas P, Gurnett CA, Schreiber S, Bassuk AG, Guipponi M, Stephani U, Helbig I, Eichler EE (2010) Genome-wide copy number variation in epilepsy: novel susceptibility loci in idiopathic generalized and focal epilepsies. PLoS Genet 6:e1000962

Merikangas AK, Corvin AP, Gallagher L (2009) Copy-number variants in neurodevelopmental disorders: promises and challenges. Trends Genet 25:536–544

Mill J, Petronis A (2008) Pre- and peri-natal environmental risks for attention-deficit hyperactivity disorder (ADHD): the potential role of epigenetic processes in mediating susceptibility. J Child Psychol Psychiatry 49:1020–1030

Mill J, Petronis A (2007) Molecular studies of major depressive disorder: the epigenetic perspective. Mol Psychiatry 12:799–814

Mill J, Tang T, Kaminsky Z, Khare T, Yazdanpanah S, Bouchard L, Jia P, Assadzadeh A, Flanagan J, Schumacher A, Wang SC, Petronis A (2008) Epigenomic profiling reveals DNA-methylation changes associated with major psychosis. Am J Hum Genet 82:696–711

Moore JH, Williams SM (2009) Epistasis and its implications for personal genetics. Am J Hum Genet 85:309–320

Morris AP, Zeggini E (2010) An evaluation of statistical approaches to rare variant analysis in genetic association studies. Genet Epidemiol 34:188–193

Nachman MW, Crowell SL (2000) Estimate of the mutation rate per nucleotide in humans. Genetics 156:297–304

Nakabayashi K, Komaki G, Tajima A, Ando T, Ishikawa M, Nomoto J, Hata K, Oka A, Inoko H, Sasazuki T, Shirasawa S (2009) Identification of novel candidate loci for anorexia nervosa at 1q41 and 11q22 in Japanese by a genome-wide association analysis with microsatellite markers. J Hum Genet 54:531–537

Neale BM, Lasky-Su J, Anney R, Franke B, Zhou K, Maller JB, Vasquez AA, Asherson P, Chen W, Banaschewski T, Buitelaar J, Ebstein R, Gill M, Miranda A, Oades RD, Roeyers H, Rothenberger A, Sergeant J, Steinhausen HC, Sonuga-Barke E, Mulas F, Taylor E, Laird N, Lange C, Daly M, Faraone SV (2008) Genome-wide association scan of attention deficit hyperactivity disorder. Am J Med Genet B Neuropsychiatr Genet 147B:1337–1344

Ng PC, Levy S, Huang J, Stockwell TB, Walenz BP, Li K, Axelrod N, Busam DA, Strausberg RL, Venter JC (2008) Genetic variation in an individual human exome. PLoS Genet 4:e1000160

O'Rahilly S (2009) Human genetics illuminates the paths to metabolic disease. Nature 462:307–314

Pembrey M (1996) Imprinting and transgenerational modulation of gene expression; human growth as a model. Acta Genet Med Gemellol (Roma) 45:111–125

Pembrey ME, Bygren LO, Kaati G, Edvinsson S, Northstone K, Sjostrom M, Golding J (2006) Sex-specific, male-line transgenerational responses in humans. Eur J Hum Genet 14:159–166

Pinheiro AP, Bulik CM, Thornton LM, Sullivan PF, Root TL, Bloss CS, Berrettini WH, Schork NJ, Kaye WH, Bergen AW, Magistretti P, Brandt H, Crawford S, Crow S, Fichter MM, Goldman D, Halmi KA, Johnson C, Kaplan AS, Keel PK, Klump KL, La VM, Mitchell JE, Strober M, Rotondo A, Treasure J, Woodside DB (2010) Association study of 182 candidate genes in anorexia nervosa. Am J Med Genet B Neuropsychiatr Genet 153B:1070–1080

Prokopenko I, McCarthy MI, Lindgren CM (2008) Type 2 diabetes: new genes, new understanding. Trends Genet 24:613–621

Rask-Andersen M, Olszewski PK, Levine AS, Schioth HB (2010) Molecular mechanisms underlying anorexia nervosa: focus on human gene association studies and systems controlling food intake. Brain Res Rev 62:147–164

Rasmussen HB, Werge TM (2007) Novel procedure for genotyping of the human serotonin transporter gene-linked polymorphic region (5-HTTLPR)–a region with a high level of allele diversity. Psychiatr Genet 17:287–291

Renthal W, Nestler EJ (2008) Epigenetic mechanisms in drug addiction. Trends Mol Med 14:341–350

Ribases M, Gratacos M, Armengol L, de Cid R, Badia A, Jimenez L, Solano R, Vallejo J, Fernandez F, Estivill X (2003) Met66 in the brain-derived neurotrophic factor (BDNF) precursor is associated with anorexia nervosa restrictive type. Mol Psychiatry 8:745–751

Ribases M, Gratacos M, Fernandez-Aranda F, Bellodi L, Boni C, Anderluh M, Cavallini MC, Cellini E, Di BD, Erzegovesi S, Foulon C, Gabrovsek M, Gorwood P, Hebebrand J, Hinney A, Holliday J, Hu X, Karwautz A, Kipman A, Komel R, Nacmias B, Remschmidt H, Ricca V, Sorbi S, Wagner G, Treasure J, Collier DA, Estivill X (2004) Association of BDNF with anorexia, bulimia and age of onset of weight loss in six European populations. Hum Mol Genet 13:1205–1212

Risch N, Herrell R, Lehner T, Liang KY, Eaves L, Hoh J, Griem A, Kovacs M, Ott J, Merikangas KR (2009) Interaction between the serotonin transporter gene (5-HTTLPR), stressful life events, and risk of depression: a meta-analysis. JAMA 301:2462–2471

Risch N, Merikangas K (1996) The future of genetic studies of complex human diseases. Science 273:1516–1517

Rujescu D, Ingason A, Cichon S, Pietilainen OP, Barnes MR, Toulopoulou T, Picchioni M, Vassos E, Ettinger U, Bramon E, Murray R, Ruggeri M, Tosato S, Bonetto C, Steinberg S, Sigurdsson E, Sigmundsson T, Petursson H, Gylfason A, Olason PI, Hardarsson G, Jonsdottir GA, Gustafsson O, Fossdal R, Giegling I, Moller HJ, Hartmann AM, Hoffmann P, Crombie C, Fraser G, Walker N, Lonnqvist J, Suvisaari J, Tuulio-Henriksson A, Djurovic S, Melle I, Andreassen OA, Hansen T, Werge T, Kiemeney LA, Franke B, Veltman J, Buizer-Voskamp JE, Sabatti C, Ophoff RA, Rietschel M, Nothen MM, Stefansson K, Peltonen L, St CD, Stefansson H, Collier DA (2009) Disruption of the neurexin 1 gene is associated with schizophrenia. Hum Mol Genet 18:988–996

Rutter M, Moffitt TE, Caspi A (2006) Gene-environment interplay and psychopathology: multiple varieties but real effects. J Child Psychol Psychiatry 47:226–261

Schanen NC (2006) Epigenetics of autism spectrum disorders. Hum Mol Genet 15 Spec No 2: R138–150

Scherag S, Hebebrand J, Hinney A (2010) Eating disorders: the current status of molecular genetic research. Eur Child Adolesc Psychiatry 19:211–226

Schork NJ, Murray SS, Frazer KA, Topol EJ (2009) Common vs. rare allele hypotheses for complex diseases. Curr Opin Genet Dev 19:212–219

Sebat J, Levy DL, McCarthy SE (2009) Rare structural variants in schizophrenia: one disorder, multiple mutations; one mutation, multiple disorders. Trends Genet 25:528–535

Stefansson H, Rujescu D, Cichon S, Pietilainen OP, Ingason A, Steinberg S, Fossdal R, Sigurdsson E, Sigmundsson T, Buizer-Voskamp JE, Hansen T, Jakobsen KD, Muglia P, Francks C, Matthews PM, Gylfason A, Halldorsson BV, Gudbjartsson D, Thorgeirsson TE, Sigurdsson A, Jonasdottir A, Jonasdottir A, Bjornsson A, Mattiasdottir S, Blondal T, Haraldsson M, Magnusdottir BB, Giegling I, Moller HJ, Hartmann A, Shianna KV, Ge D, Need AC, Crombie C, Fraser G, Walker N, Lonnqvist J, Suvisaari J, Tuulio-Henriksson A, Paunio T, Toulopoulou T, Bramon E, Di FM, Murray R, Ruggeri M, Vassos E, Tosato S, Walshe M, Li T, Vasilescu C, Muhleisen TW, Wang AG, Ullum H, Djurovic S, Melle I, Olesen J, Kiemeney LA, Franke B, Sabatti C, Freimer NB, Gulcher JR, Thorsteinsdottir U, Kong A, Andreassen OA, Ophoff RA, Georgi A, Rietschel M, Werge T, Petursson H, Goldstein DB, Nothen MM, Peltonen L, Collier DA, St CD, Stefansson K (2008) Large recurrent microdeletions associated with schizophrenia. Nature 455:232–236

Steiger H, Richardson J, Schmitz N, Joober R, Israel M, Bruce KR, Gauvin L, Dandurand C, Anestin A (2009) Association of trait-defined, eating-disorder sub-phenotypes with (biallelic and triallelic) 5HTTLPR variations. J Psychiatr Res 43:1086–1094

Stein AD, Zybert PA, van de Bor M, Lumey LH (2004) Intrauterine famine exposure and body proportions at birth: the Dutch hunger winter. Int J Epidemiol 33:831–836

Susser ES, Lin SP (1992) Schizophrenia after prenatal exposure to the Dutch hunger winter of 1944–1945. Arch Gen Psychiatry 49:983–988

Tabor HK, Risch NJ, Myers RM (2002) Candidate-gene approaches for studying complex genetic traits: practical considerations. Nat Rev Genet 3:391–397

Tamiya G, Shinya M, Imanishi T, Ikuta T, Makino S, Okamoto K, Furugaki K, Matsumoto T, Mano S, Ando S, Nozaki Y, Yukawa W, Nakashige R, Yamaguchi D, Ishibashi H, Yonekura M, Nakami Y, Takayama S, Endo T, Saruwatari T, Yagura M, Yoshikawa Y, Fujimoto K, Oka A, Chiku S, Linsen SE, Giphart MJ, Kulski JK, Fukazawa T, Hashimoto H, Kimura M, Hoshina Y, Suzuki Y, Hotta T, Mochida J, Minezaki T, Komai K, Shiozawa S, Taniguchi A, Yamanaka H, Kamatani N, Gojobori T, Bahram S, Inoko H (2005) Whole genome association study of rheumatoid arthritis using 27 039 microsatellites. Hum Mol Genet 14:2305–2321

Thapar A, Langley K, Owen MJ, O'Donovan MC (2007) Advances in genetic findings on attention deficit hyperactivity disorder. Psychol Med 37:1681–1692

The International HapMap Consortium (2007) A second generation map of over 3.1 million SNPS Nature, 449: 851–861

Thorleifsson G, Walters GB, Gudbjartsson DF, Steinthorsdottir V, Sulem P, Helgadottir A, Styrkarsdottir U, Gretarsdottir S, Thorlacius S, Jonsdottir I, Jonsdottir T, Olafsdottir EJ, Olafsdottir GH, Jonsson T, Jonsson F, Borch-Johnsen K, Hansen T, Andersen G, Jorgensen T, Lauritzen T, Aben KK, Verbeek AL, Roeleveld N, Kampman E, Yanek LR, Becker LC, Tryggvadottir L, Rafnar T, Becker DM, Gulcher J, Kiemeney LA, Pedersen O, Kong A, Thorsteinsdottir U, Stefansson K (2009) Genome-wide association yields new sequence variants at seven loci that associate with measures of obesity. Nat Genet 41:18–24

Uher R (2009) The role of genetic variation in the causation of mental illness: an evolution-informed framework. Mol Psychiatry 14:1072–1082

Vink T, Hinney A, van Elburg AA, van Goozen SH, Sandkuijl LA, Sinke RJ, Herpertz-Dahlmann BM, Hebebrand J, Remschmidt H, van Engeland H, Adan RA (2001) Association between an agouti-related protein gene polymorphism and anorexia nervosa. Mol Psychiatry 6:325–328

Weber JL, Wong C (1993) Mutation of human short tandem repeats. Hum Mol Genet 2:1123–1128

Wendland JR, Martin BJ, Kruse MR, Lesch KP, Murphy DL (2006) Simultaneous genotyping of four functional loci of human SLC6A4, with a reappraisal of 5-HTTLPR and rs25531. Mol Psychiatry 11:224–226

Wendland JR, Moya PR, Kruse MR, Ren-Patterson RF, Jensen CL, Timpano KR, Murphy DL (2008) A novel, putative gain-of-function haplotype at SLC6A4 associates with obsessive-compulsive disorder. Hum Mol Genet 17:717–723

Wong CC, Caspi A, Williams B, Craig IW, Houts R, Ambler A, Moffitt TE, Mill J (2010) A longitudinal study of epigenetic variation in twins. Epigenetics 5-epub ahead of print

Wray NR, Visscher PM (2010) Narrowing the boundaries of the genetic architecture of schizophrenia. Schizophr Bull 36:14–23

Xu MQ, Sun WS, Liu BX, Feng GY, Yu L, Yang L, He G, Sham P, Susser E, St CD, He L (2009) Prenatal malnutrition and adult schizophrenia: further evidence from the 1959–1961 Chinese famine. Schizophr Bull 35:568–576

Ziegler A, Hebebrand J, Gorg T, Rosenkranz K, Fichter M, Herpertz-Dahlmann B, Remschmidt H, Hinney A (1999) Further lack of association between the 5-HT2A gene promoter polymorphism and susceptibility to eating disorders and a meta-analysis pertaining to anorexia nervosa. Mol Psychiatry 4:410–412

The Influence of Gender and Puberty on the Heritability of Disordered Eating Symptoms

Kristen M. Culbert, Sarah E. Racine, and Kelly L. Klump

Contents

Abstract Eating disorders and disordered eating symptoms are clearly heritable, but recent research has suggested that genetic and environmental influences on disordered eating symptoms vary as a function of gender and development. Data are limited, but evidence suggests that gender may moderate the type of genetic risk, rather than the magnitude of genetic effects, on disordered eating symptoms. Only a moderate proportion of the genetic influences on disordered eating symptoms are shared between males and females. In females, at least some of the unique genetic risk may be related to puberty and ovarian hormone regulation. The heritability of disordered eating symptoms in females increases with both pubertal development and increasing levels of estradiol. Although much more research is needed to elucidate specific mechanisms, gonadal hormones may be promising candidates for understanding sex and developmental effects and the ways in which genes exert their influence on disordered eating.

K.M. Culbert and S.E. Racine
Department of Psychology, Michigan State University, 43 Psychology Building, East Lansing, MI 48824-1116, USA
e-mail: culbertk@msu.edu, racinesa@msu.edu

K.L. Klump (✉)
Department of Psychology, Michigan State University, 107B Psychology Building, East Lansing, MI 48824-1116, USA
e-mail: klump@msu.edu

R.A.H. Adan and W.H. Kaye (eds.), *Behavioral Neurobiology of Eating Disorders*, Current Topics in Behavioral Neurosciences 6, DOI 10.1007/7854_2010_80

Keywords Development · Disordered eating · Eating disorders · Environment · Gender · Genes · Hormones · Puberty · Sex difference

1 Introduction

Eating disorders, anorexia nervosa and bulimia nervosa, and disordered eating symptoms are heritable (see Thornton et al. 2010), with approximately 50% of variance accounted for by genetic factors. Nevertheless, genetic and environmental influences on eating pathology appear to be moderated by both gender and puberty. This moderation is perhaps not surprising given robust phenotypic sex differences in prevalence (i.e., female-to-male ratio 4:1–10:1; American Psychiatric Association 2000) and significant increases in eating pathology in females during puberty. The purpose of this chapter is to review studies that have investigated the role of gender and puberty in etiologic effects on disordered eating symptoms. We first discuss data exploring sex differences in the genetic basis of eating pathology in adulthood. We then review findings suggesting developmental changes, particularly during puberty, in etiologic influences on disordered eating. Finally, we review recent data suggesting that gonadal hormones may underlie developmental changes in etiologic risk for disordered eating symptoms.

2 Gender and the Heritability of Disordered Eating Symptoms

Sex differences in eating disorder prevalence are among the most pronounced of any psychiatric disorder, with the female-to-male ratio estimated at 4:1–10:1 *in adulthood* (American Psychiatric Association 2000). Identifying mechanisms underlying sex differences is important for understanding differential risk between genders. Twin studies provide a powerful design for investigating these potential mechanisms, as these studies are able to determine the extent to which genetic and environmental influences on disordered eating symptoms vary between men and women.

No twin studies have examined sex differences in genetic and environmental effects for clinical eating disorders, likely due to low prevalence of these disorders in males. Nonetheless, investigations examining disordered eating symptoms have been conducted and have the potential to increase the understanding of a range of eating pathology. Table 1 includes a summary of current results. Only a small number of these studies directly tested sex differences using structural equation models, as many studies simply reported the estimates in each gender without testing whether etiologic effects significantly differed across sex. When such differences were tested statistically, there were no significant differences in genetic or environmental effects across males and females in *any of the studies*. Therefore, current research suggests a lack of sex differences in the magnitude of genetic and environmental influences on all types of disordered eating.

Table 1 Genetic and environmental influences on disordered eating symptoms in males and females

Disordered eating symptoms		Males			Females			Significant sex difference in magnitude
Phenotype	Measure	a^2	c^2	e^2	a^2	c^2	e^2	Yes/No
Binge eating								
Baker et al. (2009)	EDI-2, bulimia	0.33 (0.00–0.44)	0.00 (0.00–0.30)	0.67 (0.56–0.80)	0.16 (0.00–0.42)	0.16 (0.00–0.36)	0.69 (0.57–0.81)	Not examined
Reichborn-Kjennerud et al. (2003)	One item, binge eating	0.51 (0.43–0.58)		0.49 (0.42–0.57)	0.51 (0.43–0.58)	–	0.49 (0.42–0.57)	No
Reichborn-Kjennerud et al. (2004b)	One item, binge eating with no compensatory behavior	0.41 (0.31–0.50)	–	0.59 (0.50–0.69)	0.41 (0.31–0.50)	–	0.59 (0.50–0.69)	No
Tholin et al. (2005)	TFEQ-R21, emotional eating	0.60 (0.52–0.67)	–	0.40 (0.33–0.48)	Not examined			Not examined
Tholin et al. (2005)	TFEQ-R21, uncontrolled eating	0.45 (0.36–0.53)	–	0.55 (0.67–0.64)	Not examined			Not examined
Body shape concerns								
Keski-Rahkonen et al. (2005a)	EDI-2, body dissatisfaction	–	0.85 (0.83–0.87)	0.15 (0.13–0.17)	0.59 (0.53–0.65)	–	0.41 (0.35–0.47)	Not examined
Baker et al. (2009)	EDI-2, body dissatisfaction	0.40 (0.06–0.57)	0.07 (0.00–0.35)	0.53 (0.44–0.64)	0.57 (0.30–0.70)	0.07 (0.00–0.31)	0.36 (0.30–0.44)	Not examined
Body weight concerns								
Baker et al. (2009)	EDI-2, drive for thinness	0.20 (0.33–0.68)	0.11 (0.00–0.35)	0.69 (0.57–0.82)	0.61 (0.33–0.68)	0.01 (0.00–0.25)	0.38 (0.31–0.46)	Not examined
Keski-Rahkonen et al. (2005a)	EDI-2, drive for thinness	–	0.86 (0.84–0.88)	0.14 (0.12–0.16)	0.51 (0.44–0.58)	–	0.49 (0.43–0.56)	Not examined
Keski-Rahkonen et al. (2005)	One item, intentional weight loss	0.38 (0.19–0.55)	–	0.62 (0.45–0.81)	0.66 (0.55–0.75)	–	0.34 (0.25–0.45)	Not examined
Reichborn-Kjennerud et al. (2004a)	One item, undue influence of body weight on self-evaluation	–	0.31 (0.24–0.38)	0.69 (0.68–0.76)	–	0.31 (0.24–0.38)	0.69 (0.68–0.76)	No

(*continued*)

Table 1 (continued)

Disordered eating symptoms		Males			Females			Significant sex difference in magnitude
Phenotype	Measure	a^2	c^2	e^2	a^2	c^2	e^2	Yes/No
Sloft-Op't Landt et al. (2008)	Four items, i.e., dieting to lose weight, fear of weight gain, undue importance of body shape/weight, binge eating	0.39 (0.28–0.49)	–	0.35 (0.29–0.42)	0.65 (0.58–0.71)	–	0.35 (0.29–0.42)	Not examined
Cognitive restraint								
Tholin et al. (2005)	TFEQ-R21, cognitive restraint	0.59 (0.52–0.66)	–	0.41 (0.34–0.48)	Not examined			Not examined
Overall disordered eating								
Slane et al. (2009)	MEBS, total score, i.e., binge eating, compensatory behavior, body dissatisfaction, weight preoccupation	0.53 (0.27–0.64)	0.00 (0.00–0.21)	0.47 (0.36–0.60)	0.53 (0.27–0.64)	0.00 (0.00–0.21)	0.47 (0.36–0.60)	No

Note. a^2 additive or nonadditive genetic effects; c^2 shared environmental effects; e^2 nonshared environmental effects; *EDI-2* eating disorder inventory 2 (Garner 1991); *TFEQ*-R21 three-factor eating questionnaire-revised 21 item version (de Lauzon et al. 2004; Karlsson et al. 2000; Stunkard & Messick 1985; Tholin et al. 2005); *MEBS* Minnesota eating behaviors survey (von Ranson 2005). Ninety-five percent confidence intervals are in parentheses. Parameter estimates are from the best-fitting model reported in each study. Dashes indicate that the parameter was dropped in the best-fitting model. Estimates of genetic effects represent additive genetic effects for all studies except Tholin et al. (2005) who found that the genetic effects on cognitive restraint and uncontrolled eating were nonadditive in nature. Parameter estimates reported for Sloft-Op't Landt et al. (2008) are from a bivariate Cholesky analysis of body mass index and disordered eating symptoms, and notably, heritability estimates unique to disordered eating symptoms were 0.53 for women and 0.29 for men

However, because so few studies directly tested sex differences, and larger sample sizes are needed to detect sex differences than sex similarities (Plomin et al. 2008), we tentatively interpret differences in the magnitude of reported estimates across sex from all studies. The pattern of estimates suggests that sex differences might be present for the body weight and shape concerns that are common in individuals with eating disorders. In several studies, genetic factors appeared to be more influential in females (heritability estimates: 51–61%) than in males (heritability estimates: 0–40%) for intentional weight loss (Keski-Rahkonen et al. 2005b), body dissatisfaction (Baker et al. 2009; Keski-Rahkonen et al. 2005a), drive for thinness (Baker et al. 2009; Keski-Rahkonen et al. 2005a), and weight/shape concerns (Sloft-Op't Landt et al. 2008).

Sex differences may also exist in the types of genetic factors underlying risk for disordered eating in females versus males. Identifying sex differences in genetic factor *types* requires data from both same-sex and opposite-sex male and female twin pairs, so that the male–female "genetic correlation" (r_g) can be calculated. The "genetic correlation" indexes the degree to which the genetic influences that are important for disordered eating symptoms in females overlap with those relevant in males. Evidence thus far is consistent in suggesting moderate overlap in the genetic factors ($r_g = 0.49–0.67$) influencing disordered eating attitudes (i.e., body dissatisfaction, drive for thinness; Baker et al. 2009) and behaviors (i.e., binge eating; Reichborn-Kjennerud et al. 2003, 2004b) between genders. These preliminary results are intriguing and suggest that differential causal mechanisms may underlie a substantial portion of genetic risk for disordered eating symptoms in males versus females.

Taken together, initial findings indicate that there are no sex differences in the magnitude of genetic/environmental influences on most forms of eating pathology. By contrast, sex differences appear to exist for the type of genetic factors contributing to heritability in each sex. Studies have been few in number, sample sizes have been relatively small, and statistical comparisons of sex effects have not frequently been conducted. In addition, some studies used measures that lacked variability or symptom endorsement in males (i.e., Baker et al. 2009; Keski-Rahkonen et al. 2005a) or only included one item of each construct (see Table 1). Since measurement error loads onto the nonshared environmental parameter, genetic and shared environmental effects may be underestimated in these studies. Additional research that improves upon these limitations is needed to confirm our initial impressions and further clarify sex differences in genetic and environmental influences on disordered eating symptoms.

3 Puberty and the Heritability of Disordered Eating Symptoms

Adolescence is a key developmental period for increases in disordered eating symptoms and changes in etiologic effects. Cross-sectional (Klump et al. 2000) and longitudinal (Klump et al. 2007a; Silberg and Bulik 2005) twin studies have indicated age-related differences in genetic and environmental effects on disordered eating symptoms in females during adolescence. Little-to-no genetic effects (~0–5%) on disordered eating have been observed in preadolescent female twins, as shared and nonshared environmental factors have predominated (Klump et al. 2000, 2007a). In contrast, similar and substantial genetic effects (>50%) were observed from mid- to late adolescence, with the remaining variance accounted for by nonshared environmental influences (Klump et al. 2000, 2007a). Developmental changes in genetic and environmental effects on disordered eating symptoms have also been reported in other samples of twins. Silberg and Bulik (2005) found that genetic influences on eating pathology increased from early- to late adolescence. Although heritability estimates were somewhat higher in early adolescent twins (~25%; Silberg and Bulik 2005) than those observed by Klump and colleagues, such differences could have resulted from differences in age categorizations. Silberg and Bulik (2005) included somewhat older twins (i.e., 13 years) in the early adolescent group which may have elevated heritability estimates, as Klump et al. (2007a) categorized twins ages 13–16 in the mid-adolescent group (where genetic effects were observed). Overall, results from these studies are important in suggesting that early- to mid-adolescence is a critical period for the genetic diathesis of disordered eating.

Researchers have attempted to identify mechanisms responsible for these age-related changes. Puberty has been a primary focus of this research, as puberty occurs between pre- to mid-adolescence, puberty is related to a cascade of biological and psychosocial transitions, and puberty is associated with increases in the incidence of disordered eating in females (Bulik 2002; Graber et al. 1994). Developmental twin studies have confirmed that puberty increases or "moderates" the heritability of disordered eating symptoms, even after controlling for age (Culbert et al. 2009; Klump et al. 2003; Klump et al. 2007b). Little-to-no genetic influences on disordered eating were found in prepubertal 11-year-old twins, whereas genetic effects accounted for approximately 50% of the variance in 11-year-old twins who were in mid- to late puberty (Klump et al. 2003). The magnitude of genetic effects in the 11-year-old twins who had begun puberty was equal to estimates observed in young adult twins (Klump et al. 2003). These results have now been replicated in two independent twin samples. Specifically, Klump et al. (2007b) demonstrated that puberty moderates genetic effects on disordered eating symptoms, as genetic influences on disordered eating increased linearly with advancing pubertal development. In addition, Culbert et al. (2009) found nominal genetic influences on disordered eating symptoms in prepubertal adolescent twins, whereas substantial and equal genetic effects were found in the pubertal and young adult twin groups (Culbert et al. 2009).

Findings have also demonstrated that the shift in etiologic influences will be missed if a late indicator (i.e., menarche) of pubertal development is used as the sole marker of puberty. No significant differences in the heritability of disordered eating symptoms were found between premenarcheal and postmenarcheal twins; genetic effects were moderate and equal in these groups (Culbert et al. 2009; Rowe et al. 2002). Menarche occurs during the last stage of puberty, and thus, elevated genetic effects in the premenarche group may be due to the inclusion of twins in mid-puberty in this group. Importantly, a recent study (i.e., Culbert et al. 2009) confirmed this hypothesis by showing that developmental differences in genetic effects are only observed when using early rather than late (i.e., menarche) indicators of pubertal development. These results highlight early- to mid-puberty as the key developmental period for changes in genetic risk (Culbert et al. 2009). Taken together, findings indicate that puberty accounts for age-related changes in genetic effects on disordered eating symptoms in females and that early indicators of pubertal status are important for the detection of developmental effects.

Recent research has begun to try to identify the factors that underlie puberty's effects. Estradiol has been of particular interest since levels of estradiol substantially increase during puberty, and estradiol is known to directly regulate gene transcription within the central nervous system (Ostlund et al. 2003). Klump et al. (2010) divided pre and early adolescent twins into two groups based on levels of estradiol (i.e., low levels versus high levels) and found no genetic influences on disordered eating in the low estradiol group, but substantial genetic effects in the high estradiol group. Estradiol's effects were independent of several other factors that are associated with puberty including age, body mass index, and pubertal physical changes (e.g., breast development) (Klump et al. 2010). These findings suggest that the increase in estradiol at puberty may "activate" genetic risk for disordered eating and account for pubertal increases in genetic influences on disordered eating in girls.

Although results from developmental twin studies have been largely consistent, there are several directions for future research. Longitudinal data are limited as only two studies (i.e., Klump et al. 2007a; Silberg and Bulik 2005) have been conducted to date. Thus, additional longitudinal studies are needed to confirm that developmental shifts in etiologic influences on disordered eating symptoms reflect within-twin changes in genetic/environmental influences during puberty. Further, only one small study examined the moderating effects of estradiol on the heritability of disordered eating during adolescence. Replicating these findings within larger samples will be important for understanding the magnitude of moderation. Researchers have also not yet examined other potential mechanisms that could account for the developmental shift in the genetic diathesis of disordered eating symptoms during puberty. For example, other factors that increase during the pubertal period (e.g., internalization of the thin ideal, weight based teasing; Hermes & Keel 2003) could also contribute to developmental shifts in etiology and would be important to examine in future research.

4 Conclusions

Research into the effects of sex and developmental stage on genetic risk for eating disorders is a relatively new yet promising area of research. Initial data indicate that gender and puberty may moderate the genetic and environmental influences on disordered eating symptoms. Gender may moderate the type (rather than magnitude) of genetic risk, while age and puberty moderate the timing and magnitude of genetic influences. Much more research is needed to elucidate the mechanisms underlying these effects. Interestingly, estradiol may contribute to both types of effects, as estradiol activation at puberty is specific to girls and appears to account for puberty's effects on genetic risk in this gender. In order to further investigate this possibility, future studies should examine developmental changes in the heritability of disordered eating symptoms in both males and females using biological indices (e.g., gonadal hormones) of risk in large samples of twins. These studies will have the strongest power for determining whether and how gender, puberty, and reproductive hormones impact the genetic diathesis of eating disorders across development.

References

American Psychiatric Association (2000) Diagnostic and statistical manual of mental disorders, fourth edition – text revision (DSM-IV-TR). American Psychiatric Association, Washington, DC

Baker JH, Maes HH, Lissner L, Aggen SH, Lichtenstein P, Kendler KS (2009) Genetic risk factors for disordered eating in adolescent males and females. J Abnorm Psychol 118:576–586

Bulik CM (2002) Eating disorders in adolescents and young adults. Child Adolesc Psychiatr Clin N Am 11:201–218

Culbert KM, Burt SA, McGue M, Iacono WG, Klump KL (2009) Puberty and the genetic diathesis of disordered eating attitudes and behaviors. J Abnorm Psychol 118:788–796

de Lauzon B, Romon M, Deschamps V, Lafay L, Borys JM, Karlsson J et al. (2004) The Three-Factor Eating Questionnaire-R18 is able to distinguish between different eating patterns in a general population. J Nutr 134:2372–2380.

Garner DM (1991) Eating Disorder Inventory-2: Professional manual. Odessa, FL: Psychological Assessment Resources.

Graber J, Brooks-Gunn J, Paikoff R, Warren M (1994) Prediction of eating problems: an 8-year study of adolescent girls. Dev Psychol 30:823–834

Hermes SF, Keel PK (2003) The influences of puberty and ethnicity on awareness and internalization of the thin ideal. Int J Eat Disord 33:465–467

Karlsson J, Persson LO, Sjostrom L, Sullivan M. (2000) Psychometric properties and factor structure of the Three-Factor Eating Questionnaire (TFEQ) in obese men and women. Results from the Swedish Obese Subjects (SOS) study. Int J Obes Relat Metab Disord 24:1715–25.

Keski-Rahkonen A, Bulik CM, Neale BM, Rose RJ, Rissanen A, Kaprio J (2005a) Body dissatisfaction and drive for thinness in young adult twins. Int J Eat Disord 37:188–199

Keski-Rahkonen A, Neale BM, Bulik CM, Pietilainen KH, Rose RJ, Kaprio J et al (2005b) Intentional weight loss in young adults: sex-specific genetic and environmental effects. Obes Res 13:745–753

Klump KL, McGue M, Iacono WG (2000) Age differences in genetic and environmental influences on eating attitudes and behaviors in preadolescent and adolescent female twins. J Abnorm Psychol 109:239–251

Klump KL, McGue M, Iacono WG (2003) Differential heritability of eating pathology in prepubertal versus pubertal twins. Int J Eat Disord 33:287–292

Klump KL, Burt SA, McGue M, Iacono WG (2007a) Changes in genetic and environmental influences on disordered eating across adolescence: a longitudinal twin study. Arch Gen Psychiatry 64:1409–1415

Klump KL, Perkins PS, Burt SA, McGue M, Iacono WG (2007b) Puberty moderates genetic influences on disordered eating. Psychol Med 37:627–634

Klump KL, Keel PK, Sisk C, Burt SA (2010). Preliminary evidence that estradiol moderates genetic influences on disordered eating attitudes and behaviors during puberty. Psychol Med 40:1745–1753

Ostlund H, Keller E, Hurd YL (2003) Estrogen receptor gene expression in relation to neuropsychiatric disorders. Ann N Y Acad Sci 1007:54–63

Plomin R, DeFries JC, McClearn GE, McGuffin P (2008) Behavioral genetics, 5th edn. Worth Publishers, New York

Reichborn-Kjennerud T, Bulik CM, Kendler KS, Roysamb E, Maes HHM, Tambs K et al (2003) Gender differences in binge-eating: a population-based twin study. Acta Psychiatr Scand 108:196–202

Reichborn-Kjennerud T, Bulik CM, Kendler KS, Roysamb E, Tambs K, Torgersen S et al (2004a) Undue influence of weight on self-evaluation: a population-based twin study of gender differences. Int J Eat Disord 35:123–132

Reichborn-Kjennerud T, Bulik CM, Tambs K, Harris JR (2004b) Genetic and environmental influences on binge eating in the absence of compensatory behaviors: a population-based twin study. Int J Eat Disord 36:307–314

Rowe R, Pickles A, Simonoff E, Bulik CM, Silberg JL (2002) Bulimic symptoms in the Virginia twin study of adolescent behavioral development: correlates, comorbidity, and genetics. Biol Psychiatry 51:172–182

Silberg JL, Bulik CM (2005) The developmental association between eating disorders symptoms and symptoms of depression and anxiety in juvenile twin girls. J Child Psychol Psychiatry 46:1317–1326

Slane JD, Burt SA, Klump KL (2009). Gender differences in genetic relationships between disordered eating and alcohol use. Paper presentation at the 2009 International Conference on Eating Disorders, Cancun, Mexico.

Sloft-Op't Landt MCT, Bartels M, Van Furth EF, van Beijsterveldt CEM, Meulenbelt I, Slagboom PE et al (2008) Genetic influences on disordered eating behaviour are largely independent of body mass index. Acta Psychiatr Scand 117:348–356

Stunkard AJ & Messick S (1985) The three-factor eating questionnaire to measure dietary restraint, disinhibition and hunger. J Psychosom Res 29:71–83.

Tholin S, Rasmussen F, Tynelius P, Karlsson J (2005) Genetic and environmental influences on eating behavior: the Swedish young male twins study. Am J Clin Nutr 81:564–569

Thornton LM, Mazzeo SE, Bulik CM (2010) The heritability of eating disorders: methods and current findings. doi:10.1007/7854_2010_91

von Ranson KM, Klump KL, Iacono WG, McGue M (2005) Development and validation of Minnesota Eating Behaviors Survey: a brief measure of disordered eating attitudes and behaviors. *Eat Behav* 6:373–392.

Part IV
Neuroendocrinology and Animal Research

New Frontiers in Endocrinology of Eating Disorders

Palmiero Monteleone

Contents

Abstract Alterations of both central and peripheral feeding regulatory substances occur in the acute phases of anorexia nervosa (AN) and bulimia nervosa (BN) and, generally, reverse after recovery. Some of these alterations are believed not only to sustain the altered eating behavior but also to contribute to certain psychopathological aspects and/or etiopathogenetic processes of eating disorders (EDs). It has been suggested that EDs are clinical conditions linked to reward-related mechanisms leading to a kind of addiction to self-starvation and/or overeating. Most of the feeding regulatory substances, which are dysregulated in EDs, are also implicated in the modulation of reward, emotional, and cognitive functions, thus representing possible links between altered nutritional regulation, motivated behaviors and reward

P. Monteleone
Department of Psychiatry, University of Naples SUN, Largo Madonna delle Grazie, 80138 Naples, Italy
e-mail: monteri@tin.it

R.A.H. Adan and W.H. Kaye (eds.), *Behavioral Neurobiology of Eating Disorders*,
Current Topics in Behavioral Neurosciences 6, DOI 10.1007/7854_2010_87

processes. In this chapter, the ED literature dealing with ghrelin, brain-derived neurotrophic factor, opioid peptides, and endocannabinoids, which have prominent effects on eating behavior, body weight, reward, emotional, and cognitive functions, is reviewed in view of the above suggested links. Moreover, the potential therapeutics of new medications developed on the basis of neuroendocrine aberrations found in EDs is also presented.

Keywords Anorexia nervosa · Bulimia nervosa · Eating disorders · Feeding regulators · Motivated behaviors · Neuroendocrinology · Reward

List of Abbreviations

β-EP	Beta-endorphin
2-AG	2-Arachidonoylglycerol
AEA	Anandamide
AN	Anorexia nervosa
ANBP	Anorexia nervosa binge-purging
ANR	Anorexia nervosa restricting
BDNF	Brain-derived neurotrophic factor
BN	Bulimia nervosa
BW	Body weight
CCK	Cholecystokinin
CSF	Cerebrospinal fluid
ED	Eating disorder
OLA	Opiate like activity

1 Introduction

Although anorexia nervosa (AN) and bulimia nervosa (BN) are commonly defined as disorders of eating, the altered feeding component is probably only the tip of an iceberg whose underwater section represents the still unknown main psychopathological component of the disorder leading to a secondary alteration of eating behavior. Nevertheless, the impaired feeding aspects of eating disorders (EDs) and their biochemical background may affect the development, the course, the prognosis, and the responses to therapies of these conditions. For these reasons, a great deal of research has been done on neurotransmitters, neuromodulators, neuropeptides, and peripheral peptides involved in the regulation of eating behavior and body weight (BW). Derangements in the physiology of these substances have been detected in the acute phase of AN and BN and, generally, have been shown to disappear after recovery. Therefore, at the moment, it is commonly believed, although not definitively proved, that most of the secretory alterations of feeding regulatory substances are the consequence of the nutritional changes occurring with the disorder. However, it has been suggested that even if those alterations are secondary phenomena disappearing after the recovery from the ED, they may hypothetically contribute to the maintenance of both aberrant eating behaviors and/or other symptomatic features. In this regard, it is interesting to note that some of the feeding regulatory substances intervene also in the modulation of reward-related processes, motivated behaviors, and various psychological functions, including cognition and emotions, whose alterations have been reported to contribute to the psychopathological core of AN and BN. Therefore, changes in central and peripheral components of the feeding regulatory system may not only sustain altered motivated behaviors, such as aberrant self-starvation and binge eating, but also contribute to psychopathological aspects of EDs.

It is known that sophisticated physiological mechanisms have evolved to regulate eating behavior and energy homeostasis in humans. It is beyond the scope of this review to discus and illustrate the complex physiological regulation of feeding and the physiological aberrations of feeding modulatory substances occurring in patients with EDs. To this purpose, several excellent published articles are available (Schwartz et al. 2000; Murphy and Bloom 2006; Monteleone et al. 2008a; Brambilla et al. 2008; Prince et al. 2009) and a synopsis of the most relevant alterations of peptidergic feeding regulatory substances in both acute and recovered AN and BN patients is provided in Table 1. The present review, instead, selectively focuses on clinical research dealing with ghrelin, brain-derived neurotrophic factor (BDNF), opioid peptides, and endocannabinoids, which are important signal systems with prominent effects on eating behavior, BW, reward, emotional, and cognitive functions, to highlight possible links between altered nutritional regulation, motivated behaviors, and psychopathological aspects of AN and BN. Moreover, the potential therapeutics of new medications developed on the basis of those neuroendocrine aberrations found in EDs is also discussed.

2 Ghrelin and Obestatin

Food ingestion stimulates the release of a variety of peptides from enteroendocrine cells throughout the gastrointestinal tract and the pancreas. These substances modulate food intake by signaling meal initiation, meal termination, and inhibition of subsequent meal intake. All but one of these enteropancreatic hormones have the potential to increase satiety and decrease food intake; the only known exception is ghrelin that has opposite effects. Ghrelin, initially characterized as a ligand for the growth hormone secretagogue receptor, is primarily produced by endocrine cells in the stomach. The physiological relevance of such a localization remained unclear until it was found that the hormone plays an important role in the regulation of food intake and metabolism. Ghrelin is a 28-aminoacid peptide with an acyl side chain attached to the serine residue at position 3 that is crucial for its orexigenic and

Table 1 Summary of the most relevant changes of central and peripheral peptides regulating appetite in anorexia nervosa (AN) and bulimia nervosa (BN)[a]

	AN		BN	
	Acute phase	Weight-restored	Acute phase	Recovered
Neuropeptides stimulating hunger and food intake				
Neuropeptide Y (NPY)				
CSF NPY	⇑	⇒ ⇑[b]	⇒	⇒
Plasma NPY	⇒ ⇓		⇑	
Agouti-related peptide (AGRP)				
Plasma AGRP	⇑			
Opioid peptides				
CSF β-endorphins	⇒ ⇓	⇒	⇒ ⇓	
CSF dynorphins	⇒	⇒	⇒	
Plasma β-endorphins	⇒ ⇑		⇓ ⇑	
T-lymphocyte β-endorphins	⇑		⇒	
Galanin				
CSF galanin	⇒	⇓		
Plasma galanin	⇒	⇓	⇒	
Vasopressin				
CSF vasopressin	⇒ ⇑	⇒ ⇑		
Plasma vasopressin	⇒ ⇑	⇒ ⇑	⇒	
Plasma vasopressin (osmotic response)	⇓		⇓	
Neuropeptides inhibiting hunger and food intake				
α-Melanocyte stimulating hormone (α-MSH)				
Plasma α-MSH	⇒			
Corticotropin-releasing hormone (CRH)				
CSF CRH	⇑	⇑	⇑	
Thyrotropin-releasing hormone (TRH)				
CSF TRH	⇓	⇓	⇒	
Neurotensin				
Plasma neurotensin	⇒	⇒	⇒	⇒
Somatostatin (SRIF)				
CSF SRIF	⇓		⇒	⇑
Plasma SRIF	⇑			
Plasma SRIF (response to test meal)	⇑	⇒		
Oxytocin				
CSF oxytocin	⇓			
Brain-derived neurotrophic factor (BDNF)				
Serum BDNF	⇓	⇒ ⇑ ⇓	⇓	
Plasma BDNF	⇑		⇑	
Peripheral peptides stimulating hunger and food intake				
Ghrelin				
Plasma ghrelin	⇑	⇒	⇒	
Plasma ghrelin (response to test meal)	⇓		⇓	
Peripheral peptides inhibiting hunger and food intake				
Leptin				
CSF leptin	⇓	⇒		
Plasma leptin	⇓	⇒	⇒ ⇑ ⇓	
Plasma leptin (response to test meal)			⇒	
Plasma leptin (response to acute fasting)			⇓	

(*continued*)

Table 1 (continued)

	AN		BN	
	Acute phase	Weight-restored	Acute phase	Recovered
Insulin				
Plasma insulin	⇒ ⇓	⇒	⇒	
Cholecystokinin (CCK)				
CSF CCK	⇒ ⇑	⇒	⇓	
Plasma CCK	⇒ ⇑	⇒	⇒ ⇓	
Lymphocyte CCK	⇓		⇓	
Plasma CCK (response to meal)	⇒ ⇑	⇒	⇓	
Peptide YY_{3-36} (PYY_{3-36})				
CSF PYY_{3-36}			⇒	⇒
Plasma PYY_{3-36}	⇒ ⇑		⇒	⇒
Plasma PYY_{3-36} (response to test meal)	⇑ or time-delayed	⇒[c]	⇓⇑	
Pancreatic polypeptide (PP)				
Plasma PP	⇒			⇒
Plasma PP (response to test meal)	⇒ ⇑ ⇓		⇒ ⇑	⇑
Gastrin				
Plasma gastrin	⇓⇑			
Plasma gastrin (response to test meal)	⇒			
Resistin				
Plasma resistin	⇒ ⇓		⇒	
Adipose tissue resistin	⇑			
Adiponectin				
Plasma adiponectin	⇒ ⇑ ⇓[d]	⇑ ⇓	⇒ ⇑ ⇓	

⇒ not different from healthy controls; ⇑ higher than healthy controls; ⇓ lower than healthy controls

[a]For references see: Monteleone et al. (2008a), Brambilla et al. (2008), Prince et al. (2009)

[b]Short-term weight restored patients with persistent amenorrhea still have significantly raised concentrations of CSF NPY

[c]Improved, but not completely restored

[d]Most of the studies found increased adiponectin levels in malnourished AN patients

gastric emptying effects (Kojima et al. 1999). Acylated ghrelin (active form) represents less than 10% of circulating ghrelin, which includes acylated and desacylated fragments (inactive forms). This issue is particularly relevant because of the demonstration that desacyl ghrelin decreases food intake and delays gastric emptying in mice and rats (Asakawa et al. 2005; Chen et al. 2005), thus showing opposite effects of the active form. Furthermore, ghrelin is synthesized from a precursor peptide of 117 residues, called preproghrelin, which undergoes stepwise processing to form ghrelin (Zhu et al. 2006). Recently, it has been shown that preproghrelin undergoes additional proteolytic cleavage, generating a 23-amino acid peptide, which has been named obestatin (Zhang et al. 2005). In contrast to ghrelin, obestatin has anorexigenic effects, reduces gastric emptying, inhibits jejunal contractions, and suppresses BW gain in the animal (Zhang et al. 2005). Therefore, obestatin has been postulated to antagonize ghrelin actions on energy homeostasis and gastrointestinal functions, although this has been questioned by other authors (Seoane et al. 2006; Gourcerol et al. 2007) and controversies still exist on the definite effects of obestatin on food intake/energy balance as well as on the measurements of the hormone levels in the human blood (Garg 2007).

Currently, ghrelin is considered as a "hunger hormone" that signals the brain the need to initiate food consumption. Indeed, its circulating levels increase before meals achieving concentrations sufficient to stimulate hunger and promote meal initiation (Cummings et al. 2001). After food ingestion, plasma levels of ghrelin drastically decrease. Some but not all the studies provided the evidence that in the cephalic phase of food ingestion the vagal efferent system promotes ghrelin secretion (Arosio et al. 2004; Simonian et al. 2005; Monteleone et al. 2008b; 2010). This would support the view that the ghrelin increase that physiologically occurs before a meal is due to the cephalic phase of vagal stimulation. Given the dynamic changes in ghrelin secretion in relation to meals, it is conceivable that this peptide may be implicated in the pathophysiology of AN and BN.

2.1 Ghrelin and Obestatin in Anorexia Nervosa

Fasting circulating levels of ghrelin have been consistently reported to be increased in underweight patients with AN. Some but not all the authors reported that this increase is more pronounced in patients with AN binge-purging type (ANBP) as compared to patients with AN restricted-type (ANR), suggesting that binge-purging behavior may have some influence on circulating ghrelin (Otto et al. 2001, 2004; Tanaka et al. 2003a,b; Misra et al. 2004, 2005; Soriano-Guillen et al. 2004; Troisi et al. 2005; Monteleone et al. 2008b). The enhanced ghrelin concentrations of underweight AN subjects tend to normalize with the recovery of BW, and the reduction of circulating ghrelin seems to parallel the progressive increase in BW during weight restoration treatments (Otto et al. 2001, 2005; Soriano-Guillen et al. 2004; Tanaka et al. 2004; Janas-Kozik et al. 2007). This supports the view that the enhanced ghrelin production in symptomatic AN individuals is a state-dependent phenomenon that disappears with the restoration of normal eating habits.

The dynamics of ghrelin secretion after energy intake has been extensively studied in emaciated AN patients with mixed results. The food-induced suppression of circulating ghrelin in underweight AN patients was found almost completely absent (Nedvidkova et al. 2003), whereas the suppressant effect of oral glucose administration on plasma ghrelin was significantly blunted in ANR women, and normal but delayed in ANBP patients (Tanaka et al. 2003c). Two other studies reported that in symptomatic AN individuals enhanced preprandial levels of ghrelin, although suppressed by food ingestion in percentages similar to normal subjects, remained significantly higher than in controls (Misra et al. 2004; Stock et al. 2005). Differences in the clinical characteristics of patients' samples, in the type, composition, and total calories of test meals, and in the timing of blood collection may partially explain discrepancies among the studies.

One of the major technical issues is that all the above-mentioned studies measured total ghrelin concentrations without differentiating active from inactive ghrelin. Therefore, increases in total ghrelin plasma levels in AN may not be representative of an increased active ghrelin production. When this aspect has been taken into account, conflicting results have emerged. In fact, Nakai et al. (2003) showed that, in symptomatic AN subjects, plasma active ghrelin was increased and normally suppressed after oral glucose administration. Hotta et al. (2004), instead, found that total ghrelin was enhanced in underweight AN patients and did not decrease after glucose infusion whereas active ghrelin was reduced and normally suppressed by glucose ingestion. Recently, one small study found significantly increased levels of total ghrelin but only slightly enhanced concentrations of acylated ghrelin in a mixed sample of symptomatic AN and BN patients (Uehara et al. 2005).

As for the ghrelin sibling peptide obestatin, it has been shown that underweight AN patients display significant increases in its levels with an enhancement of the ghrelin to obestatin ratio (Monteleone et al. 2008b,c; Harada et al. 2008; Nakahara et al. 2008). Therefore, it seems that parallel changes in ghrelin and obestatin secretion occur in underweight AN patients, which suggests that the dysregulated metabolic status may potentially affect the preproghrelin gene expression and/or the splicing of its products leading to an enhanced production of ghrelin and obestatin. It is likely that the enhanced secretion of both peptides does not occur on a one to one ratio and this may lead to an increase in the ghrelin to obestatin ratio. The secretion of ghrelin and obestatin in response to sham feeding (i.e., only smelling and chewing of meal without swallowing), which is a technique able to explore the cephalic phase of food ingestion, has been reported to be deranged in acute AN individuals, who showed an enhanced sham feeding-induced ghrelin secretion and a more robust obestatin drop (Monteleone et al. 2008b). If one assumes that ghrelin and obestatin have opposite effects on food intake, then the increased baseline ghrelin to obestatin ratio and the enhanced opposite changes in obestatin and ghrelin secretion during the cephalic phase of food ingestion could result in an amplification of the peripheral hunger signal likely aiming to oppose the rigid control that AN patients exert over their food intake.

2.2 Ghrelin and Obestatin in Bulimia Nervosa

In BN, it was initially reported that fasting ghrelin levels were increased in symptomatic patients (Tanaka et al. 2002, 2003b). This increase was evident in patients with frequent binge-purging episodes, but not in non-purging ones, supporting the idea that binge/purge cycles with vomiting as opposed to binge eating episodes without vomiting have an influence on fasting plasma ghrelin. However, subsequent studies did not detect any significant difference in plasma ghrelin concentrations between binge-purge BN individuals and healthy controls (Troisi et al. 2005; Nakazato et al. 2004; Monteleone et al. 2005a) and found no significant correlation between plasma ghrelin concentrations and the severity of the binge/vomiting behavior (Monteleone et al. 2005a).

The ghrelin responses to a macronutrient balanced meal and to a fat rich meal have been reported to be blunted in symptomatic binge/purge BN women as compared to healthy controls (Monteleone et al. 2003, 2005b; Kojima et al. 2005). The suppression of circulating ghrelin after food ingestion may denote a compensatory activation of a peripheral signal aimed at promoting termination of food ingestion. In this vein, it is worth mentioning that some, although not all the experimental human studies have suggested the occurrence in BN patients of diminished satiety responses to meals (Owen et al. 1985; Kissileff et al. 1996), which could be mediated, at least in part, by the impaired food-induced responses of ghrelin. The ghrelin response to sham feeding in symptomatic BN women has been found to be increased (Monteleone et al. 2010).

No significant changes in plasma levels of obestatin or in the ghrelin to obestatin ratio have been detected in symptomatic BN women (Monteleone et al. 2008c).

2.3 Ghrelin, Reward, and Cognition in Eating Disorders

It has been recently shown that ghrelin not only acts as an orexigenic signal but also intervenes in the modulation of reward and motivated behaviors as well as in the regulation of higher brain functions such as learning and memory. A functional magnetic resonance imaging study (Malik et al. 2008) shows that intravenous ghrelin administration in healthy subjects increased the neural response to food pictures in brain areas implicated in reward processing and appetitive behavior such as the amygdala, ventral striatum, anterior insula, and orbitoforntal cortex. Moreover, experimental data demonstrate that injection of ghrelin into the third ventricle of food-deprived mice significantly increased locomotor activity as well as extracellular dopamine levels in the nucleus accumbens (Jerlhag et al. 2007), a neurochemical system involved in reward and motivated behavior as well as in mediating the incentive salience of food (Berridge and Robinson 1998). Therefore, it has been suggested that the ghrelin-induced increase of locomotor activity likely enhances exploratory behavior, which in turn would facilitate food-seeking behavior, while the ghrelin-induced activation of the dopaminergic reward system increases the incentive value of signals associated with motivated behavior of importance for survival such as feeding behavior (Jerlhag et al. 2007). Furthermore, Diano et al. (2006) detected ghrelin receptors on mouse hippocampal neurons and showed that ghrelin knock-out animals had a reduced number of hippocampal spine synapses and performed worse than their wild-type littermates in memory spatial learning tests. Peripheral administration of ghrelin rapidly restored those deficits as well as promoted the formation of new spine synapses and generation of long-term potentiation in wild-type animals. These ghrelin-induced synaptic changes were paralleled by enhanced spatial learning and memory. This fits with the observation that, in the above-mentioned study of Malik et al. (2008), food pictures shown in the ghrelin condition were more easily recalled than those shown in the placebo condition, and with the demonstration that ED patients were more accurate than healthy controls on spatial executive tasks (Galderisi et al. 2003). Therefore, it could be speculated that the raised ghrelin levels of chronically fasting AN patients enhance their cognitive performances related to spatial learning and unfocused attention in order to potentiate their ability to identify and locate energy source, an effort that is, however, overwhelmed by their rigid control over food ingestion. Furthermore, since in primates mesolimbic dopaminergic neurons have been shown to be easily conditioned to stimuli predicting reward (Schultz et al. 1997), it could be that in the presence of starvation-associated cues, enhanced ghrelin levels may stimulate the dopamine reward system in the ventral striatum and this may provide a mechanism through which the patient's control over the ghrelin-induced hunger is perceived as rewarding. This would promote a starvation-dependent syndrome that has been proposed as a main pathophysiological mechanism in the development and/or the maintenance of EDs (Støving et al. 2009).

3 Brain-Derived Neurotrophic Factor

BDNF is a member of the neurotrophin family, which plays important roles in neuronal outgrowth and differentiation, synaptic connectivity, and neuronal repair in both the brain and the periphery (Lewin and Barde 1996). Several lines of evidence indicate a role of this neurotrophin also in energy homeostasis. Indeed, heterozygous mice with one functional BDNF allele and mice in which the BDNF gene has been deleted in excitatory brain neurons display hyperphagia and develop an age-dependent obesity (Kernie et al. 2000). Moreover, both central and peripheral administration of BDNF decreases food intake, increases energy expenditure, reduces BW, and ameliorates hyperinsulinemia and hyperglycemia in diabetic *db/db* mice by a central nervous system-mediated mechanism (Nakagawa et al. 2000; Tsuchida et al. 2001). Finally, BDNF and its tyrosine kinase receptor are highly expressed in the hypothalamus and the dorsal vagal complex, the two major autonomic centers implicated in the regulation of eating behavior and energy homeostasis (Nakagawa et al. 2000).

It has been demonstrated that BDNF infusion increases serotonin turnover in the hypothalamus of rats (Pelleymounter et al. 1995), and that heterozygous mice with one functional BDNF allele show reduced serotonin turnover in the hypothalamus (Lyons et al. 1999), which suggests an involvement of serotonin transmission in the feeding regulation by BDNF. Moreover, recent evidence suggests a link between the hypothalamic melanocortin system and BDNF, supporting the view that BDNF modulates long-term energy homeostasis by acting downstream of leptin signaling in the hypothalamus (Nakagawa et al. 2003; Xu et al. 2003). Finally, it has been proved that BDNF participates also in the short-term modulation of energy homeostasis. Indeed, cholecystokinin (CCK), a peripheral peptide which is released after food intake and promotes meal termination, has been shown to increase the levels of the neurotrophin in the rat brainstem dorsal vagal complex and that such an increase correlates with the rapid and short-lived CCK-mediated decrease of food intake (Beriohay et al. 2005). Collectively, these findings support a possible role for BDNF signaling in EDs.

3.1 BDNF in Anorexia and Bulimia Nervosa

Nakazato et al. (2003) found reduced serum BDNF levels in women with AN or BN and showed that this reduction was more pronounced in AN than BN. Moreover, serum BDNF levels negatively correlated with depressive symptoms, suggesting the idea that reduced circulating BDNF in ED patients might be related to concomitant depression and not to the ED. This likelihood was ruled out in subsequent studies (Monteleone et al. 2004), which confirmed that circulating BDNF was significantly reduced in drug-free underweight AN and symptomatic BN patients, and found a significant positive correlation between serum BDNF levels and the subjects' BW and body mass index but not with the severity of depressive symptomatology (Monteleone et al. 2005c). Furthermore, serum concentrations of BDNF did not significantly differ between AN and BN patients with or without comorbid depressive disorders (Monteleone et al. 2005c). Recently, increased plasma BDNF levels in both AN and BN were reported by Mercader et al. (2007), who showed also a significant association between some haplotypes of the BDNF gene, increased plasma levels of the neurotrophin and BN. The discrepancy with the previous studies could be partially explained by the fact that plasma and not serum BDNF was assayed. Finally, one small study showed that decreased circulating BDNF levels in emaciated AN patients were not restored after partial weight recovery (Nakazato et al. 2006), whereas a larger investigation found a normalization or even an elevation of serum BDNF in weight-recovered patients with AN (Ehrlich et al. 2009).

As for the pathophysiological significance of decreased circulating BDNF in ED patients, one hypothesis could be that, as BDNF exerts a satiety effect, its reduction may represent an adaptive phenomenon aiming to counteract the reduced calorie intake by increasing hunger. In addition, in animal models of anorexia, BDNF has been demonstrated to be involved in motivational and anticipatory aspects of eating behavior. Food anticipatory activity, that is the increase in locomotor activity prior to feeding in rodents undergoing restricted feeding paradigm, is considered an expression of the motivation to eat while the presence of wheel motor activity when food is available is believed to be an expression of the lack of motivation to eat. It has been recently shown that food anticipatory activity in mice was accompanied by a strong increase in BDNF expression levels in the hippocampus, whereas an increase of wheel activity in food-restricted mice when exposed to food was found in parallel with reduced hippocampal BDNF expression (Gelegen et al. 2008). Therefore, if reduced levels of circulating BDNF in AN mirror a similar decrease in hippocampal BDNF, this might sustain a reduced motivation to eat.

4 Opioids

An involvement of opioid peptides in the regulation of food intake and energy metabolism has been repeatedly suggested and specifically demonstrated for the proopiomelanocortin-derived β-endorphin (β-EP) and for dynorphin. Opioid peptides and their k and δ receptors are located in the hypothalamic paraventricular and ventromedial nuclei, the perifornical area, the amygdala, the globus pallidus, and the nucleus accumbens, forebrain regions that are critical for regulating reward-related behaviors. (Kelley et al. 2002; Will et al. 2003). As a matter of fact, these areas are believed to play a major role in controlling the hedonic and rewarding aspects of food choice and consumption, and in governing the positive emotional responses to highly palatable food, such as fat, sugar, salt, and ethanol (Morley and Blundell 1988).

4.1 Opioids in Anorexia and Bulimia Nervosa

In emaciated AN patients, opioid-like activity (OLA), which includes all the molecules expressing activity on the μ receptor, was found to be increased in the cerebrospinal fluid (CSF) (Kaye et al. 1982). However, when the opioid peptides were examined separately, baseline CSF levels of β-EP were found to be either normal (Gerner and Sharp 1982) or significantly lower than normal (Baranowska 1990; Kaye 1996). The reduced CSF β-EP concentrations persisted in short-term restored patients but normalized in long-term recovered anorexics (Kaye et al. 1987). Dynorphin concentrations in the CSF were detected to be normal in both emaciated and recovered AN patients (Lesem et al. 1991; Kaye 1996). Plasma β-EP levels in symptomatic AN patients were reported to be higher (Brambilla et al. 1991; Tepper et al. 1992) or lower than normal (Baranowska 1990).

In BN patients, β-EP concentrations in the CSF either did not differ from those of controls (Lesem et al. 1991) or were lower than normal (Brewerton et al. 1992), while dynorphin levels were in the normal range (Brewerton et al. 1992). In plasma, β-EP concentrations were found to be lower than normal in some studies (Waller et al. 1986) and higher in others (Fullerton et al. 1986).

The changes in both central and peripheral opioid substances observed in symptomatic ED individuals are consistent with the proposed dysregulation of reward processes in EDs (Kaye et al. 2009).

5 Endocannabinoids

The endocannabinoid system, consisting of two cannabinoid receptors (CB_1 and CB_2) and the endogenous ligands anandamide (arachidonoylethanolamide, AEA) and 2-arachidonoylglycerol (2-AG), is deeply involved in the modulation of energy balance by controlling food intake within the brain and periphery (Cota et al. 2003a). In the brain, endocannabinoids modulate feeding at two levels. First, they tonically reinforce the motivation to find and consume food with a high incentive value, possibly by interacting with the mesolimbic pathway involved in reward mechanisms. Second, they are released "on demand" in the hypothalamus after short-term food deprivation and then transiently regulate the levels and/or action of other orexigenic and anorectic mediators such as leptin, ghrelin, and melanocortins (Cota et al. 2003b). In particular, hypothalamic endocannabinoid levels are elevated in leptin-deficient mice and leptin administration has been shown to reduce hypothalamic levels of AEA and 2-AG, suggesting a negative modulation of leptin on the endocannabinoid system in the hypothalamus (Di Marzo et al. 2001). Therefore, endocannabinoids could play a role in the pathophysiology of those EDs associated with a deranged leptin signaling, such as AN. Moreover, CB1 receptors are particularly abundant in brain areas connected with reward mechanisms and endocannabinoids progressively increase in the limbic system following food deprivation (Kirkham et al. 2002). The fasting-induced increase of endogenous cannabinoids may drive the motivation to eat as well as the enjoyment of food during ingestion. Indeed, according to current accepted terminology (Berridge 1997), endocannabinoids modulate the "wanting" (that is incentive processes stimulating behavior toward food acquisition) and the "liking" (that is hedonic evaluation of ingested food) of eating. Evidence has been provided that the cannabinoid-induced food seeking and liking are mediated by release of dopamine in the nucleus accumbens (Solinas et al. 2006) and can be attenuated by opioid antagonists (Williams and Kirkham 2002), suggesting the existence of functional relationships between dopamine, endocannabinoids, and the endogenous opioid system in the reward-related aspects of eating behavior. All these data support the idea that deranged endocannabinoid signaling could have a role in the pathophysiology of EDs.

5.1 Endocannabinoids in Anorexia and Bulimia Nervosa

To date, only one study assessed circulating levels of endocannabinoids in patients with EDs. In this study, plasma levels of AEA were found significantly enhanced in patients with ANR or with binge eating disorder, but not in women with BN (Monteleone et al. 2005d). Circulating levels of 2-AG, instead, did not significantly differ between patients and healthy controls. An inverse correlation between plasma AEA levels and leptin concentrations was also detected, suggesting a possible involvement of the decreased leptin signaling of underweight AN patients in the enhancement of AEA levels. Recently, significant associations have been reported between AN and BN and single nucleotide polymorphisms of genes coding the CB1 receptor and the fatty acid amide hydrolase, the enzyme degrading AEA and 2-AG (Monteleone et al. 2009), although these results were not consistent with a family trios study (Muller et al. 2008).

Because of the role of endocannabinoids in food intake, enhanced levels of AEA in AN patients may represent an adaptive response aiming at counteracting their restrictive behavior by increasing the motivation to eat. Alternatively, the enhancement of AEA levels might be involved in the mediation of rewarding aspects of the aberrant eating behavior occurring in AN. Restricting anorexics starve themselves, avoid particular foods and adopt highly rigid eating patterns, which result in a sense of power over eating that is extremely rewarding. Hence, the starvation-induced chronic elevation of the endocannabinoid tone in AN may mediate in part the patients' addiction to self-starvation, enabling them to bear with the chronic hunger associated with prolonged food restriction. Similarly, endocannabinoids may favor overconsumption of food in binge eating syndromes by amplifying the orosensory or palatability of foods.

6 Potential Therapeutics of Feeding Modulators

Clinical trials and the clinical practice show that an appreciable proportion of ED patients do not respond adequately to treatments; therefore, there is an extreme need of new treatment strategies and, possibly, of new and effective drug therapies. Findings coming from the neuroendocrine research in AN and BN may help to promote the development of innovative medications. In particular, ghrelin, ghrelin agonists, and cannabinoids have been proposed and clinically tested for the treatment of AN patients.

Despite the finding of increased ghrelin levels in AN, a role for this peptide in the treatment of this disorder characterized by chronic malnutrition due to decreased food intake could be proposed because of its potential to stimulate food intake and promote motivated behavior, such as food searching in condition of negative energy balance. The therapeutic use of such a peptide could have the advantage to target only the specific feeding regulatory pathways without affecting other modulatory systems, and this will likely result in a reduction of untoward effects. To this regard, single or repeated i.v. infusions of ghrelin in both healthy humans and patients with different organic diseases characterized by loss of appetite and decreased BW, such as cachexia associated with congestive heart failure and chronic obstructive pulmonary disease, have been demonstrated to increase hunger and BW and to be safe with only minimal side effects (Nagaya et al. 2004; 2005; Neary et al. 2004).

Up to date, only a few studies have explored the potential therapeutics of ghrelin in underweight AN patients. One study reported that 1 μg/kg bolus infusion of ghrelin in emaciated AN patients increased their hunger sensation (Broglio et al. 2004). To the contrary, Miljic et al. (2006) reported that 5 h continuous infusion of 5 pmol/kg/min of ghrelin in both underweight and partially recovered AN patients did not affect appetite. Finally, a recent open-label pilot study showed that ghrelin infusion (3 μg/kg twice a day for 14 days) decreased gastrointestinal symptoms and enhanced energy intake and hunger sensation without significant adverse effects in five underweight restricted AN women who were motivated to recover BW but could not increase their food intake because of gastrointestinal discomfort after eating (Hotta et al. 2009). These data, if confirmed in double-blind randomized controlled studies, seem to support the idea that exogenous ghrelin could help motivated underweight patients in their process of BW recovery. In this vein, ghrelin receptor agonists are also available. The synthetic BIM-28131 ghrelin agonist has been shown to suppress BW loss and increase food intake when injected intraperitoneally in cachectic tumor-implanted mice (Hanada et al. 2003), which suggests a potential therapeutic effect in AN. However, it has been argued that ghrelin and/or ghrelin receptor agonists may stimulate also the growth hormone secretion and this could be a contraindication for the treatment of cachectic conditions. Recently, the ghrelin agonist TZP-101 has been shown to be able to accelerate gastric emptying and stimulate food intake without eliciting significant GH release (Fraser et al. 2008), which makes this compound a promising candidate in the development of new compounds for the treatment of conditions of anorexia and/or cachexia.

Although exogenous cannabinoids (dronabinol or nabilone) are currently licensed in the USA and some other countries as appetite stimulants for patients with cancer or HIV infection (Haney et al. 2005) and have been shown to be useful in the wasting and appetite loss of patients with dementias (Volicer et al. 1997), their potential therapeutic employment in AN has received relatively little attention. Indeed, only two small trials explored the effects of Δ9-THC in underweight AN patients (Gross et al. 1983; Berry 2006). In both studies, no significant effect of the cannabinoid on BW was observed, but an amelioration of depressive symptoms and perfectionism was reported in the second trial. The role of exogenous cannabinoid substances in the treatment of EDs necessitates further investigations. However, in the light of the above reported evidence of an increased endocannabinoid tone in chronic starving anorexic patients, endocannabinoid antagonists also deserve investigation.

7 Conclusions and Perspectives

A growing body of evidence supports the view that EDs might be linked to reward-related processes leading to a sort of addiction and dependence to self-starvation in AN and to overconsumption of food in binge eating syndromes. In line with this hypothesis, positive experiences of starvation have been long recognized among chronically fasting AN individuals, whose almost exclusive aim, as well as pleasure, is to maintain their cachectic conditions (Lasègue 1873; Pearce 2004), and people with binge eating syndromes usually report strong urges to eat with a sense of loss of control over their consummatory behavior and a transient binge-induced reduction of negative emotional states. In this vein, the mesolimbic reward system, including the amygdala, insula, ventral striatum, and ventral regions of the anterior cingulated cortex and of the orbitofrontal cortex, has been proposed to play a pivotal role in the genesis and/or the maintenance of AN (Kaye et al. 2009). Indeed, a functional magnetic resonance imaging study showed that, in underweight AN patients, the ventral striatal reward system was hyper-activated upon processing of disease-specific stimuli, such as watching female body images with underweight features (Fladung et al. 2010). This finding is consistent with the hypothesis that an altered evaluation of reward stimuli in AN individuals, represented by reduced responsiveness toward disease-nonspecific stimuli and enhanced reaction toward disease-specific stimuli, such as cues of emaciation, might result into addiction to self-starvation (Wagner et al. 2007; Friederich et al. 2006; Kaye 2008). In addition, changes in cognitive performances are widely recognized in ED patients and likely affect motivated behaviors. The literature reviewed above provides the evidence that feeding regulatory substances such as ghrelin, BDNF, opioids, and endocannabinoids is also involved in the modulation of reward processes, motivated behaviors, and cognitive performances. Therefore, changes in endocannabinoids, opioids, BDNF, and ghrelin physiology occurring in the acute state of an ED may play a pivotal role in the pathophysiology of the disorder by providing a possible link between motivated behaviors, reward processes, cognitive functions, and energy balance.

Findings are still too scanty to draw definite conclusions as to what comes first: the neuroendocrine alterations, which could represent pathogenetic causes of EDs, or ED psychopathology, which would result in nutritional impairments and, as a consequence, in neuroendocrine aberrations. Anyway, as suggested above, these and other neuroendocrine changes, even if secondary to malnutrition, may contribute to the maintenance of the EDs. Moreover, although most of the above reported neuroendocrine alterations represent state-dependent changes, some of them may persist after recovery from the ED. These persistent neuroendocrine aberrations may constitute a vulnerable biological background predisposing the patients to future relapses and/or recurrencies. In this vein, it is intuitive that medications developed on the basis of reported neuroendocrine alterations could be helpful for the treatment and/or secondary prevention of EDs. Preliminary findings support potential therapeutic effects of ghrelin and/or ghrelin agonists in AN, but studies in this area are still at an early stage, and the therapeutic implications of those findings are far from being conclusive.

References

Arosio M, Ronchi CL, Beck-Peccoz P et al (2004) Effects of modified sham feeding on ghrelin levels in healthy human subjects. J Clin Endocrinol Metab 89:5101–5104

Asakawa A, Inui A, Fujimiya M et al (2005) Stomach regulates energy balance via acylated ghrelin and desacyl ghrelin. Gut 54:18–24

Baranowska B (1990) Are disturbances in opioid and adrenergic systems involved in the hormonal dysfunction of anorexia nervosa. Psychoneuroendocrinology 15:371–379

Beriohay B, Lebrun B, Moyse E et al (2005) Brain-derived neurotrophic factor plays a role as an anorexigenic factor in the dorsal vagal complex. Endocrinology 146:5612–5620

Berridge KC (1997) Brain reward systems for food incentives hedonics in normal appetite and eating disorders. In: Kirkham TC, Cooper SJ (eds) Appetite and body weight: integrative systems and the development of anti-obesity drugs. Academic Press, Burlington
Berridge KC, Robinson TE (1998) What is the role of dopamine in reward: hedonic impact, reward learning, or incentive salience? Brain Res Brain Res Rev 28:309–369
Berry EM (2006) Pilot study of THC (2.5 mg x 2) in 9 ambulatory AN patients (abstract). In: The sixth nordic congress on eating disorders, Aarhus, Denmark
Brambilla F, Ferrari E, Petraglia F et al (1991) Peripheral opioid secretory pattern in anorexia nervosa. Psychiatry Res 39:115–127
Brambilla F, Monteleone P, Maj M (2008) Central and peripheral peptide regulation of hunger, satiety and food intake in eating disorders. In: Czerbska MT (ed) Psychoneuroendocrinology research trends. Nova Biomedical Books, New York
Brewerton TD, Lydiard RB, Laraia MT et al (1992) CSF-β-endorphin and dynorphin in bulimia nervosa. Am J Psychiatry 149:1086–1090
Broglio F, Gianotti L, Destefanis S et al (2004) The endocrine response to acute ghrelin administration is blunted in patients with anorexia nervosa, a ghrelin hypersecretory state. Clin Endocrinol (Oxf) 60:592–599
Chen CY, Inui A, Asakawa A et al (2005) Des-acyl ghrelin acts by CRF type 2 receptors to disrupt fasted stomach motility in conscious rats. Gastroenterology 129:8–25
Cota D, Marsicano G, Tschop M et al (2003a) The endogenous cannabinoid system affects energy balance via central orexigenic drive and peripheral lipogenesis. J Clin Invest 112:423–431
Cota D, Marsicano G, Lutz B et al (2003b) Endogenous cannabinoid system as a modulator of food intake. Int J Obes Relat Metab Disord 27:289–301
Cummings DE, Purnell JQ, Frayo RS et al (2001) A preprandial rise in plasma ghrelin levels suggests a role in meal initiation in humans. Diabetes 50:1714–1719
Di Marzo V, Goparaju SK, Wang L et al (2001) Leptin-regulated endocannabinoids are involved in maintaining food intake. Nature 410:822–825
Diano S, Farr SA, Benoit SC et al (2006) Ghrelin controls hippocampal spine synapse density and memory performance. Nat Neurosci 9:381–388
Ehrlich S, Salbach-Andrae H, Eckart S et al (2009) Serum brain-derived neurotrophic factor and peripheral indicators of the serotonin system in underweight and weight-recovered adolescent girls and women with anorexia nervosa. J Psychiatry Neurosci 34(4):323–329
Fladung AK, Grön G, Grammer K et al (2010) A neural signature of anorexia nervosa in the ventral striatal reward system. Am J Psychiatry 167(2):206–212
Fraser GL, Hoveyda HR, Tannenbaum GS (2008) Pharmacological demarcation of the growth hormone, gut motility and feeding effects of ghrelin using a novel ghrelin receptor agonist. Endocrinology 149:6280–6288
Friederich HC, Kumari V, Uher R et al (2006) Differential motivational responses to food and pleasurable cues in anorexia and bulimia nervosa: a startle reflex paradigm. Psychol Med 36:1327–1335
Fullerton DT, Swift WJ, Getto CJ et al (1986) Plasma immunoreactive beta-endorphin in bulimics. Psychol Med 16:56–63
Galderisi S, Mucci A, Monteleone P et al (2003) Neurocognitive functioning in subjects with eating disorders: the influence of neuroactive steroids. Biol Psychiatry 53:921–927
Garg A (2007) Commentary: the ongoing saga of obestatin: is it a hormone? J Clin Endocrinol Metab 92:3396–3398
Gelegen C, van den Heuvel J, Collier DA et al (2008) Dopaminergic and brain-derived neurotrophic factor signalling in inbred mice exposed to a restricted feeding schedule. Genes Brain Behav 7:552–559
Gerner RH, Sharp S (1982) B-endorphin immunoreactivity in normal, schizophrenic, depressed, manic, and anorexic subjects. Brain Res 237:244–247
Gourcerol G, St-Pierre DH, Taché Y (2007) Lack of obestatin effects on food intake: should obestatin be renamed ghrelin-associated peptide (GAP)? Regul Pept 141:1–7

Gross H, Ebert MH, Faden VB et al (1983) A double-blind trial of delta 9-tetrahydrocannabinol in primary anorexia nervosa before and during weight recovery. J Clin Psychopharmacol 3:165–171
Hanada T, Toshinai K, Kajimura N et al (2003) Anti-cachectic effect of ghrelin in nude mice bearing human melanoma cells. Biochem Biophys Res Commun 301:275–279
Haney M, Rabkin J, Gunderson E et al (2005) Dronabinol and marijuana in HIV+mar ijuana smokers: acute effects on caloric intake and mood. Psychopharmacology (Berl) 181:170–178
Harada T, Nakahara T, Yasuhara D et al (2008) Obestatin, acylghrelin, and des-acylghrelin responses to an oral glucose tolerance test in the restricting type of anorexia nervosa. Biol Psychiatry 63:245–247
Hotta M, Ohwada R, Katakami H et al (2004) Plasma levels of intact and degraded ghrelin and their responses to glucose infusion in anorexia nervosa. J Clin Endocrinol Metab 89:5707–5712
Hotta M, Ohwada R, Akamizu T et al (2009) Ghrelin increases hunger and food intake in patients with restricting-type anorexia nervosa: a Pilot Study. Endocr J 56:1119–1128
Janas-Kozik M, Krupka-Matuszczyk I, Malinowska-Kolodziej I et al (2007) Total ghrelin plasma level in patients with the restrictive type of anorexia nervosa. Regul Pept 140:43–46
Jerlhag E, Egecioglu E, Dickson SL et al (2007) Ghrelin administration into tegmental areas stimulates locomotor activity and increases extracellular concentration of dopamine in the nucleus accumbes. Addict Biol 12:6–16
Kaye WH (1996) Neuropeptide abnormalities in anorexia nervosa. Psychiatry Res 62:65–74
Kaye WH (2008) Neurobiology of anorexia and bulimia nervosa. Physiol Behav 94:121–135
Kaye WH, Pickar D, Naber D et al (1982) Cerebrospinal fluid opioid activity in anorexia nervosa. Am J Psychiatry 139:643–645
Kaye WH, Berrettini WH, Gwirtsman HE et al (1987) Reduced cerebrospinal fluid levels of immunoreactive pro-opiomelanocortin related peptides (including β-endorphin) in anorexia nervosa. Life Sci 41:2147–2155
Kaye WH, Fudge JL, Paulus M (2009) New insight into symptoms and neurocircuit function of anorexia nervosa. Nature Rev Neurosci 10:573–584
Kelley AE, Balski VP, Haber SN et al (2002) Opioid modulation of taste hedonics within the ventral striatum. Physiol Behav 76:365–377
Kernie SG, Liebl DJ, Parada LF (2000) BDNF regulates eating behavior and locomotor activity in mice. EMBO J 19:1290–1300
Kirkham TC, Williams CM, Fezza F et al (2002) Endocannabinoid levels in rat limbic forebrain and hypothalamus in relation to fasting, feeding and satiation: stimulation of eating by 2-arachidonoyl glycerol. Br J Pharmacol 136:550–557
Kissileff HR, Wentzlaff TH, Guss JL et al (1996) A direct measure of satiety disturbance in patients with bulimia nervosa. Physiol Behav 60:1077–1085
Kojima M, Hosoda H, Date Y et al (1999) Ghrelin is a growth-hormone-releasing acylated peptide from stomach. Nature 402:656–660
Kojima S, Nakahara T, Nagai N et al (2005) Altered ghrelin and peptide YY responses to meals in bulimia nervosa. Clin Endocrinol (Oxf) 62:74–78
Lasègue C (1873) De l'anorexie hystèrique. Arch Gèn Mèd 1:384–403
Lesem MD, Berrettini WH, Kaye WH et al (1991) Measurement of CSF dynorphin A 1-8 immunoreactivity in anorexia nervosa and normal-weight bulimia. Biol Psychiatry 29:244–252
Lewin GR, Barde YA (1996) Physiology of neurotrophins. Annu Rev Neurosci 19:289–317
Lyons WE, Mamounas LA, Ricaurte GA et al (1999) Brain-derived neurotrophic factor-deficient mice develop aggressiveness and hyperphagia in conjunction with brain serotonergic abnormalities. Proc Natl Acad Sci USA 96:15239–15244
Malik S, McGlone F, Bedrossian D et al (2008) Ghrelin modulates brain activity in areas that control appetitive behavior. Cell Metab 7:400–409
Mercader JM, Ribasès M, Gratacòs M et al (2007) Altered brain-derived neurotrophic factor blood levels and gene variability are associated with anorexia and bulimia. Genes Brain Behav 6:706–716

Miljic D, Pekic S, Djurovic M et al (2006) Ghrelin has partial or no effect on appetite, growth hormone, prolactin, and cortisol release in patients with anorexia nervosa. J Clin Endocrinol Metab 91:1491–1495
Misra M, Miller K, Herzog D et al (2004) Growth hormone and ghrelin responses to an oral glucose load in adolescent girls with anorexia nervosa and controls. J Clin Endocrinol Metab 89:1605–1612
Misra M, Miller KK, Stewart V (2005) Ghrelin and bone metabolism in adolescent girls with anorexia nervosa and healthy adolescents. J Clin Endocrinol Metab 90:5082–5087
Monteleone P, Martiadis V, Fabrazzo M et al (2003) Ghrelin and leptin responses to food ingestion in bulimia nervosa: implications for binge-eating and compensatory behaviours. Psychol Med 33:1387–1394
Monteleone P, Tortorella A, Martiadis V et al (2004) Opposite changes in the serum brain-derived neurotrophic factor in anorexia nervosa and obesity. Psychosom Med 66:744–748
Monteleone P, Fabrazzo M, Tortorella A et al (2005a) Circulating ghrelin is decreased in non-obese and obese women with binge eating disorder as well as in obese non-binge eating women, but not in patients with bulimia nervosa. Psychoneuroendocrinology 30:243–250
Monteleone P, Martiadis V, Rigamonti AE et al (2005b) Investigation of peptide YY and ghrelin responses to a test meal in bulimia nervosa. Biol Psychiatry 57:926–931
Monteleone P, Fabrazzo M, Martiadis V et al (2005c) Circulating brain-derived neurotrophic factor is decreased in women with anorexia and bulimia nervosa but not in women with binge-eating disorder: relationships to co-morbid depression, psychopathology and hormonal variables. Psychol Med 35:897–905
Monteleone P, Matias I, Martiadis V et al (2005d) Blood levels of the endocannabinoid anandamide are increased in anorexia nervosa and in binge-eating disorder, but not in bulimia nervosa. Neuropsychopharmacology 30:1216–1221
Monteleone P, Castaldo E, Maj M (2008a) Neuroendocrine dysregulation of food intake in eating disorders. Regul Pept 149:39–50
Monteleone P, Serritella C, Martiadis V et al (2008b) Deranged secretion of ghrelin and obestatin in the cephalic phase of vagal stimulation in women with anorexia nervosa. Biol Psychiatry 64:1005–1008
Monteleone P, Serritella C, Martiadis V et al (2008c) Plasma obestatin, ghrelin, and ghrelin/obestatin ratio are increased in underweight patients with anorexia nervosa but not in symptomatic patients with bulimia nervosa. J Clin Endocrinol Metab 93:4418–4421
Monteleone P, Bifulco M, Di Filippo C et al (2009) Association of CNR1 and FAAH endocannabinoid gene polymorphisms with anorexia nervosa and bulimia nervosa: evidence for synergistic effects. Genes Brain Behav 8:728–732
Monteleone P, Serritella C, Scognamiglio P et al (2010) Enhanced ghrelin secretion in the cephalic phase of food ingestion in women with bulimia nervosa. Psychoneuroendocrinology 35:284–288. doi:10.1016/j.psyneuen.2009.07.001
Morley JE, Blundell JE (1988) The neurobiological basis of eating disorders: some formulations. Biol Psychiatry 23:53–78
Muller TD, Reichwald K, Bronner G et al (2008) Lack of association of genetic variants in genes of the endocannabinoid system with anorexia nervosa. Child Adolesc Psychiatry Ment Health 2:33–39
Murphy KG, Bloom SR (2006) Gut hormones and the regulation of energy homeostasis. Nature 444:854–859
Nagaya N, Moriya J, Yasumura Y et al (2004) Effects of ghrelin administration on left ventricular function, exercise capacity, and muscle wasting in patients with chronic heart failure. Circulation 110:3674–3679
Nagaya N, Itoh T, Murakami S et al (2005) Treatment of cachexia with ghrelin in patients with COPD. Chest 128:1187–1193
Nakagawa T, Tsuchida A, Itakura Y et al (2000) Brain-derived neurotrophic factor regulates glucose metabolism by modulating energy balance in diabetic mice. Diabetes 49:436–444

Nakagawa T, Ogawa Y, Ebihara K et al (2003) Anti-obesity and anti-diabetic effects of brain-derived neurotrophic factor in rodent models of leptin resistance. Int J Obes Relat Metab Disord 27:557–565

Nakahara T, Harada T, Yasuhara D et al (2008) Plasma obestatin concentrations are negatively correlated with body mass index, insulin resistance index, and plasma leptin concentrations in obesity and anorexia nervosa. Biol Psychiatry 64:252–255

Nakai Y, Hosoda H, Nin K et al (2003) Plasma levels of active form of ghrelin during oral glucose tolerance test in patients with anorexia nervosa. Eur J Endocrinol 149:R1–R3

Nakazato M, Hashimoto K, Shimizu E (2003) Decreased levels of serum brain-derived neurotrophic factor in female patients with eating disorders. Biol Psychiatry 54:485–490

Nakazato M, Hashimoto K, Shiina A (2004) No changes in serum ghrelin levels in female patients with bulimia nervosa. Prog Neuropsychopharmacol Biol Psychiatry 28:1181–1184

Nakazato M, Hashimoto K, Yoshimura K (2006) No change between the serum brain-derived neurotrophic factor in female patients with anorexia nervosa before and after partial weight recovery. Prog Neuropsychopharmacol Biol Psychiatry 30:1117–1121

Neary NM, Small CJ, Wren AM et al (2004) Ghrelin increases energy intake in cancer patients with impaired appetite: acute, randomized, placebo-controlled trial. J Clin Endocrinol Metab 89:2832–2836

Nedvidkova J, Krykorkova I, Bartak V (2003) Loss of meal-induced decrease in plasma ghrelin levels in patients with anorexia nervosa. J Clin Endocrinol Metab 88:1678–1682

Otto B, Cuntz U, Fruehauf E (2001) Weight gain decreases elevated plasma ghrelin concentrations of patients with anorexia nervosa. Eur J Endocrinol 145:R5–R9

Otto B, Tschop M, Cuntz U (2004) Similar fasting ghrelin levels in binge eating/purging anorexia nervosa and restrictive anorexia nervosa. Psychoneuroendocrinology 29:692–693

Otto B, Tschop M, Fruhauf E et al (2005) Postprandial ghrelin release in anorectic patients before and after weight gain. Psychoneuroendocrinology 30:577–581

Owen WP, Halmi KA, Gibbs J et al (1985) Satiety responses in eating disorders. J Psychiatr Res 19:279–284

Pearce JM (2004) Richard Morton: origins of anorexia nervosa. Eur Neurol 52:191–192

Pelleymounter MA, Cullen MJ, Wellman CL (1995) Characteristics of BDNF-induced weight loss. Exp Neurol 131:229–238

Prince AC, Brooks SJ, Stahl D et al (2009) Systematic review and meta-analysis of the baseline concentrations and physiological responses of gut hormones to food in eating disorders. Am J Clin Nutr 89:755–765

Schultz W, Dayan P, Montague PR (1997) A neural substrate of prediction and reward. Science 275:1593–1599

Schwartz MW, Woods SC, Porte D Jr et al (2000) Central nervous system control of food intake. Nature 404:661–671

Seoane LM, Al-Massadi O, Pazos Y et al (2006) Central obestatin administration does not modify either spontaneous or ghrelin-induced food intake in rats. J Endocrinol Invest 29:RC13–RC15

Simonian HP, Kresge KM, Boden GH et al (2005) Differential effects of sham feeding and meal ingestion on ghrelin and pancreatic polypeptide levels: evidence for vagal efferent stimulation mediating ghrelin release. Neurogastroenterol Motil 17:348–354

Solinas M, Justinova Z, Goldberg SR et al (2006) Anandamide administration alone and after inhibition of fatty acid amide hydrolase (FAAH) increases dopamine levels in the nuclear accumbens shell in rats. J Neurochem 98:408–419

Soriano-Guillen L, Barrios V, Campos-Barros A et al (2004) Ghrelin levels in obesity and anorexia nervosa: effect of weight reduction or recuperation. J Pediatr 144:36–42

Stock S, Leichner P, Wong AC (2005) Ghrelin, peptide YY, glucose-dependent insulinotropic polypeptide, and hunger responses to a mixed meal in anorexic, obese, and control female adolescents. J Clin Endocrinol Metab 90:2161–2168

Støving RK, Andries A, Brixen K et al (2009) Leptin, ghrelin, and endocannabinoids: potential therapeutic targets in anorexia nervosa. J Psychiatr Res 43:671–679. doi:10.1016/j.jpsychires.2008.09.007

Tanaka M, Naruo T, Muranaga T (2002) Increased fasting plasma ghrelin levels in patients with bulimia nervosa. Eur J Endocrinol 146:1–3

Tanaka M, Naruo T, Yasuhara D (2003a) Fasting plasma ghrelin levels in subtypes of anorexia nervosa. Psychoneuroendocrinology 28:829–835

Tanaka M, Narau T, Nagai N et al (2003b) Habitual binge/purge behavior influences circulating ghrelin levels in eating disorders. J Psychiatr Res 37:17–22

Tanaka M, Tatebe Y, Nakahara T et al (2003c) Eating pattern and the effect of oral glucose on ghrelin and insulin secretion in patients with anorexia nervosa. Clin Endocrinol (Oxf) 59:574–579

Tanaka M, Nakahara T, Kojima S et al (2004) Effect of nutritional rehabilitation on circulating ghrelin and growth hormone levels in patients with anorexia nervosa. Regul Pept 122:163–168

Tepper R, Weixman A, Apter A et al (1992) Elevated plasma immunoreactive β-endorphin in anorexia nervosa. Clin Neuropharmacol 15:387–391

Troisi A, Di Lorenzo G, Lega I et al (2005) Plasma ghrelin in anorexia, bulimia, and binge-eating disorder: relations with eating patterns and circulating concentrations of cortisol and thyroid hormones. Neuroendocrinology 81:259–266

Tsuchida A, Nonomura T, Ono-Kishino M (2001) Acute effects of brain-derived neurotrophic factor on energy expenditure in obese diabetic mice. Int J Obes Relat Metab Disord 25:1286–1293

Uehara T, Omori I, Nakamura K et al (2005) Plasma des-acyl and acyl ghrelin in patients with eating disorders. Eat Weight Disord 10:264–266

Volicer L, Stelly M, Morris J et al (1997) Effects of dronabinol on anorexia and disturbed behaviour in patients with Alzheimer's disease. Int J Geriatr Psychiatry 12:910–919

Wagner A, Aizenstein H, Venkatraman VK et al (2007) Altered reward processing in women recovered from anorexia nervosa. Am J Psychiatry 164:1842–1849

Waller DA, Kiser RS, Hardy BW et al (1986) Eating behavior and plasma beta-endorphin in bulimia. Am J Clin Nutr 44:20–23

Will MJ, Franzblau EB, Kelley AE (2003) Nucleus accumbens μ-opioids regulate intake of high-fat diet via activation of a distributed brain network. J Neurosci 23:2882–2888

Williams CM, Kirkham TC (2002) Observational analysis of feeding induced by Delta-9-THC and anandamide. Physiol Behav 76:241–250

Xu B, Goulding EH, Zang K et al (2003) Brain-derived neurotrophic factor regulates energy balance downstream of melanocortin-4 receptor. Nat Neurosci 6:736–742

Zhang JV, Ren PG, Avsian-Kretchmer O et al (2005) Obestatin, a peptide encoded by the ghrelin gene, opposes ghrelin's effects on food intake. Science 310:996–999

Zhu X, Cao Y, Voogd K et al (2006) On the processing of proghrelin to ghrelin. J Biol Chem 281:38867–38870

Animal Models of Eating Disorder Traits

Martien J.H. Kas and Roger A.H. Adan

Contents

Abstract Eating disorders, such as anorexia and bulimia nervosa, are psychiatric disorders that are likely determined by a complex interaction between genetic variations, developmental processes, and certain life events. Cross-species analysis of traits related to eating disorders may provide a way to functionally and systematically study neurobiological mechanisms underlying these disorders. Interspecies trait genetics may offer opportunities to identify common neurobiological mechanisms underlying eating disorder characteristics relevant to the initiation, progression, and/or maintenance of the disease, such as cognitive rigidity, increased anxiety levels, and behavioral hyperactivity. These can subsequently be tested directly by studying allelic variation in mice and human subjects and by applying methods that can modify gene expression levels in rodent models. Increasing our knowledge about these traits and their underlying neurobiological mechanisms will be relevant to develop new therapies for patients within the heterogeneous eating disorder populations. Novel mouse genetic and phenotyping tools offer a way to study these neurobehavioral traits under controlled environmental and genetic background conditions.

M.J.H. Kas (✉) and R.A.H. Adan
Department of Neuroscience and Pharmacology, Rudolf Magnus Institute of Neuroscience, University Medical Centre Utrecht, Utrecht, The Netherlands
e-mail: m.j.h.kas@umcutrecht.nl

R.A.H. Adan and W.H. Kaye (eds.), *Behavioral Neurobiology of Eating Disorders*, Current Topics in Behavioral Neurosciences 6, DOI 10.1007/7854_2010_84

Keywords Behavior · Genetics · Hyperactivity · Imaging · Mouse · Neurobiology · Phenotype · Traits

1 Why Studying Eating Disorder Traits?

The human population with eating disorders is highly diverse revealing many different phenotypes both between and within eating disorders. Anorexia nervosa, for example, is a serious psychiatric disorder characterized by severe and selective restriction of food intake. This eating disorder, with a high prevalence among young adolescent females (15–19 years), results in extreme body weight loss and has a mortality rate of up to 15% (Hoek 2006). Anorexia nervosa patients also show other symptoms that are not observed consistently throughout the anorexia nervosa patient population. For example, behavioral hyperactivity is considered an important phenotype of the disease (Bergh and Sodersten 1996; Brewerton et al. 1995b; Davis et al. 1997; Hebebrand et al. 2003), since it leads to accelerated body weight loss and is observed in a large proportion (40–80%) of the anorexia nervosa population. In addition, a large group of anorexia patients suffers an anxiety disorder that is already present pre-morbidly and that can vary greatly between patients [e.g., a general anxiety disorder, social phobia, or panic disorder (Brewerton et al. 1995a; Bulik et al. 1997; Godart et al. 2002; Toner et al. 1988)]. Furthermore, lifetime compulsion and obsession phenotypes occur in a large sample of the anorexia nervosa population (Halmi et al. 2003). Thus, the expression of an eating disorder is not uniform and is characterized by different phenotypes that are variable within the patient population.

Phenotypic heterogeneity within the eating disorder population complicates identification of disease genes and may also reflect differences in the etiology of these disorders. The current diagnostic criteria are not sensitive enough to differentiate between subgroups of anorexia nervosa patients. Further understanding these differences is highly relevant for finding the disease genes as well as in view of developing effective treatment programs that may be directed toward specific pathophysiological features of the disease. Indeed, current treatment possibilities only cure a small proportion of the eating disorder patients [for review, see (Fairburn and Harrison 2003; Treasure et al. 2010)], and high relapse rates are documented (Herzog et al. 1999; Carter et al. 2004). Furthermore, genetic studies have, thus far, not revealed many replicated candidate genes for neither anorexia nervosa nor bulimia nervosa [for review, see (Hinney et al. 2000; Kas et al. 2003b; Klump and Gobrogge 2005; Mazzeo et al. 2006; Bulik et al. 2007b)], suggesting that either different subgroups within the population may relate to differences in affected genetic pathways or major susceptibility genes have not been found. For those reasons, dissecting phenotypic variation across the patient population and targeting those behavioral deficits that are leading to the initiation, progression, and/or maintenance of the disease would be an alternative strategy in the battle against these dramatic disorders.

2 Eating Disorder Traits

Several studies have focused on the identification of heritable phenotypes that are present in the eating disorder population (Keski-Rahkonen et al. 2005; Bulik et al. 2007a; Mazzeo et al. 2009). Recently, Gottesman and Gould (2003) put forward the psychiatric endophenotype concept to facilitate the identification of genes relevant for neuropsychiatric disorders. Endophenotypes, which are expected to have lower genetic heterogeneity than clinical diagnoses, may represent simpler clues to genetic underpinnings than the disease syndrome itself. Furthermore, endophenotypes may not be specific to one psychiatric disorder but have overarching effects impacting on several different diagnoses. This promotes the view that psychiatric diagnoses can be deconstructed to facilitate more straightforward and successful genetic analysis.

As described earlier in a paper by Bulik et al. (2007a), "an endophenotype may be neurophysiological, biochemical, endocrinological, neuroanatomical, cognitive, or neuropsychological in nature. Endophenotypes are heritable, cosegregate with a psychiatric illness in the general population, are state independent, (i.e., manifest in the individual whether illness is active), and are found in nonaffected family members at a higher rate than in the general population. Other enhancements of the definition of endophenotype include being linked to the causal process, involved in plausible biological mechanisms, predictive of the disorder probabilistically, and lying closer to the site of the primary causative agent. The term subphenotype has also been used to identify more homogeneous subgroups of complex syndromes (e.g., early onset depression or BN with self-induced vomiting). Although not as clearly defined as endophenotypes, they are commonly used as a means to reduce the heterogeneity inherent in sampling based on a diagnostic category or syndrome. An example of the hierarchy would be: the phenotype of schizophrenia; the subphenotype of individuals with schizophrenia who report auditory hallucinations; and the endophenotype of P50 event-related potential suppression." Based on these definitions, Bulik et al. (2007a) have generated an overview on potential traits for eating disorders (Table 1).

By making use of neuroimaging techniques, new insights into neurobiological mechanisms of eating disorders have been obtained. As with psychological and behavioral eating disorder traits, neurocircuit wiring and their related functioning may also provide novel ways to classify subgroups within the eating disorder population. In a recent review, Kaye et al. (2009) described how altered brain activity in the insula could explain interoceptive dysfunction in anorexia nervosa. Furthermore, altered striatal brain activity may be related to differences in reward signaling in eating disorder patients (Kaye et al. 2009). Behavioral changes seen in the development of anorexia nervosa may be related and maintained by (permanent) brain circuit alterations due to developmental (e.g., puberty) and/or environmental factors (e.g., stressful life events). For those reasons, neuroimaging characteristics may provide a way complementary to behavioral traits to closer study neurobiological circuits underlying eating disorders.

Table 1 Summary of data addressing whether psychological, physical, and biological traits represent endophenotypes or subphenotypes for eating disorders (from Bulik et al. 2007a, b)

Trait	Endophenotype criteria						Endo?	Sub?
	Measurable	Heritable	Cosegregates with illness	State independent	Observed in unaffected family members	Biologically plausible causal mechanism		
Perfectionism	+	Moderate	++	+++	+	Unknown	Unknown	++
Obsessionality	+	Moderate	+++	+++	+	+	+	
Drive for thinness	+	Moderate	+++	++	+?	+	Unknown	+++
Anxiety	+	Moderate	+++	++	+	+	+	
Negative emotionality	+	Moderate	+++	++	+	+	+	
Decreased food intake	+	Moderate-large	+++	+	Unknown	+++	Unknown	+++
Low body weight (dysregulation of body weight)	+++	Moderate-large	+++	+	Unknown	Unknown	Unknown	+++
Increased physical activity	+	Moderate-large	+++	+	Unknown	Unknown	Unknown	+++
Cognitive set-shifting	++	Moderate-large	++	+	+	+++	++	
Binge eating	+	Moderate	+++	No	Unknown	+	No	+++
Self-induced vomiting	+	Moderate-large	+++	No	Unknown	Unknown	No	+++
Impulsivity	+	Moderate-large	++	++	+	+	+	
Undue influence of weight or shape	+	Small	+++	++	Unknown	No		

AN anorexia nervosa, *ANR* restricting type AN, *BN* bulimia nervosa, *ANBP* binge/purge type AN, *Endo* endophenotype, *Sub* subphenotype, *Unknown* no studies have examined that the issue or existing data are incondusive. Plus marks (+) denote the strength of data supporting each criterion. For example, in the "measurable" column, a single plus (+) denotes that only self-report measures were used to assess the trait. A double plus mark (++) indicates that observer ratings or neuropsychological data were used to assess the trait. A triple plus mark (+++) indicates that the trait can be measured objectively by an outside observer (e.g., body weight) or can be assessed with a biological assay or marker (e.g., 5HT transporter activity). In the remaining criteria columns, the plus marks indicate the strength of the data supporting the criterion in terms of the number of studies reporting positive findings (i.e., + = few studies; ++ = more studies; +++ = many studies). In the Endo? and Sub? columns, the plus marks indicate the extent to which the trait satisfies the criteria for an endophenotype or subphenotype (+ = some evidence that criteria are supported; ++ = moderate evidence; +++ = strong evidence). Traits exhibiting the strongest evidence in support of their categorization as endophenotypes are noted in bolded and outlined text

In view of these behavioral and neuro-anatomical characteristics, some considerations should also be taken into account. For instance, some of the core characteristics identified may be highly dependent on each other and/or may reflect the same pathophysiological process. For example, do neurocircuits that underlie disturbed set-shifting in anorexia nervosa (partially) overlap with those of compulsive behavior and/or of perfectionism observed in these patients, or are these truly separate components of the disease? In addition, recent studies have indicated that behavioral hyperactivity seen in anorexia nervosa is related to the levels of anxiety and food restriction (Holtkamp et al. 2004), indicating that some behavioral characteristics are highly dependent on others. This requires practical considerations about study design to assess these potentially related eating disorder characteristics properly. Once we understand better the mechanisms underlying the traits that contribute to the development and maintenance of an eating disorder, new horizons arise for novel treatments. Future research is necessary to study the relation between these eating disorder characteristics, and the genetic pathways and neural circuits underlying the pathophysiological neuroprocesses that drive these behavioral characteristics.

3 Rational for Animal Models of Eating Disorders

Understanding the biology of behavioral disorders, including eating disorders, requires identifying and functionally testing biological substrates in relation to these disorders. Interference with, for example, pharmacological agents or genetic manipulations is a standard tool in animal research to test their involvement in physiological processes. Animal studies allow systematic studies in which environmental and genetic factors can largely be controlled for. The challenge is, of course, how to develop an animal model for eating disorders, such as anorexia nervosa. Indeed, validity of translational animal models is widely accepted for common physiological processes (such as blood pressure regulation); however, cross-species comparison for psychiatric disorders offers a challenging opportunity for biomedical research (Dennis 2005).

Recently, a proof of concept of the confluence between mouse and human for psychiatric traits was presented by Chen (Chen et al. 2006). A common genetic variant of the brain-derived neurotrophic factor (BDNF) gene in humans is associated with alterations in brain anatomy, memory and has been associated with psychiatric disorders, such as eating disorders, depression, and schizophrenia (Egan et al. 2003; Ribases et al. 2004; Lang et al. 2005; Lohoff et al. 2005; Neves-Pereira et al. 2005; Ribases et al. 2005; Schumacher et al. 2005). BDNF has an important function in neuronal survival, differentiation, and synaptic plasticity. Chen and co-workers showed that when the human Val66Met variant is genetically introduced in mice, it exhibits phenotypic characteristics in humans with the variant allele, including anxiety-related behaviors. This finding illustrates the potential of comparative neurobehavioral genetic studies between mouse and human.

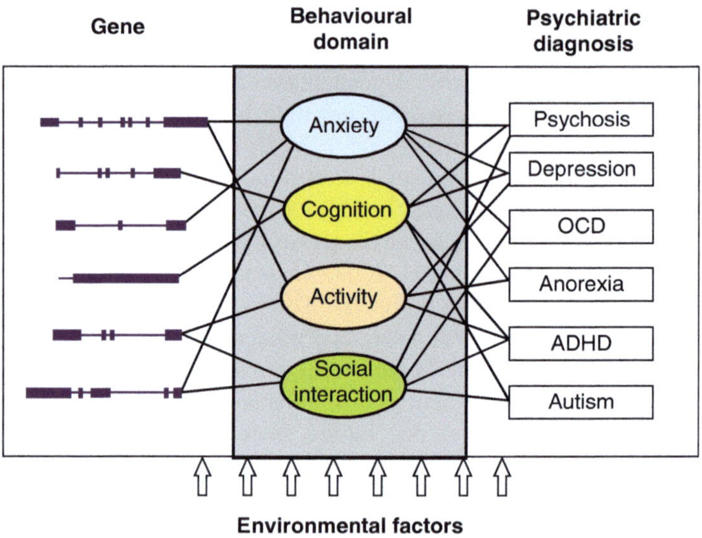

Fig. 1 The behavioral domain concept across psychiatric diagnosis. The relationship of behavioral domains to susceptibility genes will be more direct than the relationship with clinical diagnosis, since the disease will be a more heterogeneous composite of behavioral traits, which are modulated by protective and adverse life events. The contribution of these life events can also be modeled in mouse where environment can be controlled and manipulated. Note that this schematic diagram is intended to be illustrative of a behavioral domain concept, rather than demonstrate associations, which are proven by genetic epidemiology (from Kas et al. 2007)

Interestingly, certain behavioral domains that are affected in eating disorders may be relevant across the psychiatric spectrum. For example, set-shifting difficulties are observed in anorexia and bulimia nervosa (Tchanturia et al. 2004; Holliday et al. 2005), but are also observed in obsessive–compulsive disorders, suggesting that there may be common neurobiological mechanisms underlying certain aspects across diagnosis. Moreover, certain behavioral domains, such as intra- and extra-dimensional set-shifting can be studied in animals as well (Birrell and Brown 2000; Brigman et al. 2005; Bissonette et al. 2008). These behavioral domains may provide a good starting point for understanding the neurobiological mechanisms underlying these behavioral disorders (Fig. 1).

4 How These Traits Are Mimicked in Animals

As indicated with the Val66Met variant in the BDNF gene, studying gene function in relation to behavioral deficits across species may be one way to go in view of translational research for psychiatric disorders. In addition, novel gene identification methodologies for behavioral traits are another option and have been largely developed. For instance, with the current availability of a large variety of inbred mouse strains and their known genome sequences (Frazer et al. 2007; Yang et al. 2007), mouse genetics offer a challenging way to study complex neurobehavioral traits. In contrast to patient populations, mouse strains can be used to control for phenotypic and genetic heterogeneity as well as for studies aiming at understanding complex gene–environment interactions. With the recent generation of genetic reference populations, such as recombinant inbred strains (RIS) (Plomin et al. 1991) and chromosome substitution strains (CSS) (Singer et al. 2004), quantitative trait loci (QTL) analysis can be performed for complex neurobiological traits in mice. Additional approaches for mice, such as haplotype mapping (Grupe et al. 2001; Wade et al. 2002; Wade and Daly 2005), genome-wide gene expression (Sandberg et al. 2000; Jansen and Nap 2001; Fernandes et al. 2004; Letwin et al. 2006; Hovatta et al. 2007), and quantitative complementation studies (Yalcin et al. 2004) provide a complementary technological platform for gene identification. With the technology in place, the true challenge for this translational approach is the development of appropriate animal models for eating disorders and exposure of genetically defined mice to these models.

The activity-based anorexia (ABA) or semi-starvation-induced hyperactivity model is, among others (Siegfried et al. 2003), a known animal model to study pathophysiological processes in anorexia nervosa. The ABA model is around since the 1960s (Routtenberg and Kuznesof 1967) and ABA is induced in rodents with voluntary access to running wheels and that are exposed to daily scheduled restricted food availability. Reminiscent of anorexia nervosa, certain rat and mouse inbred strains exposed to this daily scheduled feeding paradigm exhibit a paradoxical behavioral hyperactivity with reduced food availability and a subsequent accelerated body weight loss (Kas et al. 2003a; Gelegen et al. 2007). Excessive behavioral hyperactivity may be a core trait of anorexia nervosa (Brewerton et al. 1995b; Bergh and Sodersten 1996; Davis et al. 1997, 1999; Hebebrand et al. 2003). In addition, it has been shown that Olanzapine, an antipsychotic drug, can suppress behavioral hyperactivity in both anorexia nervosa patients and rodents exposed to the ABA model (Hillebrand et al. 2005b). Furthermore, plasma leptin levels are correlated with physical activity levels in anorexia nervosa patients during the acute phase of the illness (Holtkamp et al. 2003; van Elburg et al. 2007), and chronic leptin infusion suppresses behavioral activity in rats exposed to the ABA model (Exner et al. 2000; Hillebrand et al. 2005a). Moreover, reminiscent of the high incidence of anorexia nervosa in young females (in the age of 15–19 years) (Hoek 2006), young adolescent rodents are more susceptible to ABA than older rodents (Barbarich-Marsteller et al. 2007). Taken together, these findings provide some face- and predictive validity of the model for pathophysiological processes observed in anorexia nervosa.

By screening a panel of CSS in the ABA model, we have recently found that different behavioral characteristics of the model can be genetically dissociated. In a CSS panel, each of the 21 mouse substrains has a C57BL/6J genetic background with a single A/J chromosome being substituted (e.g., CSS 1 carries A/J chromosome 1 in a C57BL/6J genetic background). This allows studying the contribution of single A/J chromosomes to phenotypes and providing a starting point for genetically fine mapping loci on those chromosomes that contribute to the phenotype of interest. When this particular CSS panel was tested in the ABA model, some substrains showed disorganized behavioral hyperactivity with accelerated body weight loss (Gelegen et al. 2010), while other substrains exhibited behavioral hyperactivity during the hours of limited food access (Kas et al. 2010). Some of the substrains did not contribute to ABA phenotypes. Together, these data showed that behavioral components within the ABA model can be dissociated using these CSS and that the dissected components are regulated by mechanisms of different genetic origin. Interestingly, these dissociable ABA phenotypes affect different aspects relevant to the progression and maintenance of eating disorder characteristics. For example, the disorganized behavioral hyperactivity phenotype observed in certain CSS led to accelerated body weight loss (Gelegen et al. 2010), whereas the high levels of running during the daily hours that food is present directly interferes with the eating behavior itself (Kas et al. 2010).

At the genetic level, these findings using CSS may contribute to identify novel mechanisms underlying these eating disorder traits. For example, genomic regions on A/J chromosomes that contributed to disorganized behavioral hyperactivity and subsequent accelerated body way loss display homology with regions on human chromosomes linked with genetic linkage regions in anorexia nervosa cohorts. For example, a region on human chromosome 1 (1p34.2) has been linked to the restricting subtype of AN (characterized by a severe limitation in food intake) (Grice et al. 2002), and this region shows complete overlap with a long region of mouse chromosome 4 that was identified in our genetic screen (Gelegen et al. 2010). For the other dissociable phenotype, behavioral hyperactivity during food access (Kas et al. 2010), the identified mouse chromosomes overlap with two previously observed human OCD chromosomal regions 7p and 15q (Shugart et al. 2006), suggesting that there may be overlap in compulsive wheel running during food access (while being food restricted) and compulsivity in OCD. In this way, our data open new roads for interspecies genetic studies for these neurobehavioral traits that may be relevant to anorexia nervosa and OCD.

In addition to studies that focus on behavioral traits within animal models for eating disorders, such as behavioral hyperactivity in the ABA model, one could also consider modeling susceptibility traits of eating disorders. As described above, anxiety disorders are highly comorbid with eating disorders (Fornari et al. 1999; Godart et al. 2006). Anxiety disorders, such as social phobia, may represent risk factors for eating disorders and could share common mechanisms that are relevant to the development of anorexia and bulimia nervosa. Animal models for these susceptibility traits may reveal new insights into the mechanisms underlying eating disorder development. Table 2 presents a list of characteristics that are relevant to eating disorders and approaches to their measurement in humans together with potential analogous rodent models and their measurement approaches (from Kas et al. 2009).

Table 2 Examples of how eating disorder characteristics may be modeled/tested in rodents and measured in humans (from Kas et al. 2009)

Domain	Characteristic	Human	Rodents
Anxiety and anxiety disorders	Generalized anxiety	Clinical interview	Light-dark box
	Social phobia	Laboratory measures of anxiety and arousal	Open field test
		State Trait Anxiety Inventory	Elevated plus maze
		Social Phobia and Anxiety Inventory	Novelty-suppressed feeding
	Social threat perception	Internet-based programs to assess social threat perception	Social approach behavior
Depression	Dysphoria	Clinical interview	Cocaine withdrawal
	Anhedonia	Beck Depression Inventory	Sucrose preference test
			Anticipatory activity
Weight	Weight dysregulation low	Low BMI	Strains that do not gain weight even with increased consumption
	Weight dysregulation high	High BMI	Strains that gain weight in the absence of increased caloric intake
Motor activity	Behavioral activity	Actiwatch	Activity-based anorexia model
		Observation Questionnaires	Home cage activity monitoring
			Open field testing
Cognition	Set-shifting	Trail Making Test (TMT)	Multidimensional visual stimuli task
		Wisconsin Card Sort Test (WCST)	
		Brixton task	
		Haptic Illusion CatBat task	
		Set-shifting subset of the Cambridge Neuropsychological Test Automated Battery (CANTAB)	
Obsessionality and compulsivity	OCD traits	Yale-Brown Obsessive-Compulsive Scale	Quinpirole-induced compulsive checking
	Symmetry and exactness	EatAte Life	Barbering
	Flaw detection	Internet delivered tasks	Drug seeking behavior
Hormonal	Amenorrhea in response to food deprivation/low BMI	Plasma hormone levels	Plasma hormone levels
			Vaginal cytology

(*continued*)

Table 2 (continued)

Domain	Characteristic	Human	Rodents
Eating behavior	Binge eating	Self-report/laboratory observation In response to short term food deprivation (disinhibition)	Restriction/refeeding and stress-induced eating Intermittent access to palatable foods Deprivation-induced binge eating Novelty-suppressed feeding
Impulsivity	Impulsive behavior	Barratt Impulsivity Scale (BIS) Go/No-go task	Go/No-go task
Brain activity	D2/D3 receptor activity in striatum	SPECT, fMRI, PET	SPECT, fMRI, PET Gene expression analysis
Physiology	Body temperature	Hypothermia	Hypothermia
Perfectionism	Concern over mistakes	Multidimensional Perfectionism Scale (MPS)	?
Drive for thinness	Dieting Fear of weight gain	Eating disorder inventory (EDI)	?
Body image distortion	Body dissatisfaction	Various self-report or IT-delivered measures	?

While animal models cannot mimic all eating disorder traits, such as perfectionism, body dissatisfaction, or drive for thinness, behavioral scientists have been working for decades on animal models for other behavioral characteristics relevant to anorexia and bulimia nervosa, such as, e.g., depressive symptoms [for review, see (Redei et al. 2001; Nestler et al. 2002; Cryan and Mombereau 2004; Cryan and Holmes 2005; Dranovsky and Hen 2006)], for compulsive behavior [for review, see (Joel 2006; Korff and Harvey 2006)], for impulsivity [for review, see (Evenden 1999; Jentsch and Taylor 1999; Lesch and Merschdorf 2000)], for set-shifting (Brigman et al. 2005; Brooks et al. 2006), and for body weight regulation [for review, see (Rohner-Jeanrenaud and Jeanrenaud 1997; Barsh et al. 2000; Mercer and Tups 2003; Adan et al. 2006; Buettner et al. 2007)]. Insights into mechanisms underlying these separate components may contribute to understanding the development of heterogeneity within eating disorder populations. Nevertheless, novel advances to refine assessments of these behavioral components in rodents are needed to optimize animal research for eating disorder traits.

As indicated above, there is substantial comorbidity of eating disorders and anxiety disorders [for review, see (Swinbourne and Touyz 2007)]. Studies have consistently shown that a significant number of patients with anorexia nervosa or bulimia nervosa experience one or more anxiety disorders (Kaye et al. 2004). Lifetime prevalence of at least one anxiety disorder in individuals with eating disorders varies from 25% (Keck et al. 1990) to 75% (Schwalberg et al. 1992) in bulimia nervosa and from 23% (Laessle et al. 1991) to 75% (Deep et al. 1995) in anorexia nervosa. Several studies have shown that anxiety disorders are premorbid to the development of an eating disorder (Schwalberg et al. 1992; Brewerton et al. 1995b; Deep et al. 1995; Bulik 2002; Godart et al. 2003; Brewerton et al. 1995a), indicating that studies unveiling mechanisms underlying anxiety disorders may provide insights into susceptibility factors for eating disorders.

In rodents, considerable information exists on the determination of anxiety levels. Standard laboratory tests, such as the open field, elevated plus maze, and the light–dark box test, are generally used to measure novelty-induced anxiety levels in rodents. In general, rodent species have an innate preference for sheltered places that have lower light intensities than the outside world and that provide a sense of safety via body contact with the shelter area surface (thigmotaxis). The open field test was one of the first behavioral tests developed for emotionality and that was based on the assessment of these behavioral expressions (Hall 1936). These relatively brief tests provide insights in novelty-responsiveness of the animal, but are confounded by strain differences in locomotor activity and do not provide baseline measures of anxiety levels. For these reasons, the field will benefit from novel measures that assess baseline anxiety levels and control for strain differences in locomotor activity (Kas and Van Ree 2004; Kas et al. 2008). By means of interspecies genetic analysis, we have recently found an association at the genetic level between increased baseline sheltering preference (using longitudinal automated home cage observations) and a human mood disorder (de Mooij-van Malsen et al. 2009). Furthermore, in addition to measures of anxiety levels in relation to novel environments with a nonsocial context, animal models for social phobia have also been introduced and may be relevant to eating disorder development. For instance, behavioral tests have been developed in which rodents can be tested for their preference for social approach or avoidance (Nadler et al. 2004; Moy et al. 2007). These refinements in rodent behavioral testing paradigms will contribute to face, predictive, and construct validity of animal models for eating disorder traits.

In addition to the development of behavioral testing paradigms to assess eating disorder characteristics in both mouse and human, neuroimaging approaches have recently been initiated across species to picture brain activities in relation to eating disorder development. For example, a recent study (Wagner et al. 2007) showed that individuals who have recovered from restricting-type anorexia nervosa had altered patterns of response in the ventral and dorsal striatum to positive and negative feedback. That is, an anterior ventral striatum response that distinguished between winning and losing was seen in the comparison women but not in the anorexia nervosa group. These findings suggest that individuals with anorexia nervosa may have difficulty discriminating between positive and negative feedback, relative to healthy comparison subjects. Similarly, in a study using a startle reflex paradigm (Friederich et al. 2006), a generalized failure to activate the appetitive motivational system was observed in individuals with anorexia nervosa. Interestingly, Barbarich-Marsteller et al. (2005) found changes in the striatum, hippocampus, and thalamus in rodents exposed to the ABA model. Similarly, van Kuyck et al. (2007) found altered activity in the ventral striatum, insula, thalamus, and ventral pontine nuclei, as well as a positive correlation between body weight loss and metabolism in the anterior cingulate and related regions in ABA rodents. When considered together, these human and rodent studies suggest the possibility of involvement of common pathways, but differences in imaging techniques and the effects of nutritional status make direct comparisons problematic.

Integration of genetic, behavioral, and neuroimaging findings may, in the end, provide more complete insight into the mechanisms underlying complex disorders. For example, increased anterior ventral striatum dopamine D2/D3 receptor binding in recovered anorexia nervosa patients could be contributing to the above-mentioned alteration in anterior ventral striatum function (Frank et al. 2005). Disturbed dopamine D2 receptor binding observed by these brain imaging techniques would be consistent with recently observed genetic linkage with the dopamine D2 receptor and anorexia nervosa (Nisoli et al. 2007; Bergen et al. 2005). Furthermore, recent animal studies have shown that mice with high susceptibility to develop ABA have increased striatal dopamine D2 receptor mRNA levels when compared to mice that do not develop behavioral hyperactivity in this animal model (Gelegen et al. 2008). In view of the relation between dopamine D2 receptor regulation and eating disorders and reward seeking in humans (Blum et al. 1995), it is interesting to note that mice with a genetic deletion of this receptor have deficits in reward processes (Maldonado et al. 1997; Cunningham et al. 2000; Elmer et al. 2002; Tran et al. 2002; Drew et al. 2007) that may also translate to certain aspects observed in eating disorders (Bergh and Sodersten 1996).

5 Future Directions

Eating disorders are multifactorial psychiatric disorders with unknown etiology and (at this point) relatively low success rate with current treatment programs. Understanding the complex interactions of genetic background and environmental factors will become crucial in unraveling the biology of this disease. More and more, the role of the environment, gender, and critical time periods during development have raised awareness in the development of this disease. Hand-in-hand with upcoming genetic findings, systematic and controlled animal studies will, therefore, play an important role to study their relationship in the development of eating disorders.

Extensive human genetic studies are underway and will provide novel candidate genes for eating disorders, such as anorexia nervosa. To unravel the contribution of these candidate genes in a neurobiological mechanism underlying eating disorders, translational research will be needed. Gene knockout technology in mice has provided a tremendous contribution to gene function research in all biomedical disciplines. This has evolved in novel and refined applications, such as conditional knockout technology, vector-directed gene expression, and short-hairpin interference methodologies to study gene function relationships over time [e.g., genetic deletion during development or adulthood (Gross et al. 2002) and in a tissue-specific manner]. Thus, sufficient gene manipulation technology is available to functionally test human candidate genes for anorexia nervosa. Furthermore, novel mouse genetic mapping panels allow the identification of genomic regions for mouse phenotypes that are homologous with human genomic regions identified for eating disorder characteristics (Gelegen et al. 2010; Kas et al. 2010). Together, in combination with relevant animal paradigms at the behavioral and neuro-anatomical level for eating disorder traits, these methods will contribute to unravel gene functions in neurobiological mechanisms underlying the pathophysiology of self-starvation. Better understanding of the pathophysiology of eating disorders will pave new roads for directed treatment development.

References

Adan RA, Tiesjema B, Hillebrand JJ, la Fleur SE, Kas MJ, de Krom M (2006) The MC4 receptor and control of appetite. Br J Pharmacol 149:815–827

Barbarich-Marsteller NC, Marsteller DA, Alexoff DL, Fowler JS, Dewey SL (2005) MicroPET imaging in an animal model of anorexia nervosa. Synapse 57:85–90

Barbarich-Marsteller N, Pike K, Underwood M, Foltin R, Walsh B (2007) Vulnerability of activity-based anorexia in adolescent female rats. Abstr Annu Meet Eat Disord Res Soc P8

Barsh GS, Farooqi IS, O'rahilly S (2000) Genetics of body-weight regulation. Nature 404:644–651

Bergen AW, Yeager M, Welch RA, Haque K, Ganjei JK, van den Bree MB, Mazzanti C, Nardi I, Fichter MM, Halmi KA, Kaplan AS, Strober M, Treasure J, Woodside DB, Bulik CM, Bacanu SA, Devlin B, Berrettini WH, Goldman D, Kaye WH (2005) Association of multiple DRD2 polymorphisms with anorexia nervosa. Neuropsychopharmacology 30:1703–1710

Bergh C, Sodersten P (1996) Anorexia nervosa, self-starvation and the reward of stress. Nat Med 2:21–22
Birrell JM, Brown VJ (2000) Medial frontal cortex mediates perceptual attentional set shifting in the rat. J Neurosci 20:4320–4324
Bissonette GB, Martins GJ, Franz TM, Harper ES, Schoenbaum G, Powell EM (2008) Double dissociation of the effects of medial and orbital prefrontal cortical lesions on attentional and affective shifts in mice. J Neurosci 28:11124–11130
Blum K, Sheridan PJ, Wood RC, Braverman ER, Chen TJ, Comings DE (1995) Dopamine D2 receptor gene variants: association and linkage studies in impulsive-addictive-compulsive behaviour. Pharmacogenetics 5:121–141
Brewerton TD, Lydiard RB, Herzog DB, Brotman AW, O'Neil PM, Ballenger JC (1995a) Comorbidity of axis I psychiatric disorders in bulimia nervosa. J Clin Psychiatry 56:77–80
Brewerton TD, Stellefson EJ, Hibbs N, Hodges EL, Cochrane CE (1995b) Comparison of eating disorder patients with and without compulsive exercising. Int J Eat Disord 17:413–416
Brigman JL, Bussey TJ, Saksida LM, Rothblat LA (2005) Discrimination of multidimensional visual stimuli by mice: intra- and extradimensional shifts. Behav Neurosci 119:839–842
Brooks SP, Betteridge H, Trueman RC, Jones L, Dunnett SB (2006) Selective extra-dimensional set shifting deficit in a knock-in mouse model of Huntington's disease. Brain Res Bull 69:452–457
Buettner R, Scholmerich J, Bollheimer LC (2007) High-fat diets: modeling the metabolic disorders of human obesity in rodents. Obesity (Silver Spring) 15:798–808
Bulik CM (2002) Eating disorders in adolescents and young adults. Child Adolesc Psychiatr Clin N Am 11:201–218
Bulik CM, Sullivan PF, Fear JL, Joyce PR (1997) Eating disorders and antecedent anxiety disorders: a controlled study. Acta Psychiatr Scand 96:101–107
Bulik CM, Hebebrand J, Keski-Rahkonen A, Klump KL, Reichborn-Kjennerud T, Mazzeo SE, Wade TD (2007a) Genetic epidemiology, endophenotypes, and eating disorder classification. Int J Eat Disord 40(Suppl):S52–S60
Bulik CM, Slof-Op't Landt MC, van Furth EF, Sullivan PF (2007b) The genetics of anorexia nervosa. Annu Rev Nutr 27:263–275
Carter FA, McIntosh VV, Joyce PR, Gendall KA, Frampton CM, Bulik CM (2004) Patterns of weight change after treatment for bulimia nervosa. Int J Eat Disord 36:12–21
Chen ZY, Jing D, Bath KG, Ieraci A, Khan T, Siao CJ, Herrera DG, Toth M, Yang C, McEwen BS, Hempstead BL, Lee FS (2006) Genetic variant BDNF (Val66Met) polymorphism alters anxiety-related behavior. Science 314:140–143
Cryan JF, Holmes A (2005) The ascent of mouse: advances in modelling human depression and anxiety. Nat Rev Drug Discov 4:775–790
Cryan JF, Mombereau C (2004) In search of a depressed mouse: utility of models for studying depression-related behavior in genetically modified mice. Mol Psychiatry 9:326–357
Cunningham CL, Howard MA, Gill SJ, Rubinstein M, Low MJ, Grandy DK (2000) Ethanol-conditioned place preference is reduced in dopamine D2 receptor-deficient mice. Pharmacol Biochem Behav 67:693–699
Davis C, Katzman DK, Kaptein S, Kirsh C, Brewer H, Kalmbach K, Olmsted MP, Woodside DB, Kaplan AS (1997) The prevalence of high-level exercise in the eating disorders: etiological implications. Compr Psychiatry 38:321–326
Davis C, Katzman DK, Kirsh C (1999) Compulsive physical activity in adolescents with anorexia nervosa: a psychobehavioral spiral of pathology. J Nerv Ment Dis 187:336–342
de Mooij-van Malsen AJ, van Lith HA, Oppelaar H, Hendriks J, de Wit M, Kostrzewa E, Breen G, Collier DA, Olivier B, Kas MJ (2009) Interspecies trait genetics reveals association of Adcy8 with mouse avoidance behavior and a human mood disorder. Biol Psychiatry 66:1123–1130
Deep AL, Nagy LM, Weltzin TE, Rao R, Kaye WH (1995) Premorbid onset of psychopathology in long-term recovered anorexia nervosa. Int J Eat Disord 17:291–297
Dennis C (2005) Psychiatric disease: all in the mind of a mouse. Nature 438:151–152
Dranovsky A, Hen R (2006) Hippocampal neurogenesis: regulation by stress and antidepressants. Biol Psychiatry 59:1136–1143

Drew MR, Simpson EH, Kellendonk C, Herzberg WG, Lipatova O, Fairhurst S, Kandel ER, Malapani C, Balsam PD (2007) Transient overexpression of striatal D2 receptors impairs operant motivation and interval timing. J Neurosci 27:7731–7739

Egan MF, Kojima M, Callicott JH, Goldberg TE, Kolachana BS, Bertolino A, Zaitsev E, Gold B, Goldman D, Dean M, Lu B, Weinberger DR (2003) The BDNF val66met polymorphism affects activity-dependent secretion of BDNF and human memory and hippocampal function. Cell 112:257–269

Elmer GI, Pieper JO, Rubinstein M, Low MJ, Grandy DK, Wise RA (2002) Failure of intravenous morphine to serve as an effective instrumental reinforcer in dopamine D2 receptor knock-out mice. J Neurosci 22:RC224

Evenden J (1999) Impulsivity: a discussion of clinical and experimental findings. J Psychopharmacol 13:180–192

Exner C, Hebebrand J, Remschmidt H, Wewetzer C, Ziegler A, Herpertz S, Schweiger U, Blum WF, Preibisch G, Heldmaier G, Klingenspor M (2000) Leptin suppresses semi-starvation induced hyperactivity in rats: implications for anorexia nervosa. Mol Psychiatry 5:476–481

Fairburn CG, Harrison PJ (2003) Eating disorders. Lancet 361:407–416

Fernandes C, Paya-Cano JL, Sluyter F, D'Souza U, Plomin R, Schalkwyk LC (2004) Hippocampal gene expression profiling across eight mouse inbred strains: towards understanding the molecular basis for behaviour. Eur J Neurosci 19:2576–2582

Fornari V, Wlodarczyk-Bisaga K, Matthews M, Sandberg D, Mandel FS, Katz JL (1999) Perception of family functioning and depressive symptomatology in individuals with anorexia nervosa or bulimia nervosa. Compr Psychiatry 40:434–441

Frank GK, Bailer UF, Henry SE, Drevets W, Meltzer CC, Price JC, Mathis CA, Wagner A, Hoge J, Ziolko S, Barbarich-Marsteller N, Weissfeld L, Kaye WH (2005) Increased dopamine D2/D3 receptor binding after recovery from anorexia nervosa measured by positron emission tomography and [11c]raclopride. Biol Psychiatry 58:908–912

Frazer KA, Eskin E, Kang HM, Bogue MA, Hinds DA, Beilharz EJ, Gupta RV, Montgomery J, Morenzoni MM, Nilsen GB, Pethiyagoda CL, Stuve LL, Johnson FM, Daly MJ, Wade CM, Cox DR (2007) A sequence-based variation map of 8.27 million SNPs in inbred mouse strains. Nature 448:1050–1053

Friederich HC, Kumari V, Uher R, Riga M, Schmidt U, Campbell IC, Herzog W, Treasure J (2006) Differential motivational responses to food and pleasurable cues in anorexia and bulimia nervosa: a startle reflex paradigm. Psychol Med 36:1327–1335

Gelegen C, Collier DA, Campbell IC, Oppelaar H, van den HJ A, RA KMJ (2007) Difference in susceptibility to activity-based anorexia in two inbred strains of mice. Eur Neuropsychopharmacol 17:199–205

Gelegen C, van den Heuvel J, Collier DA, Campbell IC, Oppelaar H, Hessel E, Kas MJ (2008) Dopaminergic and BDNF signalling in inbred mice exposed to a restricted feeding schedule. Genes Brain Behav 7:552–559

Gelegen C, Pjetri E, Campbell IC, Collier DA, Oppelaar H, Kas MJ (2010) Chromosomal mapping of excessive physical activity in mice in response to a restricted feeding schedule. Eur Neuropsychopharmacol 20(5): 317–326

Godart NT, Flament MF, Perdereau F, Jeammet P (2002) Comorbidity between eating disorders and anxiety disorders: a review. Int J Eat Disord 32:253–270

Godart NT, Flament MF, Curt F, Perdereau F, Lang F, Venisse JL, Halfon O, Bizouard P, Loas G, Corcos M, Jeammet P, Fermanian J (2003) Anxiety disorders in subjects seeking treatment for eating disorders: a DSM-IV controlled study. Psychiatry Res 117:245–258

Godart N, Berthoz S, Perdereau F, Jeammet P (2006) Comorbidity of anxiety with eating disorders and OCD. Am J Psychiatry 163:326–329

Gottesman II, Gould TD (2003) The endophenotype concept in psychiatry: etymology and strategic intentions. Am J Psychiatry 160:636–645

Grice DE, Halmi KA, Fichter MM, Strober M, Woodside DB, Treasure JT, Kaplan AS, Magistretti PJ, Goldman D, Bulik CM, Kaye WH, Berrettini WH (2002) Evidence for a susceptibility gene for anorexia nervosa on chromosome 1. Am J Hum Genet 70:787–792

Gross C, Zhuang X, Stark K, Ramboz S, Oosting R, Kirby L, Santarelli L, Beck S, Hen R (2002) Serotonin1A receptor acts during development to establish normal anxiety-like behaviour in the adult. Nature 416:396–400

Grupe A, Germer S, Usuka J, Aud D, Belknap JK, Klein RF, Ahluwalia MK, Higuchi R, Peltz G (2001) In silico mapping of complex disease-related traits in mice. Science 292:1915–1918

Hall CS (1936) Emotional behavior in the rat. III The relationship between emotionality and ambulatory activity. J Comp Psychol 22:345–352

Halmi KA, Sunday SR, Klump KL, Strober M, Leckman JF, Fichter M, Kaplan A, Woodside B, Treasure J, Berrettini WH, Al Shabboat M, Bulik CM, Kaye WH (2003) Obsessions and compulsions in anorexia nervosa subtypes. Int J Eat Disord 33:308–319

Hebebrand J, Exner C, Hebebrand K, Holtkamp C, Casper RC, Remschmidt H, Herpertz-Dahlmann B, Klingenspor M (2003) Hyperactivity in patients with anorexia nervosa and in semistarved rats: evidence for a pivotal role of hypoleptinemia. Physiol Behav 79:25–37

Herzog DB, Dorer DJ, Keel PK, Selwyn SE, Ekeblad ER, Flores AT, Greenwood DN, Burwell RA, Keller MB (1999) Recovery and relapse in anorexia and bulimia nervosa: a 7.5-year follow-up study. J Am Acad Child Adolesc Psychiatry 38:829–837

Hillebrand JJ, Koeners MP, de Rijke CE, Kas MJ, Adan RA (2005a) Leptin treatment in activity-based anorexia. Biol Psychiatry 58:165–171

Hillebrand JJ, van Elburg AA, Kas MJ, van Engeland H, Adan RA (2005b) Olanzapine reduces physical activity in rats exposed to activity-based anorexia: possible implications for treatment of anorexia nervosa? Biol Psychiatry 58:651–657

Hinney A, Remschmidt H, Hebebrand J (2000) Candidate gene polymorphisms in eating disorders. Eur J Pharmacol 410:147–159

Hoek HW (2006) Incidence, prevalence and mortality of anorexia nervosa and other eating disorders. Curr Opin Psychiatry 19:389–394

Holliday J, Tchanturia K, Landau S, Collier D, Treasure J (2005) Is impaired set-shifting an endophenotype of anorexia nervosa? Am J Psychiatry 162:2269–2275

Holtkamp K, Herpertz-Dahlmann B, Mika C, Heer M, Heussen N, Fichter M, Herpertz S, Senf W, Blum WF, Schweiger U, Warnke A, Ballauff A, Remschmidt H, Hebebrand J (2003) Elevated physical activity and low leptin levels co-occur in patients with anorexia nervosa. J Clin Endocrinol Metab 88:5169–5174

Holtkamp K, Hebebrand J, Herpertz-Dahlmann B (2004) The contribution of anxiety and food restriction on physical activity levels in acute anorexia nervosa. Int J Eat Disord 36:163–171

Hovatta I, Zapala MA, Broide RS, Schadt EE, Libiger O, Schork NJ, Lockhart DJ, Barlow C (2007) DNA variation and brain region-specific expression profiles exhibit different relationships between inbred mouse strains: implications for eQTL mapping studies. Genome Biol 8:R25

Jansen RC, Nap JP (2001) Genetical genomics: the added value from segregation. Trends Genet 17:388–391

Jentsch JD, Taylor JR (1999) Impulsivity resulting from frontostriatal dysfunction in drug abuse: implications for the control of behavior by reward-related stimuli. Psychopharmacology (Berl) 146:373–390

Joel D (2006) Current animal models of obsessive compulsive disorder: a critical review. Prog Neuropsychopharmacol Biol Psychiatry 30:374–388

Kas MJ, Van Ree JM (2004) Dissecting complex behaviours in the post-genomic era. Trends Neurosci 27:366–369

Kas MJ, van Dijk G, Scheurink AJ, Adan RA (2003a) Agouti-related protein prevents self-starvation. Mol Psychiatry 8:235–240

Kas MJ, van Elburg AA, van Engeland H, Adan RA (2003b) Refinement of behavioural traits in animals for the genetic dissection of eating disorders. Eur J Pharmacol 480:13–20

Kas MJ, Fernandes C, Schalkwyk LC, Collier DA (2007) Genetics of behavioural domains across the neuropsychiatric spectrum; of mice and men. Mol Psychiatry 12:324–330

Kas MJ, de Mooij-van Malsen JG, Olivier B, Spruijt BM, Van Ree JM (2008) Differential genetic regulation of motor activity and anxiety-related behaviors in mice using an automated home cage task. Behav Neurosci 122(4):769–76

Kas MJ, Kaye WH, Foulds MW, Bulik CM (2009) Interspecies genetics of eating disorder traits. Am J Med Genet B Neuropsychiatr Genet 150B:318–327

Kas MJ, Gelegen C, van Nieuwerburgh F, Westenberg HG, Deforce D, Denys D (2010) Compulsivity in mouse strains homologous with chromosomes 7p and 15q linked to obsessive-compulsive disorder. Am J Med Genet B Neuropsychiatr Genet 153B:252–259

Kaye WH, Bulik CM, Thornton L, Barbarich N, Masters K (2004) Comorbidity of anxiety disorders with anorexia and bulimia nervosa. Am J Psychiatry 161:2215–2221

Kaye WH, Fudge JL, Paulus M (2009) New insights into symptoms and neurocircuit function of anorexia nervosa. Nat Rev Neurosci 10:573–584

Keck P, Pope H, Hudson JL, McElroy S, Yurgelung-Todd D, Hundert E (1990) A controlled study of phenomenology and family history in outpatients with bulimia nervosa. Compr Psychiatry 31:275–283

Keski-Rahkonen A, Neale BM, Bulik CM, Pietilainen KH, Rose RJ, Kaprio J, Rissanen A (2005) Intentional weight loss in young adults: sex-specific genetic and environmental effects. Obes Res 13:745–753

Klump KL, Gobrogge KL (2005) A review and primer of molecular genetic studies of anorexia nervosa. Int J Eat Disord 37(Suppl):S43–S48

Korff S, Harvey BH (2006) Animal models of obsessive-compulsive disorder: rationale to understanding psychobiology and pharmacology. Psychiatr Clin North Am 29:371–390

Laessle RG, Beumont PJ, Butow P, Lennerts W, O'Connor M, Pirke KM, Touyz SW, Waadt S (1991) A comparison of nutritional management with stress management in the treatment of bulimia nervosa. Br J Psychiatry 159:250–261

Lang UE, Hellweg R, Kalus P, Bajbouj M, Lenzen KP, Sander T, Kunz D, Gallinat J (2005) Association of a functional BDNF polymorphism and anxiety-related personality traits. Psychopharmacology (Berl) 180:95–99

Lesch KP, Merschdorf U (2000) Impulsivity, aggression, and serotonin: a molecular psychobiological perspective. Behav Sci Law 18:581–604

Letwin NE, Kafkafi N, Benjamini Y, Mayo C, Frank BC, Luu T, Lee NH, Elmer GI (2006) Combined application of behavior genetics and microarray analysis to identify regional expression themes and gene-behavior associations. J Neurosci 26:5277–5287

Lohoff FW, Sander T, Ferraro TN, Dahl JP, Gallinat J, Berrettini WH (2005) Confirmation of association between the Val66Met polymorphism in the brain-derived neurotrophic factor (BDNF) gene and bipolar I disorder. Am J Med Genet B Neuropsychiatr Genet 139:51–53

Maldonado R, Saiardi A, Valverde O, Samad TA, Roques BP, Borrelli E (1997) Absence of opiate rewarding effects in mice lacking dopamine D2 receptors. Nature 388:586–589

Mazzeo SE, Slof-Op't Landt MC, Jones I, Mitchell K, Kendler KS, Neale MC, Aggen SH, Bulik CM (2006) Associations among postpartum depression, eating disorders, and perfectionism in a population-based sample of adult women. Int J Eat Disord 39:202–211

Mazzeo SE, Mitchell KS, Bulik CM, Reichborn-Kjennerud T, Kendler KS, Neale MC (2009) Assessing the heritability of anorexia nervosa symptoms using a marginal maximal likelihood approach. Psychol Med 39:463–473

Mercer JG, Tups A (2003) Neuropeptides and anticipatory changes in behaviour and physiology: seasonal body weight regulation in the Siberian hamster. Eur J Pharmacol 480:43–50

Moy SS, Nadler JJ, Young NB, Perez A, Holloway LP, Barbaro RP, Barbaro JR, Wilson LM, Threadgill DW, Lauder JM, Magnuson TR, Crawley JN (2007) Mouse behavioral tasks relevant to autism: phenotypes of 10 inbred strains. Behav Brain Res 176:4–20

Nadler JJ, Moy SS, Dold G, Trang D, Simmons N, Perez A, Young NB, Barbaro RP, Piven J, Magnuson TR, Crawley JN (2004) Automated apparatus for quantitation of social approach behaviors in mice. Genes Brain Behav 3:303–314

Nestler EJ, Gould E, Manji H, Buncan M, Duman RS, Greshenfeld HK, Hen R, Koester S, Lederhendler I, Meaney M, Robbins T, Winsky L, Zalcman S (2002) Preclinical models: status of basic research in depression. Biol Psychiatry 52:503–528

Neves-Pereira M, Cheung JK, Pasdar A, Zhang F, Breen G, Yates P, Sinclair M, Crombie C, Walker N, St Clair DM (2005) BDNF gene is a risk factor for schizophrenia in a Scottish population. Mol Psychiatry 10:208–212

Nisoli E, Brunani A, Borgomainerio E, Tonello C, Dioni L, Briscini L, Redaelli G, Molinari E, Cavagnini F, Carruba MO (2007) D2 dopamine receptor (DRD2) gene Taq1A polymorphism and the eating-related psychological traits in eating disorders (anorexia nervosa and bulimia) and obesity. Eat Weight Disord 12:91–96

Plomin R, McClearn GE, Gora-Maslak G, Neiderhiser JM (1991) Use of recombinant inbred strains to detect quantitative trait loci associated with behavior. Behav Genet 21:99–116

Redei EE, Ahmadiyeh N, Baum AE, Sasso DA, Slone JL, Solberg LC, Will CC, Volenec A (2001) Novel animal models of affective disorders. Semin Clin Neuropsychiatry 6:43–67

Ribases M, Gratacos M, Fernandez-Aranda F, Bellodi L, Boni C, Anderluh M, Cavallini MC, Cellini E, Di Bella D, Erzegovesi S, Foulon C, Gabrovsek M, Gorwood P, Hebebrand J, Hinney A, Holliday J, Hu X, Karwautz A, Kipman A, Komel R, Nacmias B, Remschmidt H, Ricca V, Sorbi S, Wagner G, Treasure J, Collier DA, Estivill X (2004) Association of BDNF with anorexia, bulimia and age of onset of weight loss in six European populations. Hum Mol Genet 13:1205–1212

Ribases M, Gratacos M, Fernandez-Aranda F, Bellodi L, Boni C, Anderluh M, Cristina CM, Cellini E, Di Bella D, Erzegovesi S, Foulon C, Gabrovsek M, Gorwood P, Hebebrand J, Hinney A, Holliday J, Hu X, Karwautz A, Kipman A, Komel R, Nacmias B, Remschmidt H, Ricca V, Sorbi S, Tomori M, Wagner G, Treasure J, Collier DA, Estivill X (2005) Association of BDNF with restricting anorexia nervosa and minimum body mass index: a family-based association study of eight European populations. Eur J Hum Genet 13:428–434

Rohner-Jeanrenaud E, Jeanrenaud B (1997) Central nervous system and body weight regulation. Ann Endocrinol (Paris) 58:137–142

Routtenberg A, Kuznesof AW (1967) Self-starvation of rats living in activity wheels on a restricted feeding schedule. J Comp Physiol Psychol 64:414–421

Sandberg R, Yasuda R, Pankratz DG, Carter TA, Del Rio JA, Wodicka L, Mayford M, Lockhart DJ, Barlow C (2000) Regional and strain-specific gene expression mapping in the adult mouse brain. Proc Natl Acad Sci USA 97:11038–11043

Schumacher J, Jamra RA, Becker T, Ohlraun S, Klopp N, Binder EB, Schulze TG, Deschner M, Schmal C, Hofels S, Zobel A, Illig T, Propping P, Holsboer F, Rietschel M, Nothen MM, Cichon S (2005) Evidence for a relationship between genetic variants at the brain-derived neurotrophic factor (BDNF) locus and major depression. Biol Psychiatry 58:307–314

Schwalberg MD, Barlow DH, Alger SA, Howard LJ (1992) Comparison of bulimics, obese binge eaters, social phobics, and individuals with panic disorder on comorbidity across DSM-III-R anxiety disorders. J Abnorm Psychol 101:675–681

Shugart YY, Samuels J, Willour VL, Grados MA, Greenberg BD, Knowles JA, McCracken JT, Rauch SL, Murphy DL, Wang Y, Pinto A, Fyer AJ, Piacentini J, Pauls DL, Cullen B, Page J, Rasmussen SA, Bienvenu OJ, Hoehn-Saric R, Valle D, Liang KY, Riddle MA, Nestadt G (2006) Genomewide linkage scan for obsessive-compulsive disorder: evidence for susceptibility loci on chromosomes 3q, 7p, 1q, 15q, and 6q. Mol Psychiatry 11:763–770

Siegfried Z, Berry EM, Hao S, Avraham Y (2003) Animal models in the investigation of anorexia. Physiol Behav 79:39–45

Singer JB, Hill AE, Burrage LC, Olszens KR, Song J, Justice M, O'Brien WE, Conti DV, Witte JS, Lander ES, Nadeau JH (2004) Genetic dissection of complex traits with chromosome substitution strains of mice. Science 304:445–448

Swinbourne JM, Touyz SW (2007) The co-morbidity of eating disorders and anxiety disorders: a review. Eur Eat Disord Rev 15:253–274

Tchanturia K, Morris RG, Anderluh MB, Collier DA, Nikolaou V, Treasure J (2004) Set shifting in anorexia nervosa: an examination before and after weight gain, in full recovery and relationship to childhood and adult OCPD traits. J Psychiatr Res 38:545–552

Toner BB, Garfinkel PE, Garner DM (1988) Affective and anxiety disorders in the long-term follow-up of anorexia nervosa. Int J Psychiatry Med 18:357–364

Tran AH, Tamura R, Uwano T, Kobayashi T, Katsuki M, Matsumoto G, Ono T (2002) Altered accumbens neural response to prediction of reward associated with place in dopamine D2 receptor knockout mice. Proc Natl Acad Sci USA 99:8986–8991

Treasure J, Claudino AM, Zucker N (2010) Eating disorders. Lancet 375(9714):583–593

van Elburg AA, Kas MJ, Hillebrand JJ, Eijkemans RJ, van Engeland H (2007) The impact of hyperactivity and leptin on recovery from anorexia nervosa. J Neural Transm 114:1233–1237

van Kuyck K, Casteels C, Vermaelen P, Bormans G, Nuttin B, Van Laere K (2007) Motor- and food-related metabolic cerebral changes in the activity-based rat model for anorexia nervosa: a voxel-based microPET study. Neuroimage 35:214–221

Wade CM, Daly MJ (2005) Genetic variation in laboratory mice. Nat Genet 37:1175–1180

Wade CM, Kulbokas EJ III, Kirby AW, Zody MC, Mullikin JC, Lander ES, Lindblad-Toh K, Daly MJ (2002) The mosaic structure of variation in the laboratory mouse genome. Nature 420:574–578

Wagner A, Aizenstein H, Venkatraman VK, Fudge J, May JC, Mazurkewicz L, Frank GK, Bailer UF, Fischer L, Nguyen V, Carter C, Putnam K, Kaye WH (2007) Altered reward processing in women recovered from anorexia nervosa. Am J Psychiatry 164:1842–1849

Yalcin B, Willis-Owen SA, Fullerton J, Meesaq A, Deacon RM, Rawlins JN, Copley RR, Morris AP, Flint J, Mott R (2004) Genetic dissection of a behavioral quantitative trait locus shows that Rgs2 modulates anxiety in mice. Nat Genet 36:1197–1202

Yang H, Bell TA, Churchill GA, Pardo-Manuel dV (2007) On the subspecific origin of the laboratory mouse. Nat Genet 39:1100–1107

Neurobiology Driving Hyperactivity in Activity-Based Anorexia

R.A.H. Adan, J.J.G. Hillebrand, U.N. Danner, S. Cardona Cano, M.J.H. Kas, and L.A.W. Verhagen

Contents

Abstract Hyperactivity in anorexia nervosa is difficult to control and negatively impacts outcome. Hyperactivity is a key driving force to starvation in an animal model named activity-based anorexia (ABA). Recent research has started unraveling what mechanisms underlie this hyperactivity. Besides a general increase in locomotor activity that may be an expression of foraging behavior and involves frontal brain regions, the increased locomotor activity expressed before food is presented (food anticipatory behavior or FAA) involves hypothalamic neural circuits. Ghrelin plays a role in FAA, whereas decreased leptin signaling is involved

S. Cardona Cano, M.J.H. Kas, and L.A.W. Verhagen
Department of Neuroscience and Pharmacology, Rudolf Magnus Institute of Neuroscience, University Medical Centre, Str. 5.203, P.O.B. 85060, 3508 AB Utrecht, The Netherlands

R.A.H. Adan (✉) and J.J.G. Hillebrand
Department of Neuroscience and Pharmacology, Rudolf Magnus Institute of Neuroscience, University Medical Centre, Str. 5.203, P.O.B. 85060 3508 AB Utrecht, The Netherlands
Altrecht Eating Disorders Rintveld, Altrecht Mental Health Institute, Zeist, The Netherlands
e-mail: r.a.h.adan@umcutrecht.nl

U.N. Danner
Altrecht Eating Disorders Rintveld, Altrecht Mental Health Institute, Zeist, The Netherlands
Clinical and Health Psychology, Utrecht University, Utrecht, The Netherlands

R.A.H. Adan and W.H. Kaye (eds.), *Behavioral Neurobiology of Eating Disorders*, Current Topics in Behavioral Neurosciences 6, DOI 10.1007/7854_2010_77

in both aspects of increased locomotor activity. We hypothesize that increased ghrelin and decreased leptin signaling drive the activity of dopamine neurons in the ventral tegmental area. In anorexia nervosa patients, this altered activity of the dopamine system may be involved not only in hyperactivity but also in aberrant cognitive processing related to food.

Keywords Activity-based anorexia · Dopamine · Ghrelin · Hyperactivity · Leptin

1 Introduction

The etiology of anorexia nervosa (AN) is poorly understood. As with many other psychiatric disorders, a role for monoamine systems in the etiology of AN has been proposed (see elsewhere is this book), but how it is involved is currently unknown.

In the last decade, genetics has played an important role in unraveling the pathways underlying disorders. In contrast to most other methods used to unravel the neurobiological pathways underlying disorders, genetics (linkage and whole genome association studies) provides an unbiased approach with no a priori hypothesis. These studies have led to the discovery of a gene that is causally related to the disorder; however, these studies do not resolve how that gene is causing the disorder. In order to understand the relationship between the gene and the disorder, two major (complementary) approaches can be taken. One is to compare phenotypes of human subjects carrying certain (disorder-associated) alleles carefully with those of subjects not carrying such alleles. Heterogeneity not related to the gene of interest and environmental differences complicate this kind of analysis. Therefore, the other approach is to investigate the role of the gene of interest in animals, preferably in a validated disease model. Here, we first discuss some of these models before describing how these were used in our studies toward unraveling the neurobiology underlying AN. As soon as new genes are discovered that are causally related to AN, these models will be helpful to unravel how these genes are involved in increasing the susceptibility for AN.

2 Animal Models for Anorexia Nervosa

The anorexia mouse (anx/anx) is a spontaneous mouse mutant characterized by poor appetite and hyperactivity. Anx/anx mice die at the age of 3–5 weeks depending on the genetic background. Several studies have revealed alterations in the dopaminergic, serotonergic, and noradrenergic systems. In the hypothalamus, neuropeptide Y (NPY) and agouti-related protein (AgRP) show aberrant expression patterns, and pro-opiomelanocortin (POMC) expression is decreased. There is evidence for neurodegenerative changes in the mediobasal hypothalamus (Johansen et al. 2007). The gene causing the anx/anx phenotype has not been identified yet, and therefore its relevance for AN is unresolved.

Very similar to anx/anx mice are contactin-deficient mice. These mice show weight loss starting at postnatal day 10 and they die before 3 weeks of postnatal life in a malnourished state. Contactin is a neural cell adhesion/recognition molecule involved in formation of specific axon projections in the developing nervous system (Fetissov et al. 2005), but whether contactin is relevant in the etiology of AN is unknown. There have been several other mouse mutants generated displaying an anorectic phenotype. Dopamine-deficient mice, generated by deleting the tyrosine hydroxylase gene (but reexpressing it in noradrenaline neurons), have an emaciated appearance and are hypoactive. Administration of L-dopa or restoration of dopamine signaling in the dorsal striatum of these mice restores feeding (Szczypka et al. 2001). These data may underscore the importance of dopamine for motivated behaviors in general and do not prove that dopamine is specific for feeding behavior or that a deficiency in dopamine is causal to developing an eating disorder. Deficiency of either (prepro) melanin-concentrating hormone (MCH) or the MCH receptor 1 also results in a lean and hyperactive phenotype in mice (Shimada et al. 1998; Marsh et al. 2002; Zhou et al. 2005). Except for dopamine-deficient mice [in the dopamine system, genetic variation has been associated with AN (Bergen 2005)], none of the pathways in these genetic models with anorectic phenotypes have been thoroughly investigated for their involvement in AN.

Besides in these genetically altered mice, anorectic behaviors can be induced by exposing wild-type animals to restricted feeding schedules. The most widely used rodent model mimicking features of human anorexia is the activity-based anorexia (ABA) model. Here, we describe results obtained from this model.

2.1 The Activity-Based Anorexia Model

In the ABA model (also referred to as semistarvation-induced hyperactivity, activity stress, or activity anorexia) (Epling and Pierce 1983), animals have voluntary access to a running wheel in their home cage. When first exposed to running wheels, running wheel activity (RWA) increases in the first 2 weeks after which it stabilizes. Animals will increase their ad libitum food intake to compensate for their increased energy expenditure. After RWA has been stabilized, food access is restricted and given only 1–2 h per day (at fixed times) for 1 week. Upon exposure to the scheduled feeding, nocturnal animals like rats and mice will increase their RWA in the dark phase and will also start to run in the light phase. The combination of reduced food intake and increased RWA will lead to substantial body weight loss (>20%) within a week (Fig. 1).

The development of excessive hyperactivity in the ABA model resembles hyperactivity as seen in patients with AN, which is difficult to control and has a negative effect on outcome (American Psychiatric Association 1994). Hyperactivity occurs in up to 80% of patients with AN (Hebebrand et al. 2003). In these patients, it

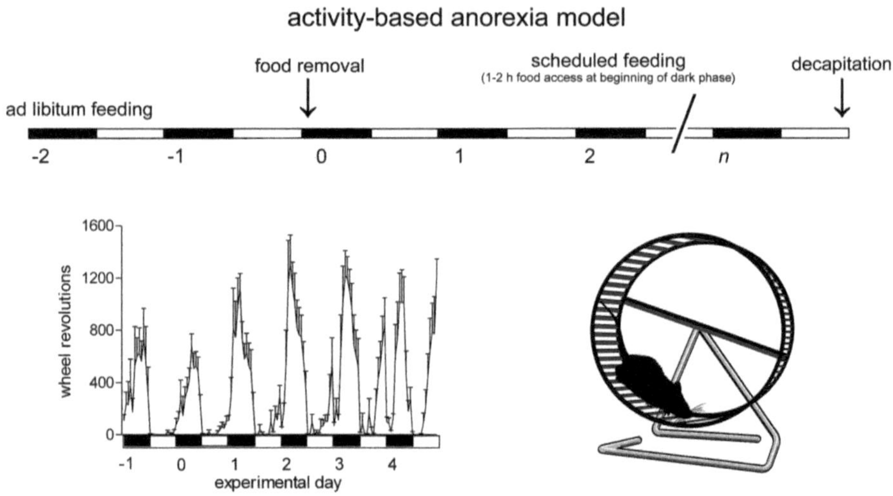

Fig. 1 The activity-based anorexia model rodents are housed in cages with running wheels. Once these animals get food once a day at a fixed time point for 1–2 h, they will increase their locomotor activity during the dark phase, but they also start to run in the light phase in the hours preceding food delivery. This latter activity is referred to as food anticipatory activity

has been regarded as a conscious attempt to lose body weight but may also be explained by a subconscious biological drive (Davis 1997). Many species display increased locomotor activity and travel large distances in periods of food scarcity to find new food sources. If at least in some animals of a group this foraging behavior is rewarding in itself, then the chance that it will occur increases. As a result, the chance of survival of that species increases as well. This may be a biological explanation of why hyperactivity may be rewarding by at least some individuals with AN.

Increased RWA upon exposure to the ABA model is displayed throughout the dark phase, and in the light phase in particular preceding food access. This latter activity is also referred to as food anticipatory activity (FAA) (Mistlberger 1994). Although in most ABA studies food is given at dark onset, the time of food access does not influence the presence of FAA. Thus, FAA will precede food access at any fixed time point of the day. The neural substrate that underlies FAA may be the food-entrainable oscillator (FEO), which has been investigated using a similar experimental setup but with milder food restriction schedules than ABA. The normal circadian rhythm, which is necessary to adapt physiological processes to time of day, can be entrained by food. Whereas noctural species are usually active in the dark phase, upon exposure to restricted feeding paradigms, locomotor activity can be shifted toward (entrained) the moment of food delivery. There are numerous reviews on the location of the FEO in the brain, many of which put forward the dorsomedial hypothalamus (DMH) as an important node, whereas others believe that the FEO is distributed over a neural network (Mistlberger et al. 2009; Gooley et al. 2006). The FEO probably contributes to FAA in the ABA model and acts to prepare the animals for the upcoming meal. In the ABA model, RWA is also increased in the dark phase (without upcoming food access). This latter increase in RWA may reflect foraging behavior, which is a goal-driven behavior that generally occurs upon food restriction. This also accounts when food is given at unpredictable times or when animals get a fixed, limited amount of food per day. Thus, the neurobiological mechanism that drives hyperactivity in the ABA model consists of at least two components: one food entrained (getting ready for eating involving the FEO) and one driving foraging behavior (searching for food).

In our attempt to unravel the brain areas involved in FAA in the ABA model, we compared wheel running in ABA rats (with 1 h food access at dark onset) with wheel running in rats that had 1 h access to food at a random time of day. The latter group of rats was not able to anticipate food access. At the end of the experiment, the random-fed rats had lower body weights because their total RWA was higher and since they could not anticipate a meal, they were not prepared to eat a lot in a short time period and therefore ate less. However, ABA rats had higher c-Fos expression levels in several hypothalamic nuclei [DMH, arcuate nucleus (Arc), lateral hypothalamus (LH)] as compared to the random-fed rats. Surprisingly, the neuronal activity in the DMH of ABA rats (measured by c-Fos expression levels) was correlated with FAA, but this was not observed in rats on the random feeding schedule. The hypothalamus thus seems to play a role in organizing the anticipatory response to upcoming food access. The specific hypothalamic neuronal populations and coupled pathways involved in food anticipation still need to be identified.

The hypothalamus, in particular the Arc, strongly responds to peripheral cues informing the brain on the status of energy balance. The Arc is in close proximity to the peripheral blood stream and senses glucose and other metabolic fuels such as free fatty acids, but also hormones that are released from the gastrointestinal system, the pancreas and adipose tissue. A recent meta-analysis on gut hormones in AN revealed increased basal plasma levels of ghrelin, PYY and CCK (Prince et al. 2009). All three hormones act in the mediobasal hypothalamus, which is also a major target site for the adipose tissue-derived hormone leptin. Plasma leptin levels are often extremely low, whereas plasma ghrelin levels are often increased in patients with AN, reflecting their low adipose tissue mass and hungry state. Since both plasma leptin and ghrelin have previously been implicated in regulating locomotor activity (Holtkamp et al. 2006; Jerlhag et al. 2007), these two hormones and their putative roles in the two components of hyperlocomotion during starvation are discussed below in more detail.

3 Leptin and Ghrelin Physiology

Leptin conveys information on adiposity, or rather starvation to the brain, whereas ghrelin is thought to be the peripheral hunger signal to the brain. Leptin injections decrease food intake by reducing meal size, decrease body weight and adipose tissue mass, and increase energy expenditure in rodents (Campfield et al. 1995;

Pelleymounter et al. 1995; Grill et al. 2002). The absence of leptin or leptin receptor (lepRb) leads to hyperphagia, obesity, hypoactivity, and neuroendocrine and metabolic malfunction (Campfield et al. 1995; Pelleymounter et al. 1995; Halaas et al. 1995; Ahima et al. 1996). During starvation, plasma leptin levels fall rapidly (even faster than adipose tissue mass), and adaptive responses, e.g., reduction of energy expenditure, suppression of the gonadal and thyroid axis, and activation of the adrenal axis, are observed. These effects of starvation are blunted by leptin treatment in rodents and humans (Ahima et al. 1996; Rosenbaum et al. 2002).

Ghrelin has emerged as an important gut-brain signal in the control of energy balance (Hosoda et al. 2002; Kojima et al. 1999) and is the only orexigenic gut hormone known so far. In contrast to many endocrine signals, plasma ghrelin levels are elevated prior to meal initiation, and they decrease again during the postprandial period (Cummings et al. 2001). Caloric restriction increases ghrelin secretion and subsequent activation of the central ghrelin signaling system via the ghrelin receptor (growth hormone secretagogue receptor 1A, GHS-R1A) in the hypothalamus [e.g., Arc, DMH, ventromedial hypothalamus (VMH) and paraventricular nucleus (PVN)] (Zigman et al. 2006; Mondal et al. 2005) increases food intake (Hosoda et al. 2002). Acute central or peripheral ghrelin injections stimulate food intake in rats (Horvath et al. 2001; Naleid et al. 2005) by increasing meal frequency but not meal size and promote fat storage leading to an increased body weight (Faulconbridge et al. 2003). Peripheral ghrelin injection also induces appetite in healthy subjects (Wren et al. 2001) and lead to similar effects on body weight and adiposity in humans (Druce et al. 2006; Wren et al. 2001). Collectively, these data are indicative of a physiological role for ghrelin in hunger and meal initiation.

Not only does plasma ghrelin levels fluctuate in relation to meal initiation (when levels are high) and meal termination (when levels are low), recent data also support the involvement of ghrelin in anticipation to meals. For instance, rhythmic ghrelin release from stomach oxyntic cells is synchronized to the time of food delivery, and GHS-R receptor knock-out mice show less FAA compared to wild-type mice when food access is restricted to 4–6 h in the light phase (LeSauter et al. 2009; Blum et al. 2009). GHS-R-deficient mice also show less c-Fos expression in the Arc, DMH, PVN, and LH following restricted feeding (Blum et al. 2009). It remains to be resolved via which neural circuit ghrelin affects FAA. Ribeiro et al. (2007) showed that the first ghrelin-responsive brain region to show c-Fos expression upon scheduled feeding is the VMH. Recent data also identified GHS-R on dopaminergic neurons of the ventral tegmental area (VTA) in the mesolimbic midbrain (Zigman et al. 2006; Guan et al. 1997; Howard et al. 1996), which have been implicated in reward-seeking behavior. Several studies have indeed demonstrated the ability of central and peripheral injected ghrelin to influence motivation and reward (Cummings et al. 2001). Moreover, Jerlhag and colleagues have demonstrated that ghrelin injection into the VTA increased dopamine release in the nucleus accumbens (NAc) and increased locomotor activity (Jerlhag et al. 2007). Direct effects of ghrelin on the electrical activity of VTA dopaminergic neurons have also been reported (Abizaid et al. 2006). Taken together, these data indicate that ghrelin may act upon the neural circuits constituting the FEO as well as at the level of the VTA to modulate the activity of dopaminergic neurons. We hypothesize that high plasma ghrelin levels in patients with AN contribute to trigger their hyperactivity.

Circulating leptin and ghrelin enter the brain at the level of the hypothalamus which has long been known as a central regulator of feeding behavior. In the Arc, two populations of neurons exist that seem to play antagonistic roles in control of feeding behavior and energy expenditure: (1) neurons expressing agouti-related protein (AgRP) and neuropeptide Y (NPY) that are both orexigenic neuropeptides and (2) neurons expressing pro-opiomelanocortin (POMC) that encodes the anorexigenic α-melanocyte-stimulating hormone (α-MSH) and cocaine- and amphetamine-regulated transcript (CART). Both types of neurons express lepRb (Munzberg and Myers 2005), whereas GHS-R are expressed only in AgRP/NPY neurons (Willesen et al. 1999). Binding of leptin to lepRb inhibits NPY/AgRP neuronal activity (Schwartz et al. 2000), whereas activation of GHS-R by ghrelin stimulates these neurons (Cowley et al. 2003). Binding of leptin to lepRb on POMC neurons directly activates POMC neurons, whereas ghrelin inhibits POMC neuronal activity indirectly via activation of GABA-ergic AgRP neurons that project to POMC neurons (Riediger et al. 2003). LepRb is also expressed in other hypothalamic nuclei, like the DMH and LH as well as in the cortex, hippocampus, midbrain, and caudal brainstem (Grill et al. 2002; Elmquist et al. 1998).

3.1 Leptin and Ghrelin in Patients with AN

Plasma and cerebrospinal fluid (csf) leptin levels are low in patients with AN and reflect reduced body weight and reduced subcutaneous fat (Hebebrand et al. 1997; Mantzoros et al. 1997; Mayo-Smith et al. 1989). Rosenbaum et al. (1997) described that body weight gain in healthy subjects leads to increased circulating leptin levels and variable leptin secretion, depending on the rate and amount of body weight gain. Plasma leptin levels were high relative to adipose tissue mass following weight gain, and low relative to adipose tissue mass following weight loss in healthy subjects (Rosenbaum et al. 1997). When the patient with AN reaches target body weight, plasma leptin levels may thus be disproportionally high (upon adjustment for BMI and adipose tissue mass) compared to healthy controls. This relative hyperleptinemia may imply a risk for renewed body weight loss since it may reduce caloric intake and increase energy expenditure resulting in poor outcome (Hebebrand et al. 2007).

Subjective measurements of motor restlessness in patients with AN axe highest during admission, when leptin levels are lowest (Exner et al. 2000). Moreover, it has been shown that hyperactivity of patients with AN (as determined by the structured inventory for anorexia and bulimia, SIAB) is negatively correlated with (lg10) leptin levels, with leptin (but not BMI) explaining 37% of the variation in hyperactivity (Holtkamp et al. 2003). Leptin levels of (adolescent) patients with AN at admission have also been found to be negatively correlated with their hyperactivity measurements using accelerometers, and with their subjective self-reports of inner restlessness (by VAS) or motor restlessness (by a 5-point Likert scale). Leptin levels have been found to predict all three hyperactivity measurements (Holtkamp et al. 2006). We found that expert ratings of hyperactivity (using a VAS examining motor restlessness and excessive exercise), which were validated with accelerometer scores, are more legitimate than patients' self-reports of hyperactivity (Van Elburg et al. 2007a). At admission, leptin levels of adolescent patients with AN axe negatively correlated to expert ratings of hyperactivity, while during treatment this relation develops into a positive one in recovering patients with AN (Van Elburg et al. 2007b).

In healthy adolescent girls, plasma ghrelin levels decrease during late pubertal stages (Whatmore et al. 2003). However, in acute patients with AN plasma, ghrelin levels are significantly elevated. The increase in plasma ghrelin levels normalizes after (partial) body weight regain (Miljic et al. 2006; Misra et al. 2004; Nedvidkova et al. 2003; Otto et al. 2001; Soriano-Guillen et al. 2004; Tolle et al. 2003; Troisi et al. 2005). One could speculate that the relative fast normalization or even further decrease in ghrelin levels contributes to inhibition of further weight gain or even to weight loss in patients with AN, since ghrelin is thought to provide orexigenic drive.

The typical postprandial decrease in plasma ghrelin levels following eating (Cummings et al. 2001; Shiiya et al. 2002) seems to remain intact in patients with AN (Misra et al. 2004; Otto et al. 2005).

Only two studies to date have measured the effect of ghrelin injection on eating and hunger in patients with AN. The group of Hotta et al. (2009) performed a pilot study on five patients with acute AN (for which conventional refeeding therapy led to weight loss) by bi-daily injecting 3 μg/kg ghrelin intravenously (iv). This resulted in increased hunger as measured by visual analog scales (VAS) and a mean increase in eating of 20%. Food intake remained higher than pretreatment once ghrelin infusion stopped; however, body weight varied from −1.5 to +2.4 kg during treatment. Miljic et al. (2006) injected nine patients with acute AN and six partially recovered patients with 1.5 ng/kg (5 pg/kg per min for 300 min) without access to food. The VAS hunger score of patients at baseline was reduced compared to controls. During ghrelin treatment, patients showed an increase in hunger according to VAS ratings, but this increase in hunger was lower than that in healthy controls. In a third study by Broglio et al. (2004), hunger was not measured but mentioned as a side effect after injection of 1 μg/kg ghrelin in six out of nine patients with AN. Unfortunately, in none of these studies, hyperactivity was measured.

4 Leptin and Ghrelin in Activity-Based Anorexia

During ABA, RWA significantly increases while plasma leptin levels drop (Kas et al. 2003; De Rijke et al. 2005; Exner et al. 2000). This observation led us and others to hypothesize that the typical increase in RWA would be prevented by increasing circulating leptin levels in ABA rats. Indeed, leptin treatment in the

ABA model suppresses hyperactivity (Hillebrand et al. 2005a; Exner et al. 2000). We showed that of hypothermia during the first 4 days of the ABA experiment. Unfortunately, reduced food intake and increased thermogenesis by treatment with leptin resulted in a rapidly worsening condition. RWA in rats fed ad libitum and locomotor activity levels in food-restricted controls were not affected by leptin treatment (Hillebrand et al. 2005a), showing that effects of leptin are dependent on the status of energy balance, with an effect of leptin on reduction of locomotor activity only during weight loss.

Using a slightly different design, it was shown before that leptin treatment (31 μg leptin/day subcutaneously via osmotic minipumps) also prevents hyperactivity in male ABA rats (body weight 230 g), both during the dark and light phase (Exner et al. 2000). Leptin-treated ABA rats in this experiment got 60% of baseline food intake/day and was not different from vehicle-treated controls (i.e., food access was not limited by time). Interestingly, the authors also showed that leptin treatment could rescue RWA when ABA had already developed.

Acute (instead of chronic) leptin treatment also reduces RWA in the ABA model. This effect is, however, only observed after a few days of exposure to the ABA model, when weight loss has already set in and the ABA rats seem to become more sensitive to exogenous leptin. The drop in endogenous leptin levels during the course of the ABA model may contribute to the higher sensitivity to exogenous leptin.

We recently found that a low dose of leptin injected directly into the VTA reduced RWA of ABA rats, suggesting that the VTA is a likely candidate region where reduced leptin signaling contributes to hyperactivity. Because leptin receptors are expressed on dopaminergic neurons and leptin has been shown to reduce the firing rate of dopamine neurons, our finding supports the hypothesis that reduced leptin signaling at the level of the VTA increases dopaminergic activity.

Other rodent models of impaired leptin signaling also show deficits in locomotor activity. For instance, mice lacking leptin or leptin receptors are hypoactive. This hypoactive behavior can be reversed by leptin treatment (Pelleymounter et al. 1995) or reexpressing leptin receptor in the Arc of leptin receptor-lacking mice (Coppari et al. 2005). Whereas we and others showed that leptin treatment (chronic 4 μg leptin in brain) in ad libitum fed rats does not influence RWA or locomotor activity (Hillebrand et al. 2005a; Surwit et al. 2000), it was also recently shown that leptin treatment increases locomotor activity in rats and that antagonizing leptin in the third ventricle reduces RWA (Choi et al. 2008). These findings seem to be in contrast with our and others findings of attenuation of hyperactivity by leptin treatment in ABA rats, and suggest that during ABA, leptin treatment affects activity levels oppositely or by different downstream targets. One explanation of these seemingly opposing results is that leptin's effects on locomotor activity depend on the status of energy balance.

In situations of a negative energy balance, like during exposure to the ABA model, plasma ghrelin levels are increased. The high plasma ghrelin levels and hyperactivity caused by food restriction in the ABA model led us to investigate whether changes in plasma ghrelin levels were associated with the development of hyperactivity in the ABA model. Indeed, while plasma leptin levels decline, plasma ghrelin levels increase dramatically over the course of the ABA model. Interestingly, the increased plasma ghrelin levels were positively correlated with FAA but not with total RWA in female ABA rats (Verhagen et al., 2010). It was also shown that FAA can be reduced by suppressing ghrelin signaling. For example, we and others showed that ghrelin receptor knock-out mice show reduced FAA. In addition, mice chronically treated with a specific ghrelin receptor do not show FAA. Furthermore, when ABA rats were given an acute central injection of ghrelin receptor antagonist just prior to the development of increased FAA, FAA was suppressed as well. In the above described models of suppressed ghrelin signaling, food intake remained affected. Thus, there is strong evidence that increased ghrelin signaling contributes to the FAA component of hyperactivity and possibly also to the general increased hyperactivity. Others showed before that ghrelin injections in Siberian hamsters peripherally (Keen-Rhinehart and Bartness 2005) increased foraging and hoarding behavior. Jerlhag et al. (2006, 2007) injected 1 μg ghrelin ilvt and directly into the VTA of mice, and these injections caused hyperactivity, most pronounced immediately after injection (0–5 min). Although increased locomotor activity following central or peripheral ghrelin treatment was not observed by others (Wellman et al. 2008; Tang-Christensen et al. 2004), these data support a role for the VTA in ghrelins effect on locomotor activity. Thus, both low leptin levels and high ghrelin levels might contribute to the hyperactivity in ABA rats and AN, and both may do so via acting on VTA dopaminergic neurons (Fig. 2).

4.1 Downstream Effector Mechanisms of Leptin and Ghrelin

As described above, peripheral leptin and ghrelin signal to (amongst other) the Arc where they influence the activity of AgRP/NPY and POMC/CART expressing neurons. In order to gain more insight in the hypothalamic response upon exposure to the ABA model and in the mechanism underlying the attenuation of hyperactivity by leptin treatment or antagonizing the ghrelin system in ABA rats, we analyzed expression levels of these neuropeptides in the Arc. AgRP and NPY mRNA levels are increased and CART mRNA levels are decreased in ABA rats compared to ad libitum fed or food-restricted controls. This suggests increased orexigenic signaling (De Rijke et al. 2005), but these changes fail to stimulate food intake since ABA rats eat less than ad libitum fed and food-restricted sedentary controls (Kas et al. 2003). We found a transient up-regulation of POMC expression during the first few days of ABA, followed by a significant down-regulation of POMC, suggesting increased activity of the melanocortin (MC) system during the first days of ABA (Kas et al. 2003; Hillebrand et al. 2006). Chronic leptin treatment reduces the up-regulation of NPY and AgRP and increases POMC mRNA levels versus vehicle-treated ABA rats and thus leads to a net result of reduced orexigenic signaling (Hillebrand et al. 2005a, b). MC binding sites in the VMH of ABA rats have also been found to be increased compared to controls (Kas et al. 2003) suggesting increased sensitivity to anorexigenic α-MSH during ABA, which can be reduced

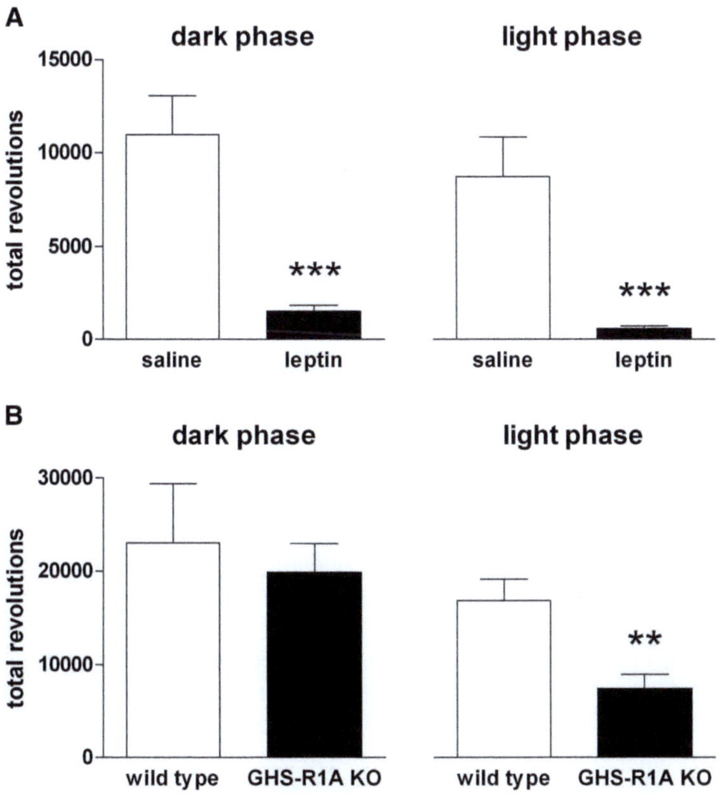

Fig. 2 Involvement of leptin and ghrelin in running wheel activity (**a**) in rats exposed to the ABA model, treatment of leptin results in decreased running wheel activity during the dark phase as well as during the light phase (**b**) in mice lacking the ghrelin receptor (GHS-R1A KO) exposed to the ABA model, running wheel activity in the dark phase is unaffected. Running wheel in the light phase (food anticipatory activity) is reduced

by binding the endogenous inverse agonist AgRP (Nijenhuis et al. 2001). Interestingly, high ambient temperatures during exposure of rats to the ABA model reduce locomotor activity and body weight loss and also decrease MC4-R expression, suggesting that heat treatment decreases the sensitivity to melanocortins (Gutierrez et al. 2009). Thus, maladaptive changes in the hypothalamic melanocortin system may contribute to the development of anorectic behaviors.

We also investigated the involvement of neural circuits downstream of ghrelin and leptin using a pharmacological approach.

4.1.1 Melanocortins and NPY

Stimulation of melanocortin receptor activity by infusion of α-MSH into the brain enhances ABA by reducing food intake, resulting in increased body weight loss.

Rats treated with α-MSH show slightly increased FAA (Hillebrand et al. 2005b). Administration of the inverse agonist AgRP(83–132) in the brain increased food intake of ABA rats, reduced hypothermia, and reduced locomotor activity, thereby attenuating ABA (Kas et al. 2003; Hillebrand et al. 2006). Administration of the competitive melanocortin receptor antagonist SHU9119 does not influence the development of ABA, suggesting that spontaneous activity of MC receptors contributes to the development of ABA. The fact that melanocortin receptor density increases upon exposure to the ABA model supports this. Another piece of evidence supporting the role of melancortin receptors in anorexia is that we found that the Ala67Thr allele of the AgRP gene occurs more frequently in AN patients than in controls (Vink et al. 2001). Thus, altered activity in AgRP and subsequently melanocortin receptor signaling may underlie anorectic behaviors in AN patients.

The increased expression of NPY upon exposure to the ABA model fails to increase food intake during exposure to food in the ABA model. However, it may contribute to increased locomotor activity as an expression of food-seeking behavior. Centrally injected NPY increases not only food intake but also food-seeking behavior in rodents, and recently it was shown that blocking the Y1 receptor by Y1 antagonist 1229U91 reverses the stimulatory effect of food deprivation on wheel running for food in hamsters (Day et al. 2005; Ammar et al. 2000; Keen-Rhinehart and Bartness 2007).

Sodersten and colleagues showed that daily injections of NPY prior to dark onset facilitated ABA in 2 h fed female rats (Nergardh et al. 2007). It was shown that NPY-treated ABA rats reduced food intake, lost more body weight, and were more hyperactive than control-treated ABA rats. In our lab, we discovered that female ABA rats with continuous infusion of NPY showed no differences in 1-h food intake, RWA during dark or light phase, or body weight loss (Hillebrand et al. 2008). The different results might be explained by differences in the restricted feeding paradigm and/or differences in administration of NPY. Daily injections rather than chronic infusion of NPY prior to food access might better reflect the physiological fluctuations due to scheduled feeding and therefore increase RWA and may reflect food-seeking behavior.

4.1.2 Opioids and Dopamine

If voluntary wheel running is rewarding for ABA rats, this may be reflected in altered activity in the opioid and dopamine systems. Hypothalamic levels of endogenous opioid β-endorphin are increased in ABA rats that lost 25% body weight (Aravich et al. 1993). This peptide, a product of the POMC gene, seems to be mainly involved in incentive motivation to acquire food reinforcers (Hayward et al. 2002). Antagonism of the opioid system with naloxone blocks RWA in rats fed ad libitum (Sisti and Lewis 2001), and μ-opioid receptor-deficient mice (β-endorphin is an agonist for this receptor) display attenuated FAA during restricted feeding (Kas et al. 2004). We attempted to attenuate the development of hyperactivity in ABA rats by chronic treatment with opioid antagonist naltrexone

(NTX). Chronic peripheral NTX treatment (0, 0.3, or 1.0 mg/kg/day) did, however, not influence RWA, food intake, or body weight loss in female ABA rats on a 1-h feeding schedule (Hillebrand et al. 2008).

Not only wheel running but also restricted feeding influences reward. For example, food restriction induces sensitization to brain stimulation reward, and this is reversed by leptin. Leptin also attenuates drug-seeking behavior after restricted feeding (Fulton et al. 2000; Shalev et al. 2001). These findings suggest the existence of a pathway between peripheral leptin that signals adiposity and the midbrain dopamine system, involved in motivational and rewarding aspects of drugs of abuse and exploration (Wise 2002). The attenuation of RWA in leptin-treated ABA rats might be explained by a direct effect of leptin on the midbrain dopamine system. As mentioned above, leptin receptors are present on dopaminergic VTA neurons projecting to the NAc (Figlewicz et al. 2003; Hommel et al. 2006), and leptin infusion into the VTA reduces firing of dopaminergic neurons, leading to decreased dopamine levels in the NAc and reduced food intake without affecting locomotor activity (Hommel et al. 2006; Fulton et al. 2006).

Long-term genetic knockdown of leptin receptor in the VTA using an adeno-associated virus containing short-hairpin leptin receptor RNA (AAV-shleptin receptor) increases food intake without affecting body weight. Furthermore, knock-down rats are more active, especially in the dark phase, compared to rats injected with the control virus and are also more sensitive to highly palatable food (Hommel et al. 2006). This suggests that decreased leptin receptor signaling increases locomotor activity followed by increased food intake (or vice versa) and, likewise, that increased leptin signaling in the VTA might underlie the attenuation of hyperactivity by central or peripheral leptin treatment in ABA rats. In fact, we showed that acute leptin injection (1 μg) bilaterally in the VTA of ABA rats suppresses RWA without affecting food intake (Verhagen et al., submitted). These studies also indicate that increased activity of the VTA dopamine system develops upon exposure to the ABA model, and future studies should therefore be aimed at direct interference with leptin signaling in the VTA in ABA rats.

Dopamine turnover in the medial basal hypothalamus is reduced during restricted feeding but seems increased or normalized in ABA rats (at 30% body weight loss) (Pirke et al. 1993). The attenuation of hyperactivity by leptin treatment compares with the effects of pimozide treatment in ABA rats, a drug with strong dopamine receptor 2 (D2) affinity (Lambert and Porter 1992). Patients with acute AN have reduced csf levels of dopamine metabolite homovanillic acid, which persist after recovery (Kaye et al. 1999). In addition, recovered patients with AN have increased D2/D3 receptor binding in the anteroventral striatum (including NAc), which may imply altered reward processing and may affect hyperactivity and eating behavior in patients with AN (Frank et al. 2005).

Blockade of dopamine signaling in the ABA model suppresses activity and increases food intake, leading to reduced body weight loss. Treatment of ABA rats with the nonselective dopaminergic antagonist *cis*-flupenthixol (0.1 mg/day subcutaneously via osmotic minipumps) resulted in higher plasma leptin levels indicative for a stimulation of the hypothalamic anorexigenic drive which is in contrast to the observed increased food intake (Verhagen et al. 2009a). This implies that antagonism of dopaminergic receptors stimulates food intake downstream of the hypothalamic circuits that respond to peripheral satiety signals.

Treatment with the atypical antipsychotic olanzapine (antagonizing both dopamine and serotonin systems), the nonselective dopamine antagonist *cis*-flupenthixol, or leptin suppresses overall hyperactive behavior in the ABA model (Verhagen et al. 2009a). Analyzing the patterns of RWA in the ABA model in more detail indicates that of these three drugs, leptin is able to reduce FAA most effectively, suggesting that changes in leptin rather than dopamine trigger FAA. Interestingly, basal dopamine release in the NAcc gradually decreased upon exposure to the ABA model, but the moment food was delivered, there was a clear peak in dopamine release (Verhagen et al. 2009b). This may reflect basic bursting activity of dopamine neurons at this time point (Stice et al. 2010). Although dopamine was not released much at the moment before food delivery (anticipation), it cannot be excluded that upon longer exposure to ABA, the dopaminergic system does play a role in FAA, since rats in the microdialysis experiments were exposed to the ABA model for only 4 days.

Taken together, the above supports the proposition that the neurobiological mechanism underlying hyperactivity (dark phase foraging behavior and FAA) in the ABA model consists of two interacting systems.

5 Discussion

We propose that hyperactivity as displayed upon exposure to the ABA model consists of two components. One component, FAA, may be mediated by activation of the FEO. Likely, ghrelin acts either directly or indirectly at the neural structures forming the FEO. But also lack of leptin signaling may contribute to FAA, since leptin strongly suppresses FAA in the ABA model (Verhagen et al. 2009a). The DMH remains a good candidate for the anticipatory hyperactivity displayed by ABA rats, since FAA correlates with c-Fos reactivity in this nucleus. The other component that drives hyperactivity in the ABA model may involve increased ghrelin and decreased leptin signaling at the level of the VTA. Ghrelin increases VTA dopamine neuronal activity, whereas leptin suppresses it. The increased activity of these neurons may mediate the increased motivation for food to frontal brain regions, such as NAcc and prefrontal cortex. The strong effect of ghrelin antagonism and leptin on RWA in ABA rats is most likely not mediated in the Arc, but in the VTA, where both can bind their respective receptors on dopaminergic neurons, consequently affecting these neurons and altering motivation to run. Future studies should investigate the precise role of these neurons during ABA, by focusing on direct effects of leptin and ghrelin as well as hypothalamic neuropeptides (e.g., orexins, CART) on these neurons in relation to hyperactivity. Molecular determinants of attenuation of hyperactivity by leptin treatment might also be further unraveled using specific mouse strains of interest, now that the mouse model of ABA is developed (Gelegen et al. 2007).

Furthermore, hyperactivity in patients with AN is also related to an anxious phenotype; therefore, neurobiological systems underlying hyperactivity and anxiety might be similar or at least connected (Holtkamp et al. 2003). Thus, studying animal models of other phenotypes, e.g., anxiety, may contribute to unraveling the molecular determinants of hyperactivity.

Leptin and ghrelin not only are related to actual eating behavior but also may have an influence on cognitive processing of food. Administration of leptin and ghrelin has been shown to alter the response of the ventral striatum to food images in humans (Malik et al. 2008; Rosenbaum et al. 2009). Interestingly, this kind of response of humans to food images is affected by genetic variation near the dopamine D2 receptor gene (Stice et al. 2008). This has also been associated with different ventral striatal responses to rewarding stimuli (Kirsch et al. 2006). It thus seems that food-related cognitive processing is strongly affected by activation of the ventral striatum and modulated by dopamine receptor activity.

The role of leptin and ghrelin and their interaction with the dopamine system may not only contribute to hyperactivity in ABA rats and patients with AN but also be a driver of cognitive processes that are disturbed in patients with AN. Food-related cognitive processing is strongly affected by activation of the ventral striatum and modulated by dopamine receptor activity (Malik et al. 2008, Rosenbaum et al. 2008; Stice et al. 2008). Interestingly AN has been associated with genetic variation in dopamine-related genes (Bergen et al. 2005) and recovered patients with AN show increased striatal dopamine D2 binding site availability (Frank et al. 2005). Thus, alterations in the dopamine system may underlie hyperactivity and food-related cognitive processing in AN, and ghrelin and leptin may affect these via their effects on the dopamine system.

Food-related cognitive processing as well as reward processing in general is disturbed in patients with AN (Wagner et al. 2007, 2008; Bailer et al. 2005, 2007; Frank et al. 2005; Kaye et al. 1999). For instance, activation of the ventral striatum by food perception is reduced in patients with AN (Wagner et al. 2008). The ventral striatum is also involved in the startle reflex which is a sort of alarm reaction (Koch et al. 1996). Exposure to food cues elicits an exaggerated startle response in patients with AN, suggesting that food primes an automatic defensive reaction (Friederich et al. 2006). Hyperactivity may be regarded as an expression of this defense reaction. Aberrations in (neuro)psychological processes like set shifting and decision making (Lopez et al. 2009; Roberts et al. 2007; Tchanturia et al. 2004, 2007) may also affect food-related cognitive processes in patients with AN. Set shifting disturbances are consistent with the perseverative obsessions with food and body shape and perfectionism that is typical for AN (Bardone-Cone et al. 2007). Interestingly, set shifting is dependent on dopaminergic activity (Avila et al. 2003). Decision making is based on the expected incentive value of an anticipated outcome and is also dependent on dopaminergic activity (Tchanturia et al. 2007; Naqvi et al. 2006). Deficits in self-regulatory control processes and frontostriatal systems are related in patients with AN (Marsh et al. 2007). The above suggests that alterations in dopaminergic signaling in patients with AN are (causally) related to their impairments in cognitive processes, including those related to food, but possibly also those related to hyperactivity. We hypothesize that impairments in cognitive processing are exaggerated by low leptin and high ghrelin levels in patients with AN.

Taken together, these data support the hypothesis that ghrelin and leptin affect the dopamine system directly via the VTA where both leptin and ghrelin receptors are expressed. Thus, the leptin and ghrelin systems are both involved in homeostatic as well as hedonic neural circuits (2). An important focus for future research is to unravel what causes the apparent leptin and ghrelin insensitivity in patients with AN, who show low and high plasma levels of leptin and ghrelin, respectively, but undereat and increase energy expenditure by physical activity.

References

Abizaid A, Liu ZW, Andrews ZB, Shanabrough M, Borok E, Elsworth JD, Roth RH, Sleeman MW, Picciotto MR, Tschop MH, Gao XB, Horvath TL (2006) Ghrelin modulates the activity and synaptic input organization of midbrain dopamine neurons while promoting appetite. J Clin Invest 116:3229–3239

Ahima RS, Prabakaran D, Mantzoros C, Qu D, Lowell B, Maratos-Flier E, Flier JS (1996) Role of leptin in the neuroendocrine response to fasting. Nature 382:250–252

American Psychiatric Association (1994) Diagnostic and statistical manual. APA, Washington DC

Ammar AA, Sederholm F, Saito TR, Scheurink AJ, Johnson AE, Sodersten P (2000) NPY-leptin: opposing effects on appetitive and consummatory ingestive behavior and sexual behavior. Am J Physiol Regul Integr Comp Physiol 278:R1627–R1633

Aravich PF, Rieg TS, Lauterio TJ, Doerries LE (1993) Beta-endorphin and dynorphin abnormalities in rats subjected to exercise and restricted feeding: relationship to anorexia nervosa? Brain Res. 622(1–2):1–8

Avila C, Barros A, Ortet G, Parcet MA, Ibanez MI (2003) Set-shiftingand sensitivity to reward: a possible dopamine mechanism for explaining disinhibitory disorders. Cogn Emot 17:951–959

Bardone-Cone AM, Wonderlich SA, Frost RO et al. (2007) Perfectionism and eating disorders: current status and future directions. Clin Psychol Rev 27:384–405

Bergen AW, Yeager M, Welch RA, Haque K, Ganjei JK, van den Bree MB, Mazzanti C, Nardi I, Fichter MM, Halmi KA, Kaplan AS, Strober M, Treasure J, Woodside DB, Bulik CM, Bacanu SA, Devlin B, Berrettini WH, Goldman D, Kaye WH (2005) Association of multiple DRD2 polymorphisms with anorexia nervosa. Neuropsychopharmacology. 30(9):1703–10

Blum ID, Patterson Z, Khazall R, Lamont EW, Sleeman MW, Horvath TL, Abizaid A (2009) Reduced anticipatory locomotor responses to scheduled meals in ghrelin receptor deficient mice. Neuroscience 164:351–359

Broglio F, Gianotti L, Destefanis S, Fassino S, Abbate DG, Mondelli V, Lanfranco F, Gottero C, Gauna C, Hofland L, Van der Lely AJ, Ghigo E (2004) The endocrine response to acute ghrelin administration is blunted in patients with anorexia nervosa, a ghrelin hypersecretory state. Clin Endocrinol 60:592–599

Campfield LA, Smith FJ, Guisez Y, Devos R, Burn P (1995) Recombinant mouse OB protein: evidence for a peripheral signal linking adiposity and central neural networks. Science 269:546–549

Choi YH, Li C, Hartzell DL, Little DE, Della-Fera MA, Baile CA (2008) ICV leptin effects on spontaneous physical activity and feeding behavior in rats. Behav Brain Res 188:100–108

Coppari R, Ichinose M, Lee CE, Pullen AE, Kenny CD, McGovern RA, Tang V, Liu SM, Ludwig T, Chua SC Jr, Lowell BB, Elmquist JK (2005) The hypothalamic arcuate nucleus: a key site for mediating leptin's effects on glucose homeostasis and locomotor activity. Cell Metab 1:63–72

Cowley MA, Smith RG, Diano S, Tschop M, Pronchuk N, Grove KL, Strasburger CJ, Bidlingmaier M, Esterman M, Heiman ML, Garcia-Segura LM, Nillni EA, Mendez P, Low MJ, Sotonyi P, Friedman JM, Liu H, Pinto S, Colmers WF, Cone RD, Horvath TL (2003) The distribution and mechanism of action of ghrelin in the CNS demonstrates a novel hypothalamic circuit regulating energy homeostasis. Neuron 20(37):649–661

Cummings DE, Purnell JQ, Frayo RS, Schmidova K, Wisse BE, Weigle DS (2001) A preprandial rise in plasma ghrelin levels suggests a role in meal initiation in humans. Diabetes 50:1714–1719

Davis C (1997) Eating disorders and hyperactivity: a psychobiological perspective. Can J Psychiatry 42:168–175

Day DE, Keen-Rhinehart E, Bartness TJ (2005) Role of NPY and its receptor subtypes in foraging, food hoarding, and food intake by Siberian hamsters. Am J Physiol Regul Integr Comp Physiol 289:R29–R36

de Rijke CE, Hillebrand JJ, Verhagen LA, Roeling TA, Adan RA (2005) Hypothalamic neuropeptide expression following chronic food restriction in sedentary and wheel-running rats. J Mol Endocrinol 35:381–390

Druce MR, Neary NM, Small CJ, Milton J, Monteiro M, Patterson M, Ghatei MA, Bloom SR (2006) Subcutaneous administration of ghrelin stimulates energy intake in healthy lean human volunteers. Int J Obes 30:293–296

Elmquist JK, Bjorbaek C, Ahima RS, Flier JS, Saper CB (1998) Distributions of leptin receptor mRNA isoforms in the rat brain. J Comp Neurol 395:535–547

Epling WF, Pierce WDSL (1983) A theory of activity-basedanorexia. Int J Eat Disord 3:27–46

Exner C, Hebebrand J, Remschmidt H, Wewetzer C, Ziegler A, Herpertz S, Schweiger U, Blum WF, Preibisch G, Heldmaier G, Klingenspor M (2000) Leptin suppresses semi-starvation induced hyperactivity in rats: implications for anorexia nervosa. Mol Psychiatry 5:476–481

Faulconbridge LF, Cummings DE, Kaplan JM, Grill HJ (2003) Hyperphagic effects of brainstem ghrelin administration. Diabetes 52:2260–2265

Fetissov SO, Bergström U, Johansen JE, Hökfelt T, Schalling M, Ranscht B (2005) Alterations of arcuate nucleus neuropeptidergic development in contactin-deficient mice: comparison with anorexia and food-deprived mice. Eur J Neurosci. 22(12):3217–28

Figlewicz DP, Evans SB, Murphy J, Hoen M, Baskin DG (2003) Expression of receptors for insulin and leptin in the ventral tegmental area/substantia nigra (VTA/SN) of the rat. Brain Res 964:107–115

Frank GK, Bailer UF, Henry SE, Drevets W, Meltzer CC, Price JC, Mathis CA, Wagner A, Hoge J, Ziolko S, Barbarich-Marsteller N, Weissfeld L, Kaye WH (2005) Increased dopamine D2/D3 receptor binding after recovery from anorexia nervosa measured by positron emission tomography and [11c]raclopride. Biol Psychiatry 58:908–912

Fulton S, Woodside B, Shizgal P (2000) Modulation of brain reward circuitry by leptin. Science 287:125–128

Fulton S, Pissios P, Manchon RP, Stiles L, Frank L, Pothos EN, Maratos-Flier E, Flier JS (2006) Leptin regulation of the mesoaccumbens dopamine pathway. Neuron 51:811–822

Gelegen C, Collier DA, Campbell IC, Oppelaar H, van den Heuvel J, Adan RA, Kas MJ (2007) Difference in susceptibility to activity-based anorexia in two inbred strains of mice. Eur Neuropsychopharmacol. 17(3):199–205. Epub 2006 Jun 2

Gooley JJ, Schomer A, Saper CB (2006) The dorsomedial hypothalamic nucleus is critical for the expression of food-entrainable circadian rhythms. Nat Neurosci 9:398–407

Grill HJ, Schwartz MW, Kaplan JM, Foxhall JS, Breininger J, Baskin DG (2002) Evidence that the caudal brainstem is a target for the inhibitory effect of leptin on food intake. Endocrinology 143:239–246

Guan XM, Yu H, Palyha OC, McKee KK, Feighner SD, Sirinathsinghji DJ, Smith RG, Van der Ploeg LH, Howard AD (1997) Distribution of mRNA encoding the growth hormone secretagogue receptor in brain and peripheral tissues. Brain Res Mol Brain Res 48:23–29

Gutierrez E, Churruca I, Zarate J, Carrera O, Portillo MP, Cerrato M, Vazquez R, Echevarria E (2009) High ambient temperature reverses hypothalamic MC4 receptor overexpression in an animal model of anorexia nervosa. Psychoneuroendocrinology 34:420–429

Halaas JL, Gajiwala KS, Maffei M, Cohen SL, Chait BT, Rabinowitz D, Lallone RL, Burley SK, Friedman JM (1995) Weight-reducing effects of the plasma protein encoded by the obese gene. Science 269:543–546

Hayward MD, Pintar JE, Low MJ (2002) Selective reward deficit in mice lacking beta-endorphin and enkephalin. J Neurosci 22:8251–8258

Hebebrand J, Blum WF, Barth N, Coners H, Englaro P, Juul A, Ziegler A, Warnke A, Rascher W, Remschmidt H (1997) Leptin levels in patients with anorexia nervosa are reduced in the acute stage and elevated upon short-term weight restoration. Mol Psychiatry 2:330–334

Hebebrand J, Exner C, Hebebrand K, Holtkamp C, Casper RC, Remschmidt H, Herpertz-Dahlmann B, Klingenspor M (2003) Hyperactivity in patients with anorexia nervosa and in semistarved rats: evidence for a pivotal role of hypoleptinemia. Physiol Behav. 79(1):25–37

Hebebrand J, Muller TD, Holtkamp K, Herpertz-Dahlmann B (2007) The role of leptin in anorexia nervosa: clinical implications. Mol Psychiatry 12:23–35

Hillebrand JJ, Koeners MP, de Rijke CE, Kas MJ, Adan RA (2005a) Leptin treatment in activity-based anorexia. Biol Psychiatry 58:165–171

Hillebrand JJ, Kas MJ, Adan RA (2005b) alpha-MSH enhances activity-based anorexia. Peptides 26:1690–1696

Hillebrand JJ, Kas MJ, Scheurink AJ, van Dijk G, Adan RA (2006) AgRP(83-132) and SHU9119 differently affect activity-based anorexia. Eur Neuropsychopharmacol 16:403–412

Hillebrand JJ, Kas MJ, van Elburg AA, Hoek HW, Adan RA (2008) Leptin's effect on hyperactivity: potential downstream effector mechanisms. Physiol Behav 94:689–695

Holtkamp K, Herpertz-Dahlmann B, Mika C, Heer M, Heussen N, Fichter M, Herpertz S, Senf W, Blum WF, Schweiger U, Warnke A, Ballauff A, Remschmidt H, Hebebrand J (2003) Elevated physical activity and low leptin levels co-occur in patients with anorexia nervosa. J Clin Endocrinol Metab 88:5169–5174

Holtkamp K, Herpertz-Dahlmann B, Hebebrand K, Mika C, Kratzsch J, Hebebrand J (2006) Physical activity and restlessness correlate with leptin levels in patients with adolescent anorexia nervosa. Biol Psychiatry 60:311–313

Hommel JD, Trinko R, Sears RM, Georgescu D, Liu ZW, Gao XB, Thurmon JJ, Marinelli M, DiLeone RJ (2006) Leptin receptor signaling in midbrain dopamine neurons regulates feeding. Neuron 51:801–810

Horvath TL, Diano S, Sotonyi P, Heiman M, Tschop M (2001) Minireview: ghrelin and the regulation of energy balance–a hypothalamic perspective. Endocrinology 142:4163–4169

Hosoda H, Kojima M, Kangawa K (2002) Ghrelin and the regulation of food intake and energy balance. Mol Interv 2:494–503

Hotta M, Ohwada R, Akamizu T, Shibasaki T, Takano K, Kangawa K (2009) Ghrelin increases hunger and food intake in patients with restricting-type anorexia nervosa: a pilot study. Endocrine J 56:1119–1128

Howard AD, Feighner SD, Cully DF, Arena JP, Liberator PA, Rosenblum CI, Hamelin M, Hreniuk DL, Palyha OC, Anderson J, Paress PS, Diaz C, Chou M, Liu KK, McKee KK, Pong SS, Chaung LY, Elbrecht A, Dashkevicz M, Heavens R, Rigby M, Sirinathsinghji DJ, Dean DC, Melillo DG, Patchett AA, Nargund R, Griffin PR, DeMartino JA, Gupta SK, Schaeffer JM, Smith RG, Van der Ploeg LH (1996) A receptor in pituitary and hypothalamus that functions in growth hormone release. Science 273:974–977

Jerlhag E, Egecioglu E, Dickson SL, Andersson M, Svensson L, Engel JA (2006) Ghrelin stimulates locomotor activity and accumbal dopamine-overflow via central cholinergic systems in mice: implications for its involvement in brain reward. Addict Biol 11:45–54

Jerlhag E, Egecioglu E, Dickson SL, Douhan A, Svensson L, Engel JA (2007) Ghrelin administration into tegmental areas stimulates locomotor activity and increases extracellular concentration of dopamine in the nucleus accumbens. Addict Biol 12:6–16

Johansen JE, Fetissov SO, Bergström U, Nilsson I, Faÿ C, Ranscht B, Hökfelt T, Schalling M (2007) Evidence for hypothalamic dysregulation in mouse models of anorexia as well as in humans. Physiol Behav. 92(1–2):278–82

Kas MJ, van Dijk G, Scheurink AJ, Adan RA (2003) Agouti-related protein prevents self-starvation. Mol Psychiatry 8:235–240

Kas MJ, van den BR B, AM LM, Lesscher HM, Hillebrand JJ, Schuller AG, Pintar JE, Spruijt BM (2004) Mu-opioid receptor knockout mice show diminished food-anticipatory activity. Eur J Neurosci 20:1624–1632

Kaye WH, Frank GK, McConaha C (1999) Altered dopamine activity after recovery from restricting-type anorexia nervosa. Neuropsychopharmacology 21:503–506

Keen-Rhinehart E, Bartness TJ (2005) Peripheral ghrelin injections stimulate food intake, foraging, and food hoarding in Siberian hamsters. Am J Physiol Regul Integr Comp Physiol 288: R716–R722

Keen-Rhinehart E, Bartness TJ (2007) NPY Y1 receptor is involved in ghrelin- and fasting-induced increases in foraging, food hoarding, and food intake. Am J Physiol Regul Integr Comp Physiol 292:R1728–R1737

Kirsch P, Reuter M, Mier D et al (2006) Imaging gene-substance interactions: the effect of the DRD2 TaqIA polymorphism and the dopamine agonist bromocriptine on the brain activation during the anticipation of reward. Neurosci Lett 405:196–201

Kojima M, Hosoda H, Date Y, Nakazato M, Matsuo H, Kangawa K (1999) Ghrelin is a growth-hormone-releasing acylated peptide from stomach. Nature 402:656–660

Lambert KG, Porter JH (1992) Pimozide mitigates excessive running in the activity-stress paradigm. Physiol Behav 52:299–304

LeSauter J, Hoque N, Weintraub M, Pfaff DW, Silver R (2009) Stomach ghrelin-secreting cells as food-entrainable circadian clocks. Proc Natl Acad Sci USA 106:13582–13587

Lopez C, Tchanturia K, Stahl D, Treasure J (2009) Weak central coherence in eating disorders: a step towards looking for an endophenotype of eating disorders. J Clin Exp Neuropsychol 31:117–25

Mantzoros C, Flier JS, Lesem MD, Brewerton TD, Jimerson DC (1997) Cerebrospinal fluid leptin in anorexia nervosa: correlation with nutritional status and potential role in resistance to weight gain. J Clin Endocrinol Metab 82:1845–1851

Malik S, McGlone F, Bedrossian D, Dagher A (2008) Ghrelin modulates brain activity in areas that control appetitive behavior. Cell Metab. 7(5):400–9

Marsh DJ, Weingarth DT, Novi DE, Chen HY, Trumbauer ME, Chen AS, Guan XM, Jiang MM, Feng Y, Camacho RE, Shen Z, Frazier EG, Yu H, Metzger JM, Kuca SJ, Shearman LP, Gopal-Truter S, MacNeil DJ, Strack AM, MacIntyre DE, Van der Ploeg LH, Qian S (2002) Melanin-concentrating hormone 1 receptor-deficient mice are lean, hyperactive, and hyperphagic and have altered metabolism. Proc Natl Acad Sci USA. 99(5):3240–5 (Epub 2002 Feb 26)

Marsh R, Steinglass JE, Graziano K, Peterson BS, Walsh BT (2007) Self-regulatory control and habit learningin the development of eating disorders. Curr Psychiatric Rev 3:73–83

Mayo-Smith W, Rosenthal DI, Goodsitt MM, Klibanski A (1989) Intravertebral fat measurement with quantitative CT in patients with Cushing disease and anorexia nervosa. Radiology 170:835–838

Miljic D, Pekic S, Djurovic M, Doknic M, Milic N, Casanueva FF, Ghatei M, Popovic V (2006) Ghrelin has partial or no effect on appetite, growth hormone, prolactin, and cortisol release in patients with anorexia nervosa. J Clin Endocrinol Metab 91:1491–1495

Misra M, Miller KK, Herzog DB, Ramaswamy K, Aggarwal A, Almazan C, Neubauer G, Breu J, Klibanski A (2004) Growth hormone and ghrelin responses to an oral glucose load in adolescent girls with anorexia nervosa and controls. J Clin Endocrinol Metab 89:1605–1612

Mistlberger RE (1994) Circadian food-anticipatory activity: formal models and physiological mechanisms. Neurosci Biobehav Rev 18:171–195
Mistlberger RE, Buijs RM, Challet E, Escobar C, Landry GJ, Kalsbeek A, Pevet P, Shibata S (2009) Standards of evidence in chronobiology: critical review of a report that restoration of Bmal1 expression in the dorsomedial hypothalamus is sufficient to restore circadian food anticipatory rhythms in Bmal1-/- mice. J Circadian Rhythms 7(3):3
Mondal MS, Date Y, Yamaguchi H, Toshinai K, Tsuruta T, Kangawa K, Nakazato M (2005) Identification of ghrelin and its receptor in neurons of the rat arcuate nucleus. Regul Pept 126:55–59
Monteiro MP, Ribeiro AH, Nunes AF, Sousa MM, Monteiro JD, Aguas AP, Cardoso MH (2007) Increase in ghrelin levels after weight loss in obese Zucker rats is prevented by gastric banding. Obes Surg. 17(12):1599–607. Epub 2007 Nov 30
Munzberg H, Myers MG Jr (2005) Molecular and anatomical determinants of central leptin resistance. Nat Neurosci 8:566–570
Naleid AM, Grace MK, Cummings DE, Levine AS (2005) Ghrelin induces feeding in the mesolimbic reward pathway between the ventral tegmental area and the nucleus accumbens. Peptides 26:2274–2279
Naqvi N, Shiv B, Bechara A (2006) The role of emotion in decision making: a cognitive neuroscience perspective. Curr Dir Psychol Sci 15:260–264
Nedvidkova J, Krykorkova I, Bartak V, Papezova H, Gold PW, Alesci S, Pacak K (2003) Loss of meal-induced decrease in plasma ghrelin levels in patients with anorexia nervosa. J Clin Endocrinol Metab 88:1678–1682
Nergardh R, Ammar A, Brodin U, Bergstrom J, Scheurink A, Sodersten P (2007) Neuropeptide Y facilitates activity-based-anorexia. Psychoneuroendocrinology 32:493–502
Nijenhuis WA, Oosterom J, Adan RA (2001) AgRP(83-132) acts as an inverse agonist on the human-melanocortin-4 receptor. Mol Endocrinol 15:164–171
Otto B, Cuntz U, Fruehauf E, Wawarta R, Folwaczny C, Riepl RL, Heiman ML, Lehnert P, Fichter M, Tschop M (2001) Weight gain decreases elevated plasma ghrelin concentrations of patients with anorexia nervosa. Eur J Endocrinol 145:669–673
Otto B, Tschöp M, Frühauf E, Heldwein W, Fichter M, Otto C, Cuntz U (2005) Postprandial ghrelin release in anorectic patients before and after weight gain. Psychoneuroendocrinology. 30(6):577–81
Pelleymounter MA, Cullen MJ, Baker MB, Hecht R, Winters D, Boone T, Collins F (1995) Effects of the obese gene product on body weight regulation in ob/ob mice. Science 269:540–543
Pirke KM, Broocks A, Wilckens T, Marquard R, Schweiger U (1993) Starvation-induced hyperactivity in the rat: the role of endocrine and neurotransmitter changes. Neurosci Biobehav Rev 17:287–294
Prince AC, Brooks SJ, Stahl D, Treasure J (2009) Systematic review and meta-analysis of the baseline concentrations and physiologic responses of gut hormones to food in eating disorders. Am J Clin Nutr 89:755–765
Riediger T, Traebert M, Schmid HA, Scheel C, Lutz TA, Scharrer E (2003) Site-specific effects of ghrelin on the neuronal activity in the hypothalamic arcuate nucleus. Neurosci Lett 341:151–155
Roberts AC, Reekie Y, Braesicke K (2007) Synergistic and regulatory effects of orbitofrontal cortex on amygdala-dependent appetitive behavior. Ann N Y Acad Sci 1121:297– 319
Rosenbaum M, Nicolson M, Hirsch J, Murphy E, Chu F, Leibel RL (1997) Effects of weight change on plasma leptin concentrations and energy expenditure. J Clin Endocrinol Metab 82:3647–3654
Rosenbaum M, Murphy EM, Heymsfield SB, Matthews DE, Leibel RL (2002) Low dose leptin administration reverses effects of sustained weight-reduction on energy expenditure and circulating concentrations of thyroid hormones. J Clin Endocrinol Metab 87:2391–2394
Rosenbaum M, Sy M, Pavlovich K, Leibel RL, Hirsch J (2008) Leptin reverses weight loss-induced changes in regional neural activity responses to visual food stimuli. J Clin Invest. 118 (7):2583–91
Schwartz MW, Woods SC, Porte D Jr, Seeley RJ, Baskin DG (2000) Central nervous system control of food intake. Nature 404:661–671

Shalev U, Yap J, Shaham Y (2001) Leptin attenuates acute food deprivation-induced relapse to heroin seeking. J Neurosci 21:RC129

Shiiya T, Nakazato M, Mizuta M, Date Y, Mondal MS, Tanaka M, Nozoe S, Hosoda H, Kangawa K, Matsukura S (2002) Plasma ghrelin levels in lean and obese humans and the effect of glucose on ghrelin secretion. J Clin Endocrinol Metab 87:240–244

Shimada M, Tritos NA, Lowell BB, Flier JS, Maratos-Flier E (1998) Mice lacking melanin-concentrating hormone are hypophagic and lean. Nature 396(6712):670–4

Sisti HM, Lewis MJ (2001) Naloxone suppression and morphine enhancement of voluntary wheel-running activity in rats. Pharmacol Biochem Behav 70:359–365

Soriano-Guillen L, Barrios V, Campos-Barros A, Argente J (2004) Ghrelin levels in obesity and anorexia nervosa: effect of weight reduction or recuperation. J Pediatr 144:36–42

Stice E, Spoor S, Bohon C, Small DM (2008) Relation between obesity and blunted striatal response to food is moderated by TaqIA A1 allele. Science. 322(5900):449–52

Stice E, Zald D, Dagher A (2010) Dopamine-based reward circuitry responsivity, genetics, and overeating. Curr Top Behav Neurosci. doi:10.1007/7854_2010_89

Surwit RS, Edwards CL, Murthy S, Petro AE (2000) Transient effects of long-term leptin supplementation in the prevention of diet-induced obesity in mice. Diabetes 49:1203–1208

Szczypka MS, Kwok K, Brot MD, Marck BT, Matsumoto AM, Donahue BA, Palmiter RD (2001) Dopamine production in the caudate putamen restores feeding in dopamine-deficient mice. Neuron. 30(3):819–28

Tang-Christensen M, Vrang N, Ortmann S, Bidlingmaier M, Horvath TL, Tschop M (2004) Central administration of ghrelin and agouti-related protein (83-132) increases food intake and decreases spontaneous locomotor activity in rats. Endocrinology 145:4645–4652

Tchanturia K, Morris RG, Anderluh MB, Collier DA, Nikolaou V, Treasure J. (2004) Set shifting in anorexia nervosa: an examination before and after weight gain, in full recovery and relationship to childhood and adult OCPD traits. J Psychiatr Res 38:545–52

Tolle V, Kadem M, Bluet-Pajot MT, Frere D, Foulon C, Bossu C, Dardennes R, Mounier C, Zizzari P, Lang F, Epelbaum J, Estour B (2003) Balance in ghrelin and leptin plasma levels in anorexia nervosa patients and constitutionally thin women. J Clin Endocrinol Metab 88:109–116

Troisi A, Di Lorenzo G, Lega I, Tesauro M, Bertoli A, Leo R, Iantorno M, Pecchioli C, Rizza S, Turriziani M, Lauro R, Siracusano A (2005) Plasma ghrelin in anorexia, bulimia, and binge-eating disorder: relations with eating patterns and circulating concentrations of cortisol and thyroid hormones. Neuroendocrinology 81:259–266

van Elburg AA, Hoek HW, Kas MJ, van Engeland H (2007a) Nurse evaluation of hyperactivity in anorexia nervosa: a comparative study. Eur Eat Disord Rev 15:425–429

van Elburg AA, Kas MJ, Hillebrand JJ, Eijkemans RJ, van Engeland H (2007b) The impact of hyperactivity and leptin on recovery from anorexia nervosa. J Neural Transm 114:1233–1237

Verhagen LA, Egecioglu E, Luijendijk MC, Hillebrand JJ, Adan RA, Dickson SL (2010) Acute and chronic suppression of the central ghrelin signaling system reveals a role in food anticipatory activity. Eur Neuropsychopharmacol. [Epub ahead of print]

Verhagen LA, Luijendijk MC, Hillebrand JJ, Adan RA (2009) Dopamine antagonism inhibits anorectic behavior in an animal model for anorexia nervosa. Eur Neuropsychopharmacol. 19(3):153–60

Vink T, Hinney A, van Elburg AA, van Goozen SH, Sandkuijl LA, Sinke RJ, Herpertz-Dahlmann BM, Hebebrand J, Remschmidt H, van Engeland H, Adan RA (2001) Association between an agouti-related protein gene polymorphism and anorexia nervosa. Mol Psychiatry. 6(3):325–8

Wellman PJ, Hollas CN, Elliott AE (2008) Systemic ghrelin sensitizes cocaine-induced hyperlocomotion in rats. Regul Pept 146:33–37

Whatmore AJ, Hall CM, Jones J, Westwood M, Clayton PE (2003) Ghrelin concentrations in healthy children and adolescents. Clin Endocrinol 59:649–654

Willesen MG, Kristensen P, Romer J (1999) Co-localization of growth hormone secretagogue receptor and NPY mRNA in the arcuate nucleus of the rat. Neuroendocrinology 70:306–316

Wise RA (2002) Brain reward circuitry: insights from unsensed incentives. Neuron 36:229–240

Wren AM, Seal LJ, Cohen MA, Brynes AE, Frost GS, Murphy KG, Dhillo WS, Ghatei MA, Bloom SR (2001) Ghrelin enhances appetite and increases food intake in humans. J Clin Endocrinol Metab 86:5992

Zhou D, Shen Z, Strack AM, Marsh DJ, Shearman LP (2005) Enhanced running wheel activity of both Mch1r- and Pmch-deficient mice. Regul Pept. 124(1-3):53–63

Zigman JM, Jones JE, Lee CE, Saper CB, Elmquist JK (2006) Expression of ghrelin receptor mRNA in the rat and the mouse brain. J Comp Neurol 494:528–548

Part V
Translational Approach to Treatment

Translating Experimental Neuroscience into Treatment of Eating Disorders: Two Examples

Ulrike Schmidt, Anna Oldershaw, and Annemarie van Elburg

Contents

Abstract Anorexia nervosa (AN) is a serious mental disorder with impaired functioning including not only the cognitive and socio-emotional but also physical domains. Improved treatments, especially for adults with AN, are urgently needed. The insights gained from basic research in experimental animal models and the advent of cognitive neuroscience have produced major advances in our understanding of the condition, but translating these into clinical research or practice remains a challenge. We describe here what the eating disorders field can gain from schizophrenia research in this area. We use the example of socio-emotional impairments in AN to describe the iterative process between basic research and intervention

U. Schmidt (✉) and A. Oldershaw
Section of Eating Disorders, King's College London, Institute of Psychiatry, De Crespigny Park, London SE5 8AF, UK
e-mail: u.schmidt@iop.kcl.ac.uk

A. van Elburg
Rintveld Center for Eating Disorders, Altrecht Mental Health Institute, Oude Arnhemseweg 260, Zeist, Utrecht, 3705BK, The Netherlands
Department of Child and Adolescent Psychiatry, University Medical Center, Utrecht, The Netherlands
e-mail: a.van.elburg@altrecht.nl

R.A.H. Adan and W.H. Kaye (eds.), *Behavioral Neurobiology of Eating Disorders*, Current Topics in Behavioral Neurosciences 6, DOI 10.1007/7854_2010_76

development for neurobiologically informed and based treatments for this condition and briefly touch on some other examples that stem from translational science.

Keywords Anorexia neruosa · Translational research · Social cognition

1 Introduction

Translational science or research can help to bridge the gap between basic and applied research. Scientists are increasingly aware that this bench-to-bedside approach is really a two-way street. Basic scientists provide clinicians with new tools for use in patients and for assessment of their impact, and clinical researchers make novel observations about the nature and progression of disease that often stimulate basic investigations.

In the field of eating disorders, two examples of this approach are cognitive neuroscience and experimental animal models that mimic Anorexia Nervosa (AN) such as the Anorexia-Based Activity model (Adan, Chap. 4.3). The advent of cognitive neuroscience has produced 'major advances in the precision with which cognitive processes can be measured and related to underlying neural mechanisms using modern electrophysiologic and neuroimaging techniques' (Cohen and Insel 2008). As a result, the last two decades have seen a renewed interest in and significant progress in our understanding of the neurobiological underpinnings of eating disorders. Much of this work has understandably focused on AN, as this is arguably the most severe form of eating disorder with high levels of mortality and disability, physical and psychological morbidity, and impaired quality of life (Klump et al. 2009). Imaging studies have shown that food-related cognitive processing as well as reward processing in general are disturbed in patients with AN (Wagner et al. 2007, 2008; Bailer et al. 2005, 2007; Frank et al. 2005). Clinically, cognitive and emotional functioning are impaired in AN (Lopez et al. 2009; Roberts et al. 2007; Oldershaw et al. submitted; Zucker et al. 2007), making engagement in treatment difficult.

Psychotherapeutic interventions are currently the treatment of choice for AN (National Collaborating Centre for Mental Health 2004), but the results of psychotherapy depend critically on the stage of the illness. While response to psychological treatment (usually family based) is excellent in adolescents with a short duration of AN (Le Grange and Eisler 2009), the treatment response in adults with a more chronic form of the illness is much less positive and drop-out from treatment is high. The evidence base for psychological treatment of adults with AN is extremely limited. Several small trials have tested a range of therapies, including cognitive behavioural therapy (CBT), interpersonal therapy (IPT), cognitive analytical therapy, and family therapy, but no clear front-runner in terms of efficacy has emerged (National Collaborating Centre for Mental Health 2004; Fairburn 2005; Bulik et al. 2007). The urgent need to develop more effective treatments for adults with AN has been highlighted (National Collaborating Centre for Mental Health 2004; Fairburn 2005; Bulik et al. 2007).

So far, across different mental disorders, 'developments in basic science have seen little translation into clinical research or practice' (Cohen and Insel 2008). Scientific and societal hurdles make translating basic research innovations into improved treatment and outcomes for patients a major challenge. To make things more complicated, there are many different definitions of the meaning of translational research. T1 translational research is the classical 'from bench-to-bedside' approach that translates findings from basic science into clinical applications and is the focus of this chapter. Ultimately, it is hoped that improved and brain-directed treatment will allow for more personalised interventions taking into account the bio-psychosocial profile of each person. T2 translational research focuses on the delivery and dissemination of treatments in real-world situations (Woolf 2008; Wang et al. 2008). While this is also an important issue, it is beyond the scope of this chapter to address this.

As yet in the field of eating disorders, only a few articles have specifically focused on outlining promising areas for T1 translational research in AN (Klein and Walsh 2005; Kas et al. 2003, 2009). This is in contrast to other disorders, specifically schizophrenia where the iterative process between basic research and intervention development and evaluation is much further advanced. Questions such as 'What are the most promising targets for intervention? How should one assess the efficacy of any such treatments? How would one go about developing neurobiologically informed treatments? Conversely, what are the neurobiological correlates of existing treatments including psychological therapies?' have all been thought of in relation to schizophrenia, and we therefore decided to use the approach employed in this field as an example for identifying the progress and gaps in our knowledge in relation to these questions in AN.

2 What Can We Learn from Schizophrenia Research?

In schizophrenia, impaired cognition is a hallmark of the illness, present at all stages, linked to functional impairment and is largely treatment refractory. Therefore, developing treatments for impaired neurological and social cognition has been thought to be the key challenge for psychiatry in the twenty first century (Barch et al. 2009a). With this in mind, a National Institute of Health sponsored collaboration between different stakeholders (academics, pharmaceutical industry and FDA) was set up in the USA in 2002, namely the Measurement and Treatment Research to Improve Cognition in Schizophrenia (MATRICS) initiative. Its aim was to identify molecular and cognitive targets with the potential to improve cognitive processing in schizophrenia and to develop mechanisms for devising and testing interventions (Cohen and Insel 2008).

One of the tasks of the MATRICS initiative was to develop a cognitive test battery that would measure empirically derived cognitive domains in schizophrenia, based largely on factor analytic studies of neuropsychological task performance. The selection process focused mostly on paper and pen measures available for these domains, based on properties that would make them suitable for use in clinical trials such as being brief, having good psychometric properties (reliability, low practice effects), and having a strong relationship to functional outcomes (Green et al. 2004; Marder and Fenton 2004; Nuechterlein et al. 2004, 2008). The MATRICS Consensus Cognitive Battery was released in 2006 and is now widely used in clinical trials focusing on cognitive functioning in schizophrenia.

A second initiative, cognitive neuroscience treatment research to improve cognition in schizophrenia (CNTRCS) (Carter et al. 2008; Barch et al. 2008, 2009a), developed out of MATRICS, was based on the recognition that in order to truly move the field forward, the experimental tools of cognitive neuroscience needed to be used. This would allow the use of tasks that could measure discrete, specific cognitive processes, would allow researchers to distinguish between specific and general cognitive deficits and most importantly would have 'the ability to link cognitive deficits to specific neural systems, using animal models, neuropsychology and functional imaging' with key benefits for translational research (Barch et al. 2009a). Expert meetings, supplemented with online consultations, involved a wide range of experts from academia, industry, and government. Six cognitive systems to be targeted were identified. These included (1) executive control (Barch et al. 2009b), (2) working memory (Barch et al. 2009c), (3) long-term learning and memory (including reinforcement learning) (Ragland et al. 2009), (4) attention (Nuechterlein et al. 2009), (5) perception (Green et al. 2009) and (6) social and emotional processing (Carter et al. 2009). Based on these, a list of measures was identified. Criteria used to select the most promising tasks included construct validity, clarity of link to neural circuits, clarity of link to cognitive mechanisms, availability of animal model, link to neural systems through neuropsychopharmacology, amenabity for use in human neuroimaging studies, evidence of impairment in schizophrenia, link to functional outcomes in schizophrenia, and good psychometric properties (Barch et al. 2009a).

Within CENTRICS, further steps then depend on where a given task is in the 'translational research pathway' (Barch et al. 2009a). For those tasks that appear promising yet have never been studied in schizophrenia, it will be important to check out whether they are able to identify cognitive deficits and how they relate to functional outcomes. For other tasks which have been studied in schizophrenia but where psychometric properties are unknown or sub-optimal, ways have to be identified in which to improve the tasks to make them more 'user-friendly' and practical for use in clinical trials.

Another important step is the use of measures of neural systems in conjunction with measures of behaviour, as a way of assessing directly the mechanisms underpinning the efficacy of cognitive enhancers or therapies. It is hoped that such biomarker measures will be early indicators of behavioural change, thereby 'providing early information about the potential utility of different pharmacological or psychosocial approaches'. Moreover, it is hoped that such measures will help identify individual differences in response to particular treatments that will ultimately allow development of more targeted and personalised treatments. Finally, collaboration with animal cognitive neuroscientists is also required to develop and test animal models of specific paradigms in different species to 'facilitate the important interplay between human and animal work in the drug discover process' (Barch et al. 2009a).

These efforts in the field of schizophrenia are likely to 'have an impact on other domains of psychiatric research by providing a new generation of methods for studying cognitive disturbances' in other psychiatric disorders (Cohen and Insel 2008).

3 Schemes for Studying Social Cognition and Socio-Emotional Processing

The area of social cognitive and affective neuroscience is an emerging field of research within mental health and a wealth of such studies have been conducted in disorders such as Autism Spectrum Disorder (ASD) and schizophrenia. The term 'social cognition' refers to the mental processes underlying human social behaviour and interaction (Adolphs 1999) and was defined by Adolphs (2001) as 'the ability to construct representations of the relation between oneself and others and to use those representations flexibly to guide social behavior'. Social cognitive and perceptual processes are thought to be distinct from other cognitive abilities (Pinkham et al. 2003). In evolutionary terms, social cognition can be understood as enabling pro-social behaviour, promoting group dynamics and bonds, as well as assisting in dealing with competitions and hierarchies in groups (Adolphs 1999). During the MATRICS initiative, a scheme for studying these domains was identified, based on recent work in the field of schizophrenia and including five different domains, namely emotional processing, social perception, social knowledge, Theory of Mind (ToM), and attributional style (Green and Leitman 2008). As part of the CNTRICS initiative, Ochsner (2008) distilled the findings emerging from the non-clinical social cognitive and affective neuroimaging literature into a set of core underpinning abilities. Although his main interest was schizophrenia, the resulting 'Social-Emotional Processing Stream' provides a useful 'roadmap' for research seeking to identify mechanisms underlying social and emotional dysfunction in different mental disorders. This Social-Emotional Processing Stream is comprised of five main constructs or abilities, each mapping onto one another (for details see Fig. 1).

4 Social Cognition and Impact on Function

Couture et al. (2006) point out that Adolphs' (2001) definition of social cognition implies a close link between this concept and functional outcomes in the social domain, such as family, peer and romantic relationships and work and school

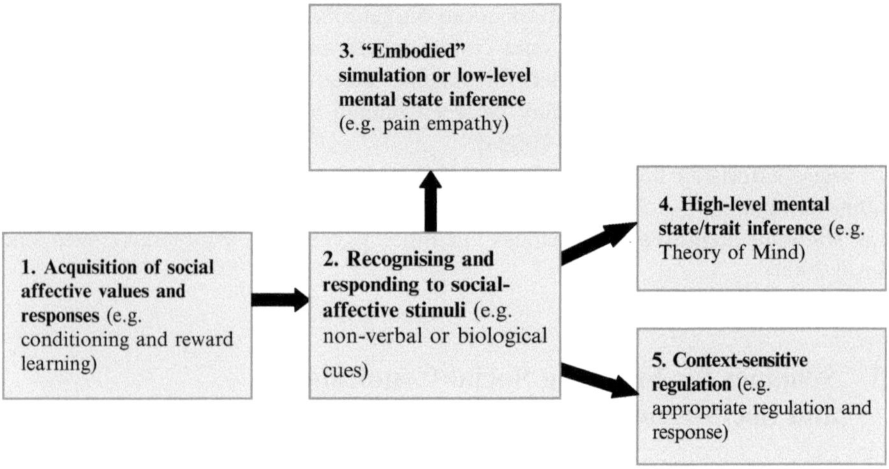

Fig. 1 The five constructs of the social-emotional processing stream (from Ochsner 2008)

interpersonal interactions. In a recent review, these authors examined the link between different social cognitive concepts including emotion perception (or emotion recognition, i.e. the ability to infer emotional information from facial expressions, vocal inflexions, or a combination of these), social perception (i.e. the ability to ascertain social cues from behaviour provided in a social context), ToM (i.e. the ability to understand and make inferences about other people's mental state and intentions), attributional style (i.e. people's tendencies for explaining the causes of events in their lives) and functional outcomes (social behaviour in the community, social skills and social problem solving) in relation to schizophrenia (Couture et al. 2006). Overall, they found clear and consistent relationships between social cognitive constructs and aspects of social functioning. These were strongest for social perception. Emotion perception too appeared to have a fairly consistent, but modest relationship with aspects of social functioning. For both these domains, there appeared to be some evidence that they mediate the relationship between neurocognition and functional outcomes. ToM and attributional style have been far less well studied in relation to their functional significance in schizophrenia.

5 Socio-Emotional Processing in Anorexia Nervosa

From the earliest descriptions of AN, the importance of emotional and social factors in its origins has been highlighted. For example, in 1873, Charles Lasègue noted that the young woman with AN 'suffer[s] from some emotions she avows or conceals' (quoted from Vandereycken and van Deth 1990). The psychoanalyst Hilde Bruch hypothesised that women with AN have an underlying impairment in their ability to identify emotional states and responses (Bruch 1962, 1977). In addition, it has been proposed that AN may be triggered by a fear of sexual maturation and adult autonomy (Crisp 1980) and that AN patients see the obtainment of thinness as a means to solve these and other psychosocial conflicts (Russell 1995).

Recent research supports these early observations of the role of socio-emotional factors in the aetiology of AN. For example, pre-morbidly, people who develop AN are commonly shy with few friends (Fairburn et al. 1999), have poor social functioning (Gillberg and Råstam 1992; Zucker et al. 2007) or social phobia (Kaye et al. 2004; Godart et al. 2003), are overanxious (Raney et al. 2008), experience high levels of negative affectivity (Pike et al. 2008) and negatively compare themselves to their unaffected sisters (Karwautz et al. 2001). Research into precipitating factors of AN found that severe interpersonal problems prior to onset were the most common form of stressor (Schmidt et al. 1997).

Whilst weight loss in AN may initially have positive social consequences, attracting the admiration and envy of peers (Branch and Eurman 1980), impairments in the social and emotional domain also arise as a consequence of AN (Schmidt et al. 1995). This is almost certainly related to the effects of starvation. These effects have been well documented in the Minnesota starvation experiment (Keys et al. 1950) and more recently in relation to caloric restriction for longevity (Vitousek et al. 2004), with starvation causing dysphoria and a withdrawal from and loss of desire for all social and sexual experiences. This is also apparent in AN where small social networks are reported (Tiller et al. 1997). In addition, AN may exacerbate pre-morbid psychosocial and emotional difficulties as the disorder causes family discord, distrust and frustration (Schmidt et al. 1995). Socio-emotional problems have been found to be associated with poor prognosis (Zipfel et al. 2000; Herpertz-Dahlmann et al. 2001). These empirical observations have led to the recognition of socio-emotional factors in AN as playing a maintaining role and thus as a focus for treatment in newer illness models (Fairburn et al. 2003; McIntosh et al. 2000; Schmidt and Treasure 2006). However, the way to identify, conceptualise and ultimately ameliorate difficulties in the socio-emotional domain remains uncertain, as research evidence in this area is piecemeal, based on different theoretical assumptions and poorly integrated. As yet, the specific social-cognitive difficulties and underlying neural mechanisms are elusive and understudied (Zucker et al. 2007).

Building on the model by Ochsner (2008) (see above under Fig. 1), we conducted a systematic review of experimental studies investigating socio-emotional factors in AN, categorising studies within Ochsner's previously established framework of the 'socio-emotional processing stream' comprising acquisition of social-affective stimuli (construct 1), recognition and response to social-affective stimuli (construct 2) low-level mental state inference (construct 3), high-level mental state inference (construct 4) and context-sensitive emotion regulation (construct 5) (Oldershaw et al. submitted). From a total of 45,617 articles identified and after removal of duplicate or irrelevant abstracts, 50 records were assessed in depth for eligibility. Of these, 43 fulfilled inclusion criteria for the review. Studies mapped on to four of the five constructs of the processing stream: Construct 1 (n = 15 studies); construct 2 (n = 21 studies); construct 3 (n = 0 studies); construct 4 (n = 7 studies); construct 5 (n = 4 studies). Socio-emotional functioning in AN was impaired across all four constructs. A meta-analysis of a subset of studies on facial emotion recognition (n = 9) was conducted and found that basic emotion recognition (n = 6 studies) is impaired in AN although effect sizes are small. More complex emotion recognition using Baron-Cohen et al. (2001) Reading the Mind in the Eyes Task (RME) is more impaired (n = 3 studies) with medium to large effect sizes. However, across all constructs studied, the majority of studies concerned current AN and were cross-sectional. Thus, more research is needed to delineate whether impairments are trait or state based. Two recent studies from our group (Oldershaw et al. 2010; Harrison et al. in press) examined RME in people who had recovered from AN but findings were inconsistent with one study finding a lack of impairment in recovered patients, and, the other finding persistence of impairments. The links between these socio-emotional constructs and social function has as yet not been assessed, although such work is currently under way.

6 Developing a Social-Cognitive Intervention for Anorexia Nervosa

As suggested above, there are many unanswered questions in relation to the role of socio-emotional processes in AN; in particular, whether they are state or trait deficits and how they impact on function. Thus, a sceptic might think that it is premature to target these processes in treatment. However, qualitative studies in patients with AN uniformly highlight that these patients experience emotions and social interactions as highly problematic (Fox 2009: Kyriacou et al. 2009), experience the perceived uncertainty of social interactions as highly stressful (Sternheim et al. 2010) and believe that overcoming these difficulties is necessary for recovery (Federici and Kaplan 2008).

Many of the psychotherapies currently used in the treatment of eating disorders have a focus on the expression and the processing of emotions and social relationships. For example, the technique of circular questioning employed in family therapy (e.g. what do you think your mother thinks about your anorexia?) invites family members to guess each other's mental state in the context of core familial relationships. In CBT, identification and challenging of patients' beliefs about their experience and emotional expression again involves a form of mentalising (thinking about one's own thoughts). IPT involves the expression of affect and some skill building in the area of social relationships. Dialectical behaviour therapy teaches emotion regulation and social skills. Thus, in future clinical trials, it will be worth assessing whether outcomes in these therapies are mediated by improved social cognitive abilities.

How then would a specifically social cognitive intervention differ from these more conventional therapies? Again, it will be useful to look at what has been done in the schizophrenia field. Here, two types of social cognitive interventions have been developed: highly targeted interventions, focusing on just one specific social cognitive domain (e.g. such as emotion perception), and interventions that are more broad based and include social-cognitive aspects in combination with other approaches such as cognitive remediation and social skills training. Both types have been found to lead to improvements in the targeted areas (for review see Couture et al. 2006). These authors go on to pose two important questions: 'can we expect the narrow focus of targeted intervention to yield improvements across social cognitive domains or to generalize to social functioning?' and 'if targeted interventions are too narrow, are broad-based interventions too burdensome?' As yet the answer to this is not known.

We have developed an outpatient treatment for people with AN (MANTRA; Maudsley Model of Anorexia Nervosa Treatment for Adults) (Schmidt 2009) which is based on a specific maintenance model of AN (Schmidt and Treasure 2006). The model and treatment are novel in several respects: (1) it is empirically based, drawing on and incorporating recent neuropsychological, social cognitive and personality trait research in AN, (2) it includes both intrapersonal and interpersonal maintaining factors and strategies to address these, (3) it is modularised with a clear hierarchy of procedures, tailored to the need of the individual. One of the key modules of this treatment is the 'Social and Emotional Mind' module which focuses on improving social-cognitive skills.

Mindful of Adolphs' (2001) definition of social cognition as 'the ability to construct representations of the relation between oneself and others and to use those representations flexibly to guide social behavior', a prominent focus of the therapeutic exercises in this module is to improve ToM. Of note, it has been recognised that ToM is a multidimensional construct with evidence supporting a distinction between affective and cognitive ToM functions with partly different neural correlates (Kalbe et al. 2009). Functional divisions in the Medial Pre-Frontal Cortex (MPFC) are associated with cognitive versus emotional processing (Amodio and Frith 2006). Neuroimaging studies suggest that the anterior rostral MPFC plays a key role in a range of more emotional mentalising tasks, where participants have to either infer the behaviour of other people in terms of their mental state or answer questions about general or specific dispositions/attitudes of others or their own perception of themselves (Frith 2007; Amodio and Frith 2006). Frith (2007) suggests that the anterior rostral MPFC has a particular role in handling communicative intentions rather than private intentions. He notes that: 'This is a more complex process than simply thinking about intentions since we have to recognize that the communicator is also thinking about our mental state'.

7 Some Other Examples

Already in 1868, Gull described several case histories in which he placed emphasis on the physical emaciation of the anorexic patient as well as the sometimes remarkable degree of energy and activity, in light of the undernourished state. A large proportion of AN patients (31–80% depending on the study methods) exhibit abnormally high activity levels or hyperactivity and over exercises (Favaro et al. 2000; Hebebrand et al. 2003; Holtkamp et al. 2003). Almost all AN patients show a constant, agitated restlessness when they are emaciated, but before they become lethargic in the final stages of starvation. In clinical practice, hyperactivity is a worrisome symptom. Hyperactivity leads to accelerated weight loss, to potentially lethal cardiovascular complications and because of its obsessive components often to drop-out of treatment programmes. The exact nature of hyperactivity remains to be clarified and although it is not included in the DSM IV criteria, several authors (Hebebrand et al. 2003; Casper 2006) have described and commented on this seemingly contradictory phenomenon. Some authors state that it might be seen as a core symptom of AN (Casper and Jabine 1996). Neurobiological factors and conscious attempts to burn calories in order to loose more weight coexist. Hyperactivity has been described as a pre-morbid feature (Davis et al. 1997), and it accelerates body weight loss during food restriction. Therefore, this behaviour is important not only for treatment but also for identifying factors for genetic and behavioural susceptibilities to AN.

Animal models mimicking AN weight loss and hyperactivity, such as the activity-based anorexia (ABA) model (Routtenberg and Kuznesof 1967; Kas et al. 2003) or the semi-starvation-induced hyperactivity (SIH) model (Pirke et al. 1993; Exner et al. 2000), have been helpful in the search for a possible biological drive or changes in physiological parameters that trigger food restriction and hyperactivity. Studies using these preclinical models showed that anorectic rats (Kas et al. 2003) and mice (Gelegen et al. 2007) have reduced plasma leptin levels and body temperature and that treating food-restricted rats with a hot plate or leptin suppresses the development of hyperactivity (Hillebrand et al. 2005a; Exner et al. 2000). Several studies in humans (Holtkamp et al. 2006; van Elburg et al. 2007) have confirmed the correlation of leptin and hyperactivity levels in acutely ill anorectic patients and the beneficial effects of external heat (Gutierrez et al. 2008).

8 Discussion

As yet the area of social-cognitive intervention development is in its infancy, as are other areas of translating experimental neuroscience into clinical interventions. A number of gaps need to be filled in order to determine whether this kind of intervention is worth pursuing further and what are potentially the most promising targets. In particular, we need more research into different aspects of social cognition in AN. This also includes use of social-cognitive tasks that are able to distinguish between domain-specific and general impairments. Moreover, studies assessing social cognitive impairments longitudinally over the course of treatment in AN and prospectively in high-risk adolescent cohorts prior to onset are needed so as to be able to distinguish between trait and state impairments and potential post-illness 'scarring' effects. Links between different social-cognitive constructs and social functioning in people with AN will need to receive much greater research attention. Additionally, the question as to how neurocognition and social cognition relate to each other in AN will need addressing.

The neural correlates of social-cognitive impairments in AN also need to be assessed. To this end, human studies of social-cognition in AN could usefully be supplemented with animal studies. The prairie vole is a model organism for understanding the social brain (McGraw and Young 2010). It is a highly affiliative animal with stable partner bonds. It has been used to study the consequences of social loss which in female voles, amongst other changes, leads to reduced sucrose intake. With the development of genomic resources and transgenic technologies comparable to those available in mice and rats, vole models of eating disorders might be possible to create.

Clearly, social-cognitive impairments are not specific to AN and are found across many other disorders. The magnitude of social-cognitive impairments in AN seems to be fairly comparable to that in high-functioning ASDs (Oldershaw et al. submitted). Moreover, recent research comparing schizophrenia and ASDs has found that whilst there is much similarity between the two in relation to their social cognitive functioning, there are also important differences (Couture et al. 2009; Sasson et al. 2007). Future research could therefore usefully focus on similarities and differences between disorders, as this might allow development of trans-diagnostic and disorder-specific interventions/intervention components.

At present, broad-based social-cognitive programmes for the treatment of AN are our 'best bet'. With this in mind, in addition to the programme described above, our group has developed a social cognitive programme for inpatients with AN (CREST, Tchanturia et al., unpublished), which combines cognitive and social cognitive training, and a group social-cognitive programme for people with bulimia nervosa (Lavender et al., unpublished). Both are currently undergoing intensive evaluation. Preliminary impressions suggest that these interventions are highly acceptable to patients.

Clearly, the translational process between basic research and treatment development is iterative and has to combine a top–down (theory-led, data-driven) and bottom–up (patient and clinician experience-led) approach. This is something that traditionally the field of psychological treatment has not been comfortable with its schools of psychotherapies where treatment procedures and components are 'invented', usually by a charismatic leader, and then become set in stone and reified over time.

One danger is that the idea of brain training through structured exercises is currently very fashionable and popular with the public, despite often exaggerated and unsubstantiated claims of its efficacy. Explanatory models for eating disorders have also undergone fashions, and the current enthusiasm for targeted brain-directed treatments should not blind us to the fact that rather humblingly the best available treatment for AN currently is a non-specific supportive treatment (McIntosh et al. 2005).

Acknowledgements We thank Mrs Hannah Broadbent, Mrs Hannah de Jong and Dr. Sebastien Guillaume and the members of URGE, the Utrecht Research Group Eating disorders, for their helpful comments on this chapter. This work was supported by the Marie Curie Research Training Network, INTACT (MRTN-CT-2006-035988), the Swiss Anorexia Foundation, the NIHR Biomedical Research Centre for Mental Health, South London and Maudsley NHS Foundation Trust and Institute of Psychiatry, King's College London and by a Department of Health NIHR Programme Grant for Applied Research (Reference number RP-PG-0606-1043). The views expressed herein are not necessarily those of DoH/NIHR.

References

Adolphs R (1999) Social cognition and the human brain. Trends Cogn Sci 3:469–479

Adolphs R (2001) The neurobiologiy of social cognition. Curr Opin Neurobiol 11:231–239

Amodio DM, Frith CD (2006) Meeting of minds: the medial frontal cortex and social cognition. Nat Rev Neurosci 7:68–77

Bailer UF, Frank GK, Henry SE, Price JC, Meltzer CC, Weissfeld L, Mathis CA, Drevets WC, Wagner A, Hoge J, Ziolko SK, McConaha CW, Kaye WH (2005) Altered brain serotonin 5-HT1A receptor binding after recovery from anorexia nervosa measured by positron emission tomography and [carbonyl11C]WAY-100635. Arch Gen Psychiatry 62(9):1032–1041

Bailer UF, Frank GK, Henry SE, Price JC, Meltzer CC, Becker C, Ziolko SK, Mathis CA, Wagner A, Barbarich-Marsteller NC, Putnam K, Kaye WH (2007) Serotonin transporter binding after recovery from eating disorders. Psychopharmacology (Berl) 195(3):315–324, Epub 2007 Aug 11

Barch DM, Carter CS, CNTRICS Executive Committee (2008) Measurement issues in the use of cognitive neuroscience tasks in drug development for impaired cognition in schizophrenia: a report of the second consensus building conference of the CNTRICS initiative. Schizophr Bull 34:613–618

Barch DM, Carter CS, Arnsten A, Buchanan RW, Cohen JD, Geyer M, Green MF, Krystal JH, Nuechterlein K, Robbins T, Silverstein S, Smith EE, Strauss M, Wykes T, Heinssen R (2009a) Selecting paradigms from cognitive neuroscience for translation into use in clinical trials: proceedings of the third CNTRICS meeting. Schizophr Bull 35:109–14

Barch DM, Braver TS, Carter CS, Poldrack RA, Robbins TW (2009b) CNTRICS final task selection: executive control. Schizophr Bull 35:115–135

Barch DM, Berman MG, Engle R, Hurdelbrink Jones J, Jonides J, MacDonald A III, Evan Nee D, Redick TS, Sponheim SR (2009c) CNTRICS final task selection: working memory. Schizophr Bull 35:136–152

Baron-Cohen S, Wheelwright S, Hill J, Raste Y, Plumb I (2001) The "Reading the Mind in the Eyes" test revised version: a study with normal adults, and adults with Asperger syndrome or high-functioning autism. J Child Psychol Psychiatry 42:241–251

Branch CH, Eurman LJ (1980) Social attitudes toward patients with anorexia nervosa. Am J Psychiatry 137:631–632

Bruch H (1962) Perceptual and conceptual disturbances in anorexia nervosa. Psychosom Med 24:187–194

Bruch H (1977) Psychotherapy in eating disorders. Can Psychiatr Assoc J 22:102–108

Bulik CM, Berkman ND, Brownley KA, Sedway JA, Lohr KN (2007) Anorexia nervosa treatment: a systematic review of randomized controlled trials. Int J Eat Disord 40:310–320

Carter CS, Barch DM, Buchanan RW, Bullmore E, Krystal JH, Cohen J, Geyer M, Green M, Nuechterlein KH, Robbins T, Silverstein S, Smith EE, Strauss M, Wykes T, Heinssen R (2008) Identifying cognitive mechanisms targeted for treatment development in schizophrenia: an overview of the first meeting of the cognitive neuroscience treatment research to improve cognition in schizophrenia (CNTRICS) inititative. Biol Psychiatry 64:4–10

Carter CS, Barch DM, Gur R, Gur R, Pinkham A, Ochsner K (2009) CNTRICS final task selection: social cognitive and affective neuroscience-based measures. Schizophr Bull 35:153–162

Casper R (2006) The 'drive for activity' and "restlessness" in anorexia nervosa: potential pathways. J Affect Disord 92:99–107

Casper R, Jabine L (1996) An eight-year follow-up: outcome from adolescent compared to adult onset anorexia nervosa. J Youth Adolesc 25:499–517

Cohen JD, Insel TR (2008) Cognitive neuroscience and schizophrenia: translational research in need of a translator. Biol Psychiatry 64:2–3

Couture SM, Penn DL, Roberts DL (2006) The functional significance of social cognition in schizophrenia: a review. Schizophr Bull 32(Suppl 1):S44–S63

Couture SM, Penn DL, Losh M, Adolphs R, Hurley R, Piven J (2009) Comparison of social cognitive functioning in schizophrenia and high functioning autism: more convergence than divergence. Psychol Med 12:1–11

Crisp AH (1980) Anorexia nervosa: let me be. Academic Press, London

Davies H, Schmidt U, Stahl D, Tchanturia K Evoked facial emotional expression and emotional experience in people with anorexia nervosa. Int J Eat Dis. In press

Exner C, Hebebrand J, Remschmidt H, Wewetzer C, Ziegler A, Herpertz S et al (2000) Leptin suppresses semi-starvation induced hyperactivity in rats: implications for anorexia nervosa. Mol Psychiatry 5:476–481

Fairburn CG (2005) Evidence-based treatment of anorexia nervosa. Int J Eat Disord 37(Suppl S26-30):discussion S41–42

Fairburn CG, Cooper Z, Doll HA, Welch SL (1999) Risk factors for anorexia nervosa: three integrated case-control comparisons. Arch Gen Psychiatry 56:468–476

Fairburn CG, Cooper Z, Shafran R (2003) Cognitive behaviour therapy for eating disorders: a "transdiagnostic" theory and treatment. Behav Res Ther 41:509–528

Favaro A, Caregaro L, Burlina AB, Santonastaso P (2000) Tryptophan levels, excessive exercise, and nutritional status in anorexia nervosa. Psychosom Med 62(4):535–538

Federici A, Kaplan AS (2008) The patient's account of relapse and recovery in anorexia nervosa: a qualitative study. Eur Eat Disord Rev 16:1–10

Fox JR (2009) A qualitative exploration of the perception of emotions in anorexia nervosa: a basic emotion and developmental perspective. Clin Psychol Psychother 16:276–302

Frank GK, Bailer UF, Henry SE, Drevets W, Meltzer CC, Price JC, Mathis CA, Wagner A, Hoge J, Ziolko S, Barbarich-Marsteller N, Weissfeld L, Kaye WH (2005) Increased dopamine D2/D3 receptor binding after recovery from anorexia nervosa measured by positron emission tomography and [11c]raclopride. Biol Psychiatry 58(11):908–912, Epub 2005 Jun 29

Frith C (2007) The social brain? Philos Trans R Soc Lond B Biol Sci 362:671–678

Gelegen C, Collier DA, Campbell IC, Oppelaar H, van den Heuvel J, Adan RA et al (2007) Difference in susceptibility to activity-based anorexia in two inbred strains of mice. Eur Neuropsychopharmacol 17(3):199–205, May 28 epub

Gillberg C, Råstam M (1992) Do some cases of anorexia-nervosa reflect underlying autistic-like conditions. Behav Neurol 5:27–32

Godart NT, Falment MF, Curt F, Perdereau F, Jeammet P (2003) Comorbidity between eating disorders and anxiety disorders. Int J Eat Disord 23:253–270

Green MF, Leitman DI (2008) Social cognition in schizophrenia. Schizophr Bull 34:670–672

Green MF, Nuechterlein KH, Gold JM, Barch DM, Cohen J, Essock S, Fenton WS, Frese F, Goldberg TE, Heaton RK, Keefe RS, Kern RS, Kraemer H, Stover E, Weinberger DR, Zalcman S, Marder SR (2004) Approaching a consensus cognitive battery for clinical trials in schizophrenia: the NIMH-MATRICS conference to select cognitive domains and test criteria. Biol Psychiatry 56:301–307

Green MF, Butler PD, Chen Y, Geyer MA, Silverstein S, Wynn JK, Yoon JH, Zemon V (2009) Perception measurement in clinical trials of schizophrenia: promising paradigms from CNTRICS. Schizophr Bull 35:163–181

Gutierrez E, Cerrato M, Carrera O, Vazquez R (2008). Heat reversal of activity-based anorexia: implications for the treatment of anorexia nervosa. Int J Eat Disord 41(7):594–601

Harrison A, Sullivan S, Tchanturia K, Treasure J (2010) Emotion recognition and regulation in anorexia nervosa: state or trait. Biol Psychiatry (in press) Biol Psychiatry Jun 28 [Epub ahead of print]
Hebebrand J, Exner C, Hebebrand K, Holtkamp K, Casper RC, Remschmidt H, Herpertz-Dahlmann B, Klingenspor M (2003) Hyperactivity in patients with anorexia nervosa and in semistarved rats: evidence for a pivotal role of hypoleptinemia. Physiol Behav 79:25–37
Herpertz-Dahlmann B, Müller B, Herpertz S, Heussen N, Hebebrand J, Remschmidt H (2001) Prospective 10-year follow-up in adolescent anorexia nervosa–course, outcome, psychiatric comorbidity, and psychosocial adaptation. J Child Psychol Psychiatry 42:603–612
Hillebrand JJG, Koeners MP, de Rijke CE, Kas MJ, Adan RA (2005a) Leptin treatment in activity-based anorexia. Biol Psychiatry 58:165–171
Hillebrand JJG, van Elburg AA, Kas MJ, van Engeland H, Adan RA (2005b) Olanzapine reduces physical activity in rats exposed to activity-based anorexia: possible implications for treatment of anorexia nervosa? Biol Psychiatry 58:651–657
Holtkamp K, Herpertz-Dahlmann B, Mika C, Heer M, Heussen N, Fichter MM, Herpertz S, Senf W, Blum WF, Schweiger U, Warnke A, Ballauff A, Remschmidt H, Hebebrand J (2003) Elevated physical activity and low leptin levels co-occur in patients with anorexia nervosa. J Clin Endocrinol Metab 88:5169–5174
Holtkamp K, Herpertz-Dahlmann B, Hebebrand K, Mika C, Kratsch J, Hebebrand J (2006) Physical activity and restlessness correlate with leptin levels in patients with adolescent anorexia nervosa. Biol Psychiatry 60(3):311–313
Kalbe E, Schlegel M, Sack AT, Nowak DA, Dafotakis M, Bangard C, Brand M, Shamay-Tsoory S, Onur OA, Kessler J (2009) Dissociating cognitive from affective theory of mind: a TMS study. Cortex 46(6):769–780, 2009 Jul 29. [Epub ahead of print]
Karwautz A, Rabe-Hesketh S, Hu X, Zhao J, Sham P, Collier DA, Treasure JL (2001) Individual-specific risk factors for anorexia nervosa: a pilot study using a discordant sister-pair design. Psychol Med 31:317–329
Kas MJ, van Elburg AA, van Engeland H, Adan RA (2003) Refinement of behavioural traits in animals for the genetic dissection of eating disorders. Eur J Pharmacol 480:13–20
Kas MJ, Kaye WH, Foulds Mathes W, Bulik CM (2009) Interspecies genetics of eating disorder traits. Am J Med Genet B Neuropsychiatr Genet 150B(3):318–327
Kaye WH, Bulik CM, Thornton L, Barbarich N, Masters K (2004) Comorbidity of anxiety disorders with anorexia and bulimia nervosa. Am J Psychiatry 161:2215–2221
Keys A, Brozek J, Henschel A, Mickelson O, Taylor H (1950) The biology of human starvation. University of Minnesota Press, Minneapolis
Klein DA, Walsh BT (2005) Translational approaches to understanding anorexia nervosa. Int J Eat Disord 37(Suppl S10–14): discussion S20–21
Klump KL, Bulik CM, Kaye WH, Treasure J, Tyson E (2009) Academy for eating disorders position paper: eating disorders are serious mental illnesses. Int J Eat Disord 42:97–103
Kyriacou O, Easter A, Tchanturia K (2009) Comparing views of patients, parents, and clinicians on emotions in anorexia: a qualitative study. J Health Psychol 14:843–854
Le Grange D, Eisler I (2009) Family interventions in adolescent anorexia nervosa. Child Adolesc Psychiatr Clin N Am 18:159–173
Lopez C, Tchanturia K, Stahl D, Treasure J (2009) Weak central coherence in eating disorders: a step towards looking for an endophenotype of eating disorders. J Clin Exp Neuropsychol 31:117–25
Marder SR, Fenton W (2004) Measurement and treatment research to improve cognition in schizophrenia: NIMH MATRICS initiative to support the development of agents for improving cognition in schizophrenia. Schizophr Res 72:5–9
McGraw LA, Young LJ (2010) The prairie vole: an emerging model organism for understanding the social brain. Trends Neurosci 33(2):103–109
McIntosh VV, Bulik CM, McKenzie JM, Luty SE, Jordan J (2000) Interpersonal psychotherapy for anorexia nervosa. Int J Eat Disord 27:125–139

McIntosh VV, Jordan J, Carter FA, Luty SE, McKenzie JM, Bulik CM, Frampton CM, Joyce PR (2005). Three psychotherapies for anorexia nervosa: a randomized, controlled trial. Am J Psychiatry 162(4):741–747

National Collaborating Centre for Mental Health (2004) Eating disorders: core interventions in the treatment and management of anorexia nervosa, bulimia nervosa and related eating disorders. British Psychological Society and Gaskell, Leicester and London

Nuechterlein KH, Barch DM, Gold JM, Goldberg TE, Green MF, Heaton RK (2004) Identification of separable cognitive factors in schizophrenia. Schizophr Res 72:29–39

Nuechterlein KH, Green MF, Kern RS, Baade LE, Barch DM, Cohen JD, Essock S, Fenton WS, Frese FJ 3rd, Gold JM, Goldberg T, Heaton RK, Keefe RS, Kraemer H, Mesholam-Gately R, Seidman LJ, Stover E, Weinberger DR, Young AS, Zalcman S, Marder SR (2008) The MATRICS consensus cognitive battery, part 1: test selection, reliability, and validity. Am J Psychiatry 165:203–213

Nuechterlein KH, Luck SJ, Lustig C, Sarter M (2009) CNTRICS final task selection: control of attention. Schizophr Bull 35:182–196

Ochsner KN (2008) The social-emotional processing stream: five core constructs and their translational potential for schizophrenia and beyond. Biol Psychiatry 64:48–61

Oldershaw A, Hambrook D, Tchanturia K, Treasure J, Schmidt U (2010) Emotional theory of mind and emotional awareness in recovered anorexia nervosa patients. Psychosom Med 72:73–79

Oldershaw A, Hambrook D; Stahl D, Tchanturia K, Treasure J, Schmidt U The socio-emotional processing stream in anorexia nervosa. Paper submitted for publication

Oldershaw A, Treasure J, Hambrook D,Tchanturia K, Schmidt U Is anorexia nervosa a female version of autism spectrum disorders? Submitted for publication

Pike KM, Hilbert A, Wilfley DE, Fairburn CG, Dohm FA, Walsh BT, Striegel-Moore R (2008) Toward an understanding of risk factors for anorexia nervosa: a case-control study. Psychol Med 38:1443–1453

Pinkham AE, Penn DL, Perkins DO, Lieberman J (2003) Implications for the neural basis of social cognition for the study of schizophrenia. Am J Psychiatry 160:815–824

Pirke KM, Broocks A, Wilckens T, Marquard R, Sweiger U (1993) Starvation-induced hyperactivity in the rat: the role of endocrine and neurotransmitter changes. Neurosci Biobehav Rev 17:287–294

Ragland JD, Cools R, Frank M, Pizzagalli DA, Preston A, Ranganath C, Wagner AD (2009) CNTRICS final task selection: long-term memory. Schizophr Bull 2009(35):197–212

Raney TJ, Thornton LM, Berrettini W, Brandt H, Crawford S, Fichter MM, Halmi KA, Johnson C, Kaplan AS, LaVia M, Mitchell J, Rotondo A, Strober M, Woodside DB, Kaye WH, Bulik CM (2008) Influence of overanxious disorder of childhood on the expression of anorexia nervosa. Int J Eat Disord 41:326–332

Roberts ME, Tchanturia K, Stahl D, Southgate L, Treasure J (2007) A systematic review and meta-analysis of set-shifting ability in eating disorders. Psychol Med 37:1075–1084

Routtenberg A, Kuznesof A (1967) Self-starvation of rats living in activity wheels on a restricted feeding schedule. J Comp Physiol Psychol 64(3):414–421

Russell G (1995) Anorexia nervosa through time. In: Dare C, Treasure J, Szmukler G (eds) Handbook of eating disorders: theory, treatment and research. Wiley, Chichester, pp 5–17

Sasson N, Tsuchiya N, Hurley R, Couture SM, Penn D, Adolephs R, Piven J (2007) Orienting to social stimuli differentiates social cognitive impairment in autism and schizophrenia. Neuropsychologia 45:2580–2588

Schmidt U (2009) Development & evaluation of the maudsley model of anorexia nervosa treatment for adults (MANTRA). Paper presented at the eating disorders research society, Brooklyn, USA, 2009

Schmidt U, Treasure J (2006) Anorexia nervosa: valued and visible. A cognitive interpersonal maintenance model and its implications for research and practice. Br J Clin Psychol 45:343–366

Schmidt U, Tiller J, Morgan H (1995) The social consequences of eating disorders. In: Szmukler G, Dare C, Treasure J (eds) Handbook of eating disorders: theory, treatment and research. Wiley, Chichester, pp 259–270

Schmidt U, Tiller J, Blanchard M, Andrews B, Treasure J (1997) Is there a specific trauma precipitating anorexia nervosa? Psychol Med 27:523–530

Sternheim L, Konstantellou A, Startup H, Schmidt U (2010) What does uncertainty mean to women with anorexia nervosa? An interpretative phenomenological analysis. Eur Eat Disord Rev Jul 28 [Epub ahead of print]

Tiller JM, Sloane G, Schmidt U, Troop N, Power M, Treasure JL (1997) Social support in patients with anorexia nervosa and bulimia nervosa. Int J Eat Disord 21:31–38

van Elburg AA, Kas MJ, Hillebrand JJ, Eijkemans RJ, van Engeland H (2007) The impact of hyperactivity and leptin on recovery from anorexia nervosa. J Neural Transm 114(9): 1233–1237, Epub 2007 May 26

Vandereycken W, van Deth R (1990) A tribute to Lasègue's description of anorexia nervosa (1873), with completion of its English translation. Br J Psychiatry 157:902–908

Vitousek KM, Manke FP, Gray JA, Vitousek MN (2004) Caloric restriction for longevity: II – The systematic neglect of behavioural and psychological outcomes in animal research. Eur Eat Disord Rev 12:338–360

Wagner A, Aizenstein H, Venkatraman VK, Fudge J, May JC, Mazurkewicz L, Frank GK, Bailer UF, Fischer L, Nguyen V, Carter C, Putnam K, Kaye WH (2007) Altered reward processing in women recovered from anorexia nervosa. Am J Psychiatry 164(12):1842–1849

Wagner A, Aizenstein H, Mazurkewicz L, Fudge J, Frank GK, Putnam K, Bailer UF, Fischer L, Kaye WH (2008) Altered insula response to taste stimuli in individuals recovered from restricting-type anorexia nervosa. Neuropsychopharmacology 33(3):513–523, Epub 2007 May 9

Wang PS, Heinssen R, Oliveri M, Wagner A, Goodman W (2008) Bridging bench and practice: translational research for schizophrenia and other psychotic disorders. Neuropsychopharmacology 34:204–212

Woolf SH (2008) The meaning of translational research and why it matters. JAMA 299:211–213

Zipfel S, Lowe B, Reas DL, Deter HC, Herzog W (2000) Long-term prognosis in anorexia nervosa: lessons from a 21-year follow-up study. Lancet 355:721–722

Zucker NL, Losh M, Bulik CM, LaBar KS, Piven J, Pelphrey KA (2007) Anorexia nervosa and autism spectrum disorders: guided investigation of social cognitive endophenotypes. Psychol Bull 133:976–1006

Cognitive Remediation Therapy for Eating Disorders: Development, Refinement and Future Directions

Kate Tchanturia and James Lock

Contents

Abstract In this chapter, we aim to address some basic conceptual and practical questions about cognitive remediation therapy (CRT) for eating disorders. We begin by providing an overall historical, conceptual, and theoretical framework for CRT. Next, we discuss the specific indications for how and why CRT might be useful for eating disorders based on existing neuropsychological research evidence. We also provide an overview of the types of tasks and stimuli used in CRT and a general protocol for a manualized version of CRT. In addition, modifications of the adult CRT manual for use with adolescents as well as preliminary acceptability of the approach with this younger age group are described. We also propose various ways to integrate CRT in a variety of inpatient and outpatient programmes.

K. Tchanturia (✉)
Department of Psychological Medicine, King's College London, London, UK
e-mail: Kate.Tchanturia@iop.kcl.ac.uk

J. Lock
Department of Psychiatry and Behavioral Sciences, Stanford University School of Medicine, Stanford, CA, USA

R.A.H. Adan and W.H. Kaye (eds.), *Behavioral Neurobiology of Eating Disorders*, Current Topics in Behavioral Neurosciences 6, DOI 10.1007/7854_2010_90

Finally, a discussion of potential future directions in research using the tools of neurocognitive assessment, imaging and treatment research is provided.

Keywords Anorexia nervosa · Attention to detail · Cognitive exercises · Neuropsychology · Set shifting · Treatment

1 What Is CRT and How Can It Be Useful in Psychiatric Disorders?

Cognitive remediation (CR) is an intervention that targets cognitive strengths and weaknesses. The focus is on encouraging patients to become more confident in using new cognitive strategies to solve problems more efficiently. In most of the cognitive remediation modules, patients are asked to do cognitive tasks that utilize a particular cognitive skill, as well as practice and reflect on their own cognitive strategies to improve this skill. There is evidence that improving cognition helps patients to improve their daily functioning. For example, six meta-analytic studies in schizophrenic populations showed that cognition/neuropsychological performance after cognitive remediation therapy (CRT) improved with moderate to large effect size. Furthermore, although the magnitude of change is less with daily activities, these are still consistent and encouraging results (Medalia and Choi 2009). Several forms of the CRT have been developed over the years. These developments include individual, group, computerized and pen and paper version approaches.

Historically, work in CRT in general started from the 1950s. Immanent Russian neuropsychologist A. Luria's experimental and clinical work into neurological disorders and rehabilitation after brain injury highly influenced the development of cognitive remediation (Das 1999). Most of the initial work in cognitive remediation was conducted in the acquired brain injury field (for review, see Rohling et al. 2009); a recent meta-analysis based on 967 articles, after careful screening and the application of meta-analytical methods, concluded that there is sufficient evidence for CR training for brain-injured patients.

Gradually CRT became a part of the treatment for older age groups, and in the late 1970s it was developed into a new treatment for schizophrenia (the paper by McGurk et al. 2007 includes 26 studies for meta-analysis). Cognitive remediation was also effectively applied to learning disabilities, mainly focusing on skills in learning and adaptation strategies (Stevenson et al. 2002). Over the last decade, several studies have made attempts to examine the efficacy of CRT in attention-deficit hyperactivity disorder (ADHD), the aim of which is to help patients with attention by focusing various attention training programmes.

In summary, to date CRT has been broadly used in the treatment of various psychiatric disorders. The field of CRT is rapidly growing and researchers and clinicians with expertise in this approach are facing more challenges with deciding what forms, which aspects and what portion of CRT to use, as well as reviewing what CRT can and cannot do. There are several forms of CRT which include computerized, individual, group and with or without a therapist. On the other hand, there are also a number of core principles that vary from programme to programme. For example, some programmes are based on the principal of "drill and practice", whereas some put more focus on motivational mechanisms, strategies and some are focused on developing meta-cognition. Before testing CRT in any particular disorder, from our point of view there are two factors that need to be considered: (1) what is the plausible research evidence and logic to bring CRT into a treatment package? (2) what focus should CRT have for specific disorders and what ingredients should be used: practice, strategies, behavioural and ecological strategies, reflection and the form of delivery? All these questions will tailor the application of CRT in any particular area.

2 Why Might CRT Be Useful in Eating Disorders?

Neuropsychology, in a broad sense, involves studying executive functions (planning, set-shifting, problem solving and decision making). Over the last three decades, the focus of neuropsychological research in eating disorders (EDs) has visibly shifted from the assessment of broad executive function to a hypothesis-driven approach targeting areas of clinical interest to understand more effectively what neuropsychological assessments/experiments can tell us about information processes. In fact, more recently these research findings are being translated into clinical practice (Tchanturia et al. 2005, 2008).

Two parallel processes can be observed in the current literature: on the one hand, interest is rapidly growing in examining the eating disorder field from a neuroscience perspective to broaden out understanding of the disorder (Kaye et al. 2009). On the other hand, there are attempts made to clarify assessments of executive function (Chan et al. 2008) and the integration of neuropsychological and cognitive theories in rehabilitation Hill et al (2001). There are a number of studies and a few systematic reviews on executive functions in ED reviewing: planning, cognitive flexibility, sustained attention, central coherence (extreme attention to detail) and working memory (Southgate et al. 2005, 2009; Lopez et al. 2008). Most of the available experimental literature sheds light on the so-called cold cognition (for neuropsychological assessment review, see Chan et al. 2008). Executive functions also involve "hot" emotional components in the processes like decision making where regulation of reward and punishment are involved (in ED, Cavedini et al. 2004; Tchanturia et al. 2007; Russell et al. 2009 provided some experimental work in this area). To date, findings from studies on "cold" executive functions in AN were translated into the treatment intervention of the CRT module. Therefore, we will focus on the findings in cognitive flexibility and central coherence which inspired the work in cognitive remediation for AN.

The most robust findings are supported with systematic reviews (Roberts et al. 2007; Lopez et al. 2008) which found that set-shifting (the ability to quickly change and adapt strategies when environmental demands change) and central coherence (extreme attention to details) are the areas where people with ED show suboptimal performance, compared to controls. It is worth noting that neuropsychological assessment tools were predominantly developed in brain-injured populations, and when it comes to ED in most cases we do not see a "deficit": it is a more suboptimal performance as a group in a range of one standard deviation below the norms (Lauer 2002). Extreme attention to detail (weak central coherence) in laboratory research was first reported from Gillberg and colleagues (for the update Gillberg et al. 2007); later several groups have also found an "eye for detail" at the expense of the bigger picture, using different experimental measures. For example, Southgate et al. (2008) reported that people with eating disorders (particularly AN) tend to spot details faster and more efficiently than people without an eating disorder (a matching familiar test was used to assess this). Lopez et al. (2008) reported a detailed strategy when completing the Rey figure in a cross-sectional study comparing people with and without ED. Tokley and Kemps (2007) concluded from their study that there is also poor abstraction in people with AN (using Object Assembly test WAIS, embedded figure test).

Poor flexibility in set-shifting was also reported across the studies (Tchanturia et al. 2003, 2004; Holliday et al. 2005; Roberts et al. 2007) using a variety of set-shifting tasks (all based on the idea to change and shift strategies with rule changes), but tapping different domains: cognition [catbat, Brixton, Wisconsin card sorting task (WCST), attention (Trail making task), and perception (perceptual illusion task)].

This cognitive profile of inefficient set-shifting and weak central coherence fits well with what clinical and personality research reports have highlighted, e.g. patients with AN having high perfectionism, obsessive-compulsive personality traits, harm avoidance, and low novelty seeking. All these characteristics are associated with poor flexibility.

In psychological treatments, the main focus of learning (therapy is learning) is on the content of the symptoms alongside a cognitive formulation of the illness and possible ways to change the beliefs and behaviours of the patients. In behavioural therapy, for example, the content of thoughts is highly important.

If cognitive characteristics such as inflexibility and an over attention to detail are addressed in therapy, this may increase the effectiveness of the therapy as patients become more aware of their thinking styles and are in a better position to work towards change. Thus, the idea of the CRT is to help patients to make their thinking process more flexible (to help patients be aware of their own inflexible thinking styles using symptom-free cognitive games and then apply this knowledge to areas of rigid behaviours they struggle with the most, e.g. eating, food shopping and exercising), to make patients "think outside the box" and see the bigger picture (e.g. how ED affects their own life, what impact it has on work, private life, relationships, and quality of life in general).

Of course, there is a big variation within AN patients as to what degree each of the cognitive characteristics are present (e.g. someone can be very flexible and yet stuck with details, or the other way round, someone may be very rigid but able to develop a bigger picture approach). In intensive studies of patient groups with eating disorders, it becomes clear that patients with the most severe clinical symptoms (e.g. very low lifetime BMI and long duration of the illness) have these cognitive style characteristics that are represented more strongly than in mild cases, adolescents or people who have recovered from the illness.

CRT for AN was developed with the aim of addressing these cognitive and information processing characteristics.

The original study of CRT for AN consisted of ten sessions of cognitive exercises and behavioural experiments specifically designed to allow patients to practise skills in cognitive flexibility and global processing (bigger picture thinking).

Case studies and exploratory case series in CRT have found improvements in participant's body mass index (BMI), performance on neuropsychological tasks and self-reported cognitive flexibility (Davies and Tchanturia 2005; Tchanturia et al. 2006, 2007). These gains have been sustained at a 6-month follow-up (Genders et al. 2008).

CRT was developed out of the continual requirement for novel treatments for AN. The National Institute of Clinical Excellence (NICE 2004) recognizes the lack of evidence-based treatments for adults with AN, and current guidelines cannot endorse a first line treatment for the disorder in this age group (for recent reviews, see Lock and Fitzpatrick 2009; Treasure et al. 2010). With current pressures for resources placed on clinicians, there is a need for cost-effective interventions. CRT, because of its motivational delivering style, has no symptom-related material, and with a relatively low intensity training requirement for the therapists, it is worth exploring as a starting point intervention. By this, we mean that CRT offers psychological input but not to the degree that cognitive behavioral therapy (CBT) and dialectical behavior therapy (DBT) or other complex psychological interventions do; in other words, CRT for ED is not a stand-alone psychological intervention: we see it as a complimentary addition which is worth researching further.

3 Why Might CRT Be Useful for Adolescent Anorexia Nervosa?

Epidemiological, biological, cognitive developmental process and treatment response data all suggest that adolescence is the critical period for onset and treatment of AN. Epidemiological studies show that AN is not randomly distributed among all populations. Young females are the most vulnerable group, with only 5–10% of clinical samples being male (Hoek and van Hoeken 2003; Van Son et al. 2006; Keski-Rahkonen et al. 2007). Studies suggest that incidence rates are increasing among 15–24-year-old females, while rates in adult women are stable (Lucas et al. 1991, 1999). Onset of AN after age 25 is relatively uncommon. Thus, adolescents with AN are the key population for whom treatments, including CRT, should be targeted.

Amenorrhea is a clinical indication of malnutrition associated with AN in females and a current diagnostic criteria for the disorder (American Psychiatric Association 1994; Arnsten and Shansky 2004). However, the importance of amenorrhea may be greater in adolescent females' developing cognitive functioning.

Rising oestradiol levels during this critical period are thought to increase susceptibility to impaired judgment, prefrontal dysfunction and neuropsychiatric disease (Arnsten and Shansky 2004). While there are only scant studies on sex hormone changes and cognitive function in adolescents, recent work has shown a link between amenorrhea or irregular menses and deficits in cognition, including recall, verbal memory, working memory, visual reproduction, reading, math and oral language (Chui et al. 2008). It is possible that suppression of normal circulating oestradiol levels may have a much greater impact on cognition during puberty in patients with AN than in chronically ill adults.

In addition, adolescence is a critical period of brain development associated with synaptic pruning, elaboration of dendritic arborisation and increased myelination (Luna and Sweeney 2004). These developmental processes support the integration of brain circuitry associated with the prefrontal cortex (PFC) and sub-cortical structures (basal ganglia and thalamus) supportive of the executive functioning area of the brain. This maturation serves to improve inhibitory and reflective processes, making them more efficient and consistent. For some adolescents, difficulties in this maturational process lead to a range of externalizing behavioural difficulties (Casey et al. 2008); while in others, difficulties may arise because of excessive inhibitory processes (Marsh et al. 2009a).

In summary, the age of onset of AN, biological and cognitive developmental processes, and common behavioural manifestations of these processes including particularly avoidance of risk taking in inhibited individuals support the idea that adolescents with AN might benefit from an intervention such as CRT aimed at making their cognitive processes more flexible and less perseverative on detail to the neglect of the whole. It might be anticipated that such an intervention has the potential for greater impact on these processes in developing adolescents with less "fixed" cognitive processing styles.

In addition to these developmental reasons for considering CRT for adolescents with AN, we have recently conducted a study examining set-shifting and central coherence in adolescents with AN (Fitzpatrick et al. submitted) to assess the utility of targeting these processes in a younger group of non-chronic patients.

Neuropsychological data on 26 adolescent females between the ages of 12 and 18 (mean age 14.9 SD = 1.92) were collected. Participants were at a per cent ideal body weight of 82.13% (SD = 8.97), with 59% of the sample having secondary amenorrhea, 14% having primary amenorrhea, 18% currently menstruating and the remainder did not report menstrual status. Eating Disorder Examination Questionnaire (Cooper and Fairburn 1987; Cooper et al. 1989; Fairburn and Cooper 1993; Carter et al. 2001): scores on the EDE-Q were greater than two standard deviations above published mean norm scores (Fairburn and Cooper 1993) on all four subscales. Subjects were within the average range on the Weschsler assessment scales (Wechsler 1997). We examined cognitive flexibility or set-shifting using the Wisconsin Card Sort Task (WCST) (Resources 2003). Overall, adolescents with AN performed within expectations for their age and IQ in terms of total performance, but their performance was characterized by a high number of perseverative errors, a finding consistent with set-shifting difficulties. We also explored set-shifting using the Trails on the Delis–Kaplin Executive Functioning Scale (DKEFS) (Delis et al. 2001). Compared to a normative sample on this subscale, participants performed within expectations for their age, but those with AN performed significantly more slowly than adults with AN on the simple letter sequencing task and the shift task.

Central coherence was assessed using the Rey Osterrieth Complex Figure (ROCF; Osterrieth 1944). On this measure, adolescents with AN closely resembled the adults with AN with no significant differences between them. However, adolescents with AN demonstrated greater weakness in central coherence compared to adolescent healthy controls. Taken together, these findings suggest that adolescents with AN demonstrate neurocognitive inefficiencies in set-shifting and central coherence compared to adolescent norms. In addition, the neuropsychological profile suggested by these findings were similar to those found in adults with chronic AN, though set-shifting, appears to less severe than in adults with AN.

4 What Is CRT in Detail for Adults?

Neuropsychological interventions have been developed, which mainly concentrate on (1) restoration of function, to improve specific skill deficits and (2) compensatory training to adapt the presence of certain behavioural or cognitive problems (from Eslinger and Oliveri 2002). In brain injury rehabilitation, the targets are individuals with (a) impairments – *loss*/abnormalities in psychological structure and function; (b) disabilities – *restriction* or lack of ability to perform activities in the range considered normal, or (c) handicaps – disadvantages that prevent fulfilment of a role that is normal.

For psychiatric conditions, neuropsychological interventions can be used in the context of the clinical picture and the research evidence drawn from experimental studies. Brain injured, ADHD, schizophrenia and ED patient groups are very diverse and have very different needs. This presents a big challenge for researchers and clinicians who have to tailor and use the principals of cognitive remediation in a clinically meaningful way for the targeted population.

In the context of ED, first of all we focused on anorexia. Our choice was determined from the fact that:

1. There is relatively wealthy research data on neuropsychology in AN (e.g. flexibility and extreme attention to detail as problematic areas).
2. Treatment in AN to date has very poor outcomes and needs further investigation.
3. Bringing a brain hypothesis to the treatment intervention could be plausible (by brain hypothesis we mean factors like childhood preterm birth, low BMI at birth, brain synaptic pruning affected by starvation and a high association between brain lesions and ED symptoms (Uher and Treasure 2005).

There are several CRT programmes (individual, group, computerized) which can be adapted for the ED cohort. The first step for such developments is to adapt materials for the ED population, test acceptability and effectiveness as well as improvements in cognitive style and treatment engagement.

For ED, much research lies ahead to explore the benefits of CRT and adjust various approaches in remediation to eating disorders. One of the first studies making an attempt to do this adapted a model developed in schizophrenia (in-house manual by Delahunty A, Reeder C, Wykes T, Morice R, Newton E). In schizophrenia, the modules for working memory, flexibility and planning are core components presented in five workbooks. Because of the research evidence (cognitive flexibility, eye for detail), we were interested in only the flexibility module for anorexia, where we have added a number of cognitive exercises (e.g. therapist and patient come up with names of countries in alphabetical order taking turns).

Letter	Name
A	
B	
C	

It could be more challenging, if it feels appropriate, to increase the difficulty of the task, e.g. switch to boys and girls names. Several exercises are based on the Stroop effect – patients are asked to name animal pictures and then switch to reading aloud the labels of the pictures.

In addition to the flexibility tasks such as the visual illusion task and the Stroop effect tasks, we have developed simple cognitive exercises which tap into the bigger picture approach, e.g. giving patient a page of meaningless or meaningful information to come up with a catchy (relevant) title; the therapist letter or assessment page can be used in later sessions to bullet point the information. After doing the exercise, the therapist further explores what the patient's observations are on cognitive style and how this relates to everyday life behaviours. The next step is to explore alternative ways to do the same cognitive task and apply this to real-life behavioural strategies.

Pilot work has been published in several papers (Davies and Tchanturia 2005; first case report, finalizing manual for pilot case series; Tchanturia et al. 2007, case series including work conducted in 23 patients with CRT manual for AN; qualitative evaluation of patients views on intervention; Whitney et al. 2008).

5 Can CRT Modified for Use in Adolescents with AN?

In order to address the cognitive inefficiencies in AN, the basic strategies of CRT similarly target cognitive flexibility and weak central coherence. However, a range of modifications are needed to make CRT more appropriate and useful for adolescents with AN. It was not initially clear, however, how best to amend the adult version of CRT; so a case series using the adult manual was undertaken to identify areas that needed to be modified. In general, adolescents with AN found the CRT tasks interesting to do, but more range and variety of tasks seemed to be needed to keep them engaged in the therapy. In addition, some adolescents felt the tasks to be not challenging enough, while at the same time some struggled with using the reflective "thinking about thinking" aspects of CRT after the tasks were completed. It is not too surprising that some of the cognitive flexibility tasks were not as challenging to adolescents because they are less compromised in this area than severely malnourished chronically ill AN adults. It is also not unexpected that metacognitive tasks were experienced as difficult in a subject pool where the mean age was a little over 14 years since self-reflection and abstract thinking are still developing skills in this age group.

In order to address these challenges, several new and modified tasks were added to the protocol. Specifically, more challenging complex geometric figures, complex line bisection tasks and modification of letters were added. In addition, the Main Idea task is a task designed to improve central coherence through summarizing key points in a written document. Feedback from pilot adolescents suggested that the language in the adult version was confusing to them and the content not interesting. To address these issues with the Main Idea task, adolescents were asked to identify the main points and themes contained in letters and short articles that were likely to be of interest to this age group. This letters varied in length, degree of detail and contained content designed to evoke feelings – including humour, anger and frustration.

Because pilot work and preliminary information related to cognitive processing suggest that adolescents are less compromised in terms of set-shifting than adults with chronic AN, it was important to make sure that the tasks used were pitched at the correct level, taking into account by age, ability and cognitive inefficiency. Participants in pilot study provided feedback on all the tasks as well as those that remained unchanged from the adult manual to help rank them in order of difficulty. In this way, the manual was arranged so that the CRT therapist could present tasks in a sequence that provided increasing cognitive processing challenges within each session and over the course of CRT. In practice, feedback from participants in terms of their reports of difficulty, interest, as well as therapist assessments of these factors, leads to an individualized course of CRT treatment for each participant.

The initial pilot study refining the treatment manual involved 20 adolescents with AN. The feasibility of the refined manual was examined in an additional 20 adolescents with AN. Of a total of 42 adolescents approached for CRT, only two refused after the treatment was described to them (i.e. 95% acceptance rate). This is noteworthy because many adolescents (as adults) with AN are highly resistant to psychotherapy and commonly actively reject individual therapy. These pilot treatments took place on an inpatient service (see below for details); so many of the patients were particularly unhappy about being in hospital making their acceptance of CRT even more surprising. Participants appeared to like the therapy because it was similar to tasks familiar to them (e.g. homework, school work) and most of them felt competent in taking on the tasks. This was the case despite the fact that the tasks in CRT are designed to become increasingly challenging. The adolescents were not usually frustrated increasing the level of difficulty in the tasks. Therapists noted that most patients appeared to proceed well through the assigned tasks. No participants asked for stopping CRT to be discontinued. The only reason for why CRT stopping was patient discharge. The patients reported liking the therapy and their therapists – not a common report for adolescents with AN in the early stages of treatment.

Parents also appeared to feel that CRT was useful, and none objected to the treatment during the hospital stay while several requested that the treatment continue in an outpatient setting.

Comparing these preliminary results with those of CRT with adults, there is much commonality. First, CRT was acceptable and feasible, perhaps even more so than with an adult population of chronically ill adults on an inpatient service. There are several possible reasons for this. The adolescents were not as underweight as those in studies of adults with AN. As some authors have noted, extremely low weight may exacerbate cognitive inefficiencies, especially set-shifting, which might also make taking on the task of CRT more difficult, thereby possibly increasing resistance. At the same time, both adolescents and adults seemed to agree that working on their thinking styles was potentially worth while, and that having a treatment that did not focus on eating and weight was helpful in some ways by distracting them from these more challenging problems, while also being beneficial. Most thought that working with the CRT therapists was a good experience. Thus, the experience of CRT for both adolescents and adults has the potential for being a good model for later work with therapists, even therapy targeting eating, shape and weight. Some therapists thought that adolescents may have had a harder time with tasks related to meta-cognition than adults. If this was the case, it would not be too surprising since such perspective taking is more challenging for adolescents in this age group even without AN. This is a potential area for further refinement of the CR intervention.

6 How Can CRT Be Integrated into Focused Treatments for Eating Disorders?

CRT is a flexible treatment that can likely be used in conjunction with a range of other more focused treatments for eating disorders. In this section, we provide examples of how CRT might be integrated into common programmes including inpatient treatment and cognitive behavioural therapy for outpatients.

6.1 The New Maudsley Model

Anorexia is one of the most difficult conditions to treat. Treatment guidelines for AN rely on expert recommendations. These recommendations emphasize the importance of a multidisciplinary approach including medical, nutritional, social and psychological components (Lock and Fitzpatrick 2009; Treasure et al. 2010).

In relation to CRT, we have offered some reflections on similarities and differences from CBT (Baldock and Tchanturia 2007; Tchanturia and Hambrook 2009). Below we will focus on CRT in the context of the new Maudsley model. Schmidt and Treasure (2006) proposed the Maudsley Model for the treatment of AN, which is based on the following four maintenance factors:

1. Obsessive-compulsive personality traits (OCPD)
2. Avoidance of emotion
3. Pro-anorexic beliefs
4. Responses to close others

This model promotes specifically tailored interventions to address these factors.

It is highlighted in the model that the core active ingredient of the therapeutic approach is the empathic, reflective style of motivational interviewing. Engagement of AN patients in any treatment is problematic, and therefore therapists' curiosity and collaborative style are highly important. CRT is based on the same principal of working "together" with the patient and taking a journey in the exploration of cognitive styles and finding alternative approaches in doing cognitive tasks.

In the Maudsley Model, therapeutic writing is proposed to help patients with taking perspectives and develop a bigger picture approach instead of an attention to detail. In CRT, direct cognitive tasks are introduced and completed during the session, and later there is a reflection on their own thinking style. In the new model, some cognitive exercises are included in the manual and in addition to the writing tasks (e.g. estimation task, illusion task, getting the gist from the bigger chunks of information). Thus the Maudsley model addresses the maintenance factors broadly, whereas CRT focuses on cognitive styles. The delivery of the modules is highly motivational, as was proposed in the Schmidt and Treasure (2006) paper.

6.2 *Use of CRT on Short-Term Medical Unit for Adolescent AN*

CRT has also been used with adolescents with AN on an inpatient service for medical stabilization. Admissions to this service are specifically for bradycardia, hypothermia, orthostasis or severe underweight (below 75% of expected weight for height by age and gender). Lengths of stay are brief, lasting on average about 12 days. The mean age of patients on the service is between 14 and 15 years, though the range is between 11 and 21 years. Participants using CRT in preliminary studies were between the ages of 12 and 18 years, with a mean age of about 14.5 years. Although the main goal of the inpatient admission is resolution of medical problems due to malnutrition, the programme has an active psychiatric consultation service that provides individual, family and group therapy aimed at diagnosis, crisis management and outpatient referral processes.

Most of the patients in the service are resistant to typical therapies, while some are cognitively impaired due to malnutrition. All patients are restricted to their beds to insure their medical safety for a large portion of their brief stays. In this context, the potential utility of CRT was considered to address the following dilemmas that this kind of inpatient context brings about:

1. Patients have limited motivation for changing eating-related behaviours.
2. There is limited time to conduct individual therapy because of other required activities (e.g. psycho-educational groups, meals and snacks, family therapy, in hospital school and medical treatments).
3. There is limited amount of time to conduct a course of therapy because of short hospital stays.
4. Patients have limited ability to focus or use insight-oriented therapy.
5. Patients have limited ability, given the constraints of the hospital setting to use behavioural therapies likely to generalize to the outpatient setting (e.g. eating with families, managing exercise without supervision, eating with friends without eating disorders).

The use of CRT appears to be overall a good match for many of the processes of the brief medical hospitalization service. Most sessions were brief (25–40 min) and could be flexibly administered. Patients are often confined to their beds, but the therapy could easily be conducted at the bedside. Materials to conduct CRT were easily condensed and portable for therapists. For this reason, nursing staff on the unit were supportive of participants using CRT and did not feel it interfered with the routines associated with medical assessment, meal times and other milieu-based activities. In addition, while the therapeutic relationship in CRT appears to be valued, the material is not emotionally charged or intimate as is the case with some other forms of in-depth therapy making the termination process less fraught with concerns about losing a confidant. In this same vein, although CRT is warm and collaborative, it does not require the participant to develop a deep interpersonal connection to use the therapy. Treatments in hospital or other intensive setting that encourage this type of relationship run the risk that the patients will feel loss when discharged and perhaps experience challenges in taking up therapy with another provider in the outpatient setting.

There were also some limitations to using CRT in an inpatient setting such as the one described above. For example, severely underweight adolescent subjects became confused and too easily distracted to conduct CRT. Lengths of stay for medical stabilization are often very short and discharge can be precipitous; thus, on average only about five sessions of CRT could be conducted. If implemented as a regular programme feature, CRT, a specific protocol with more predictable treatment length and frequency, would be helpful. The use of trained nursing staff to conduct CRT might make the intervention easier to deliver on an inpatient medical stabilization service. Evaluation of the usefulness of the approach in an inpatient setting is needed to determine the costs and benefits of CRT in medically unstable adolescents with AN.

6.3 Fitting CRT to CBT in an Outpatient Setting for Older Adolescents and Adults with AN

We have argued that CRT might be a good "preparatory" or adjunctive treatment for other outpatient therapies for AN. This might be the case because typically focused treatments for AN often do not match well with the initial motivational state of the patient, treatment collaborations are difficult to develop when therapists and patients aims are divergent in terms of focus and goals, and capacities for using specific interventions require efficient cognitive process, particularly flexibility in thinking and an ability to see the big picture. Consideration of how CRT might particularly serve as a pre-treatment or adjunctive treatment for CBT for AN deserves particular attention. CBT is a useful therapy for many psychiatric disorders and eating disorders; but to date, it has demonstrated only limited utility for AN. Small-scale studies suggested CBT might be helpful (Treasure et al. 1995; Pike et al. 2000), but two somewhat larger treatment studies were more equivocal (Pike et al. 2004; Mcintosh et al. 2005). In one of these studies (Mcintosh et al. 2005), specialist non-specific care outperformed CBT and IPT based on treatment completers. There are likely a variety of reasons why CBT for AN is less successful than it is with adult BN, including motivation, ego-syntonic nature of AN and perhaps cognitive style.

Despite limited data, CBT is a reasonable candidate for treating AN, and it appears that those who stay in treatment may achieve benefit, treatments that complement CBT may be a useful first step to address the issue of treatment retention and ultimately treatment response (Serfaty 1999; Pike et al. 2004; Mcintosh et al. 2005). As noted above, cognitive impairments potentially make it challenging to make use of the specific treatments such as CBT. A pre-treatment to improve brain and cognitive functioning is a novel way to address this dilemma. A possible solution is to use a different therapy before beginning CBT that targets related core psychopathology, though not eating-specific psychopathology, while also providing a model of a productive therapeutic relationship and practice with relevant skills (perspective taking, practice with behavioural experiments, etc.). CRT, it can be argued, may accomplish just these goals (Davies and Tchanturia 2005; Tchanturia et al. 2006; Baldock and Tchanturia 2007).

There are several reasons to consider CRT as a likely complementary and compatible treatment with CBT. A key benefit of CRT identified by clinical reports is that it is more understandable and appealing to patients with AN than other approaches, even for those at extremely low weight (Davies and Tchanturia 2005; Whitney et al. 2008). This is the case because unlike CBT, CRT does not address weight and shape or other eating disorder symptoms directly. Instead, it examines and targets neutral material effectively, i.e. cognitive styles. In addition, patients with AN are typically emotionally avoidant; so CRT's focus is comparatively non-threatening. At the same time, even emotionally laden topics would be managed in CRT by focusing on the cognitive aspects rather than elaborating the emotional context.

Furthermore, CRT is an engaging therapy that uses activities and tasks that while aimed at changing thinking styles are also a distraction from the obsessive preoccupation with weight, food and exercise associated with AN. The treatment sessions in CRT are not a strain on concentration, while nonetheless improving concentration. Completion of CRT-related tasks provide the patient with a sense of accomplishment thereby improving self-esteem (a factor associated with higher completion rates in one study (Halmi et al. 2005) and self-efficacy (Halmi et al. 2005).

The therapeutic gains from CRT lead to improvement in concentration, cognitive flexibility and central coherence processing (Tchanturia et al. 2008). Relatively high levels of these skills are needed in CBT. The ability to take differing perspectives as well as to take a more global view of problems are necessary to be able to identify and challenge contributing factors for symptom maintenance.

CRT also encourages the practice of a thinking style that precipitates the kind of self-reflection required when evaluating the impact of eating disorder-specific preoccupations and behaviours targeted in CBT. Behavioural tasks are also a part of CRT and may also provide a kind of template for the more demanding tasks of CBT. Although the behavioural tasks in CRT are not directly challenging to eating disorder behaviours, generalizing from these tasks to the more difficult ones required in CBT may be supported by success in CRT.

Taken together, CRT, though targeting core cognitive impairments, is a treatment that complements CBT by providing opportunities for practice in developing a positive therapeutic relationship, enhanced cognitive abilities related to concentration, flexibility in thinking, and perspective taking, practice in self-reflection and behavioural experiments, all of which may better prepare the patient to agree to take up CBT, to stick with the approach, and perhaps respond more quickly and better to this more specific treatment for the cognitions and behaviours associated with AN.

7 What Are the Future Research Directions in Using CRT?

CRT is not a stand-alone treatment for eating disorders; therefore, the next step in examining CRT for eating disorders is to examine any specific benefit CRT may add to existing effective treatments for eating disorders. In the area of BN, CRT appears to be compatible with CBT, which remains the first line of treatment. CBT for BN achieves about a 40% abstinence rate; so there is room for improvement. In the area of CBT for AN, it is possible that adding CRT may improve treatment retention and improve outcomes, but that possibility needs to be tested. A pilot randomized clinical study examining this possibility is underway. It is possible that CRT may be useful in inpatient treatment programmes as the preliminary study described in this chapter suggests. In adolescent treatment, the possible benefit of CRT in improving outcomes, preventing relapse and improving general prognosis in conjunction with family therapy could also be examined. CRT used in this way might help support the adolescent in tolerating weight restoration and perhaps improve the rate of psychological recovery which tends to be delayed in family-based treatment (FBT) for adolescent AN.

A first attempt to offer CRT work in group format is also in progress (Genders and Tchanturia 2010). The sessions followed the aim of practising global and flexible thinking with the support of a peer group and group facilitators. This may have secondary benefits of increasing motivation, self-esteem and reducing social isolation, all areas known to be a problem in AN. Four sessions were designed to include the following elements: psycho-education, practical exercises, reflection and discussion within the session and inter-session tasks. As with individual CRT, continual discussion relating the exercises and inter-session work to real-life thoughts and behaviours is an essential part of the learning and reflection process. Evaluation of this intervention programme may allow for an efficient strategy for delivery of CRT.

In addition to treatment studies, a better understanding for whom and how best CRT might be used in the treatment of eating disorders is needed. As the data related to neurocognitive processes in eating disorders remain quite limited, it will first be necessary to conduct more comprehensive and larger studies of these processes in the range of eating disorders. The relationship between neurocognitive process and severity of eating-related psychopathology needs to be better understood. There is a need to examine the "hot" neurocognitive processes in addition to the "cold" ones so far examined. Although preliminary data examining the relationship between age and chronicity of disorder suggest that similar cognitive inefficiencies are present early in the disorder, it is unknown whether they are a modifiable risk factor. Refinement of CRT for specific disorders, different cognitive profiles and ages is needed if CRT is likely to be a useful treatment for eating disorders.

It has been suggested that set-shifting and weak central coherence may be an endophenotype for AN (Holliday et al. 2005; Lopez et al. 2008; Roberts et al. 2010). Cognitive functioning is highly genetic (Fossella et al. 2003; Goldberg and Weinberger 2004; Buyske et al. 2006; Gosso et al. 2006; Friedman et al. 2008; Koten, Wood et al. 2009). In AN, specific executive functioning inefficiencies in the area of cognitive flexibility (set-shifting) and central coherence have been identified that appear to meet some of the criteria for an endophenotype (e.g. found in acute and recovered states, and in unaffected family members) (Tchanturia et al. 2004; Holliday et al. 2005; Lopez et al. 2008; Roberts et al. 2010). Other psychiatric disorders share some of these inefficiencies; e.g. set-shifting difficulties have been found in schizophrenia (Almasy et al. 2008), bipolar disorder (Robinson et al. 2006), OCD (Chamberlain et al. 2007) and autism (Losh et al. 2009).

Studies have identified a number of functional neural correlates of cognitive functioning in a range of disorders, e.g. autism, OCD and eating disorders (Gilbert et al. 2008; Sanders et al. 2008; Marsh et al. 2009a). Recent neuroimaging data specific to eating disorders are more limited, but support the involvement of fronto-striatal brain circuitry (Marsh et al. 2009b; Zastrow et al. 2009). Furthermore, other recent studies have identified neural correlates of cognitive inflexibility in a sample of adults with AN (Zastrow et al. 2009), suggesting that examining such correlates is reasonable and timely. If anatomical and functional neural correlates are better understood, both the pathophysiology and, ultimately, the ethiology of these deficits can be described on a biological level.

In summary, as Siegle et al. (2007) suggest, cognitive remediation techniques are those that aim to target neurobiological and neurocognitive mechanisms thought to underlie psychological disorders. While there is solid neuropsychological evidence demonstrating neurocognitive dysfunction in AN, and preliminary evidence suggesting that CRT may improve functionality to some extent, we still know very little about the links between CRT and changes in neurobiological parameters. It is clear that more basic science research is necessary before we can fully understand the specific brain mechanisms underlying AN, and the number of neuroimaging studies involving eating disordered patients is growing. However, we still know less than researchers in other mental health fields (e.g. schizophrenia) about the neurobiological correlates of AN, and even less about the specific neural processes underlying the cognitive impairments seen in this population, and how these might be affected by the treatment. In the psychosis field, for example, several well-designed studies have already documented cognitive and functional brain changes as a result of CRT (Eack et al 2010; Vinogradow et al. 2009). Ideally, future studies in AN will follow this lead and combine neuropsychology with neuroimaging methods to help elucidate and delineate the neural mechanisms that are susceptible to intervention and bring about clinically significant improvement. Thus, there are a number of questions we would like to explore in future studies, which are as follows:

Is CRT a beneficial intervention for AN? (replication studies of Tchanturia et al. 2002; pilot work is required).

Is CRT helping to engage patients in treatment?

Does CRT help improve cognition? Daily functioning?

References

Almasy L, Gur R et al (2008) A genome screen for quantitative trait loci influencing schizophrenia and neurocognitive phenotypes. Am J Psychiatry 165:1185–1192

American Psychiatric Association (1994) Diagnostic and statistical manual of mental disorders. American Psychiatric Association, Washington, DC

Arnsten A, Shansky R (2004) Adolescence: vulnerable period for stress-induced prefrontal cortical function? Introduction to part IV. Ann NY Acad Sci 1021:143–147

Baldock E, Tchanturia K (2007) Translating laboratory research into practice: foundations, functions, and future of cognitive remediation therapy for anorexia nervosa. Therapy 4:1–8

Buyske S, Bates M et al (2006) Cognitive traits link to human chromosomal regions. Behav Genet 36:65–76

Carter C, Stewart D et al (2001) Eating disorder examinatin questionnaire: norms for young adolescent girls. Behav Res Ther 39:625–632

Casey B, Jones R et al (2008) The adolescent brain. Ann NY Acad Sci 1124:111–126

Cavedini P, Bassi T, Ubbiali A, Casolari A, Giordani S, Zorzi C, Bellodi L (2004) Neuropsychological investigation of decision-making in anorexia nervosa. *Psychiatry Research* 127:259–266

Chamberlain S, FIneberg N et al (2007) Impaired cognitive flexibility and motor inhibition in unaffected first-degree relatives of patients with obsessive-compulsive disorder. Am J Psychiatry 164:335–338

Chan R, Shum D, Toulopoulou T, Chen E (2008) Assessment of executive functions: review of instruments and identification of critical issues. Arch Clin Neurospychol 23:201–216

Chui H, Christensen B et al (2008) Cognitive function and brain structure in females with a history of adolescent-onset anorexia nervosa. Pediatrics 122:e426–e437

Cooper Z, Fairburn CG (1987) The eating disorder examination: a semi-structured interview for the assessment of the specific psychopathology of eating disorders. Int J Eat Disord 6:1–8

Cooper Z, Cooper PJ et al (1989) The validity of the eating disorder examination and its subscales. Br J Psychiatry 154:807–812

Das J (1999) A neo-Lurian approach to assessment and remediation. Neuropsychol Rev 9(2): 107–116

Davies M, Tchanturia K (2005) Cognitive remediation therapy as an intervention for acute anorexia nervosa: a case report. Eur Eat Disord Rev 13:311–316

Delis D, Kaplan E et al (2001) Examiner's manual: Delis-Kaplan executive functioning systems (D-KEFS). Psychological Corporation, San Antonio, TX

Eack SM, Hogarty GE, Cho RY, Prasad KM, Greenwald DP, Hogarty SS, Keshavan MS (2010) Neuroprotective Effects of Cognitive Enhancement Therapy Against Gray Matter Loss in Early Schizophrenia Results From a 2-Year Randomized Controlled Trial. Arch Gen Psychiatry 67(7):674–82

Eslinger P, Oliveri M (2002) Approaching interventions clinically and scientifically. In: Eslinger PJ (ed) Neuropsychological interventions – clinical research and practice. Guildford University Press, New York

Fairburn CG, Cooper I (1993) The eating disorder examination. In: Fairburn CG, Wilson GT (eds) Binge eating: nature, assessment, and treatment, 12th edn. Guilford Press, New York

Fossella J, Bishop S et al (2003) Exploring genetic influences on cognition: emerging strategies for target validation and treatment optimization. Curr Drug Targets CNS Neurol Disord 2:357–362

Friedman N, Miyake A et al (2008) Individual differences in executive functions are almost entirely genetic in origin. J Exp Psychol Gen 137:201–225

Genders R, Tchanturia K (2010) Cogntitive remediation in group format: a pilot study. J Weight Eat Disord (in press)

Genders R, Davies H, StLouis L, Kyriacou O, Hambrook D, Tchanturia K (2008) Long-term benefits of CRT for anorexia. British Journal of Healthcare Management 14(12):15–19

Gilbert S, Bird G et al (2008) Atypical recruitment of medical prefrontal cortex in autism spectrum disorders: an fMRI study of two executive function tasks. Neuropsychologia 46:2281–2291

Gillberg I, Råstam M, Wentz E, Gillberg C (2007) Cognitive and executive functions in anorexia nervosa ten years after onset of eating disorder. J Clin Exp Neuropsychol 29(2):170–178

Goldberg T, Weinberger D (2004) Genes and parsing of cognitive processes. Trends Cogn Sci 8:325–335

Gosso MF, de Geus EJ, van Belzen MJ et al (2006) The SNAP-25 gene is associated with cognitive ability: evidence from a family-based study in two independent Dutch cohorts. Mol Psychiatry 11:878–886

Halmi KA, Agras WS et al (2005) Predictors of treatment acceptance and completion in anorexia nervosa: implications for future study designs. Arch Gen Psychiatry 62:776–781

Hoek HW, van Hoeken D (2003) Review of prevalence and incidence of eating disorders. Int J Eat Disord 34:383–396

Holliday J, Tchanturia K et al (2005) Is impaired set-shifting an endophenotype of anorexia nervosa? Am J Psychiatry 162:2269–2275

Kaye W, Fudge J, Paulus M (2009) New insights into symptoms and neurocircuit function of anorexia nervosa. Nat Rev Neurosci 10:573–584

Keski-Rahkonen A, Hoek H et al (2007) Epidemiology and course of anorexia nervosa in the community. Am J Psychiatry 164:1259–165

Lauer CJ (2002) Neuropsychological findings in eating disorders. In: D'haenen H, den Boer JA, Westenberg H, Willner P (eds) Biological psychiatry. Wiley, Swansey, UK, pp 1167–1172
Lock JD, Fitzpatrick KK (2009) Anorexia nervosa. Br Med J Clin Evid 01(1011):1–19
Lopez C, Tchanturia K, Stahl D, Treasure J (2008) Central coherence in eating disorders: a systematic review. Psychol Med 38(10):1393–1404
Losh M, Adolphs R et al (2009) Neuropsychological profile of autism and the broad phenotype. Arch Gen Psychiatry 66:518–526
Lucas AR, Beard CM et al (1991) 50-year trends in the incidence of anorexia nervosa in Rochester, Minn: a population-based study. Am J Psychiatry 148:917–929
Lucas AR, Crowson C et al (1999) The ups and downs of anorexia nervosa. Int J Eat Disord 26:397–405
Luna B, Sweeney J (2004) The emergence of collaborative brain function. Ann NY Acad Sci 1021:296–309
Marsh R, Maia T et al (2009a) Functional disturbances within frontostriatal circuits across multiple childhood psychopathologies. Am J Psychiatry 166:664–674
Marsh R, Steinglass J et al (2009b) Deficient activity in the neural systems that mediate self-regulatory control in bulimia nervosa. Arch Gen Psychiatry 66:51–63
McGurk S, Twamley E, Sitzer D, McHugo G, Mueser K (2007) A meta analysis of cognitive remediation in schizophrenia. Am J Psychiatry 164:179–1802
Mcintosh VW, Jordan J et al (2005) Three psychotherapies for anorexia nervosa: a randomized, controlled trial. Am J Psychiatry 162:741–747
Medalia A, Choi J (2009) Cognitive remediation in schizophrenia. Neuropsychol Rev 19:353–364
NICE (2004) National clinical practice guideline: eating disorders: core interventions in the treatment and management of anorexia nervosa, bulimia nervosa, and related eating disorders. National Institute for Clinical Excellence, London
Osterrieth P (1944) Test of copying a complex figures: contribution to the study of perception and memory. Arch Psychol 20:206–356
Pike K, Walsh BT et al (2000) Cognitive-behavioral therapy in the relapse prevention of anorexia nervosa. Third International Congress of Neuropsychology, Kyoto
Pike K, Walsh BT et al (2004) Cognitive-behavioral therapy in the posthospitalization treatment of anorexia nervosa. Am J Psychiatry 160:2046–2049
Resources PA (2003) Computerized Wisconsin Card Sort Task, (WCST), version 4
Roberts M, Tchanturia K, Stahl D, Southgate L, Treasure J (2007) A systematic review and meta-analysis of set shifting ability in eating disorders. Psychol Med 37(8):1075–1084
Roberts M, Tchanturia K, Treasure J (2010) Exploring the neurocognitive signature of poor set-shifting in anorexia and bulimia nervosa. J Psychiatr Res PMID:20398910
Robinson I, Thompson J et al (2006) A meta-analysis of cognitive deficits in euthumic patients with bipolar disorder. J Affect Disord 93:105–115
Rohling M, Faust M, Beverly B, Demakis G (2009) Effectiveness of cognitive rehabilitation following acquired brain injury: a meta – analytic re-examination of Cicerone et al. (2000–2005) systematic reviews. Neuropsychology 23:20–39
Russell T, Schmidt U, Tchanturia K (2009) Aspects of social cognition in anorexia nervosa: Affective and cognitive theory of mind. Psychiatry Research 15; 168(3):181–5
Sanders J, Johnson K et al (2008) A review of neuropsychological and neuroimagin research in autism spectrum disorders: attention, inhibition, and cognitive flexibility. Res Autism Spectr Disord 2:1–16
Schmidt U, Treasure J (2006) Anorexia nervosa: valued and visible. A cognitive interpersonal maintenance model and its implications for research and practice. Br J Clin Psychol 45:343–366
Serfaty M (1999) Cognitive therapy versus dietary counselling in the outpatient treatment of anorexia nervosa. Eur Eat Disord Rev 7:334–350

Siegle GJ, Ghinassi F, Thase ME (2007) Neurobehavioral therapies in the 21st century: summary of an emerging field and an extended example of cognitive control training for depression. Cognit Ther Res 31:235–262

Southgate L, Tchanturia K, Treasure J (2005) Building a model of the aetiology of eating disorders by translating experimental neuroscience into clinical practice. J Ment Health 14:553–566

Southgate L, Tchanturia K, Treasure J (2008) Information processing bias in anorexia nervosa. Psychiatry Res 160:221–227

Southgate L, Tchanturia K et al (2009) Neuropsychology in eating disorders. In: Wood S, Allen N, Pantelis C (eds) Handbook of neuropsychology of mental illness. Cambridge University Press, Cambridge, pp 316–325

Stevenson C, Whitmont S, Bornholt L, Livesey D, Stevenson R (2002) A cognitive remediation programme for adults with attention deficit hyperactivity disorder. Aust NZ J Psychiatry 36 (5):610–616

Tchanturia K, Hambrook D (2009) Cognitive remediation. In: Grilo C, Mitchell J (eds) The treatment of eating disorders; Clinical handbook. Guilford Press, New York, pp 130–150

Tchanturia K, Morris R, Surguladze S, Treasure J (2002) An examination of Perceptual and Cognitive Set Shifting Tasks in Acute Anorexia Nervosa and Following Recovery. Weight and Eating Disorders 7(4): 312–316

Tchanturia K, Campbell I, Morris R, Treasure J (2005) Neuropsychological studies in AN. Int J Eat Disord, Special Issue Anorexia Nervosa 37:572–576

Tchanturia K, Whitney J, Treasure J (2006) Can cognitive exercises help treat anorexia nervosa? A case report. Eat Weight Disord 11(4):112–117

Tchanturia K, Davies H, Campbell I (2007) Cognitive Remediation for patients with anorexia nervosa: preliminary findings. Ann Gen Psychiatry 14:6–14

Tchanturia K, Davies H, Lopez C, Schmidt U, Treasure J, Wykes T (2008) Neuropsychological task performance before and after cognitive remediation in anorexia nervosa: a pilot case series. Psychol Med 38(9):1371–1373

Tokley M, Kemps E (2007) Preoccupation with detail contributes to poor abstraction in women with anorexia nervosa. J Clin Exp Neuropsychol 29:734–741

Treasure JL, Todd G et al (1995) A pilot study of a randomized trial of cognitive-behavioral analytical therapy vs educational behavioral therapy for adult anorexia nervosa. Behav Res Ther 33:363–367

Treasure J, Claudino A, Zucker N (2010) Eating disorders. Lancet 375:583–593

Uher R, Treasure J (2005) Brain lesions and eating disorders. J Neurol Neurosurg Psychiatry 76(6):852–857, Review

Van Son G, van Hoeken D et al (2006) Time trends in the incidence of eating disorders: a primary care study in the Netherlands. Int J Eat Disord 39:565–569

Wechsler D (1997) WAIS-III Weschler adult intelligence scale, 3rd edn. Psychological Corporation, San Antonio, TX

Whitney J, Easter A, Tchanturia K (2008) The patients experiences in cognitive exercise intervention for anorexia nervosa: Qualitative findings. International Journal of Eating Disorders 41 (6):542–50

Zastrow A, Kaiser S, Stippich C, Walther S, Herzog W, Tchanturia K, Belger A, Weisbrod M, Treasure J, Friederich H (2009) Neural correlates of impaired cognitive-behavioral flexibility in anorexia nervosa. Am J Psychiatry 166(5):608–616

Incorporating Dispositional Traits into the Treatment of Anorexia Nervosa

Nancy L. Zucker, David Herzog, Ashley Moskovich, Rhonda Merwin, and Tammy Lin

Contents

N.L. Zucker (✉)
Duke University Medical Center, Durham, NC, USA
Duke University, Durham, NC, USA
and
e-mail: Zucke001@mc.duke.edu

D. Herzog
Massachusetts General Hospital and Harvard Medical School, Harvard University, Cambridge, MA, USA

A. Moskovich and T. Lin
Duke University, Durham, NC, USA

R. Merwin
Duke University Medical Center, Durham, NC, USA

R.A.H. Adan and W.H. Kaye (eds.), *Behavioral Neurobiology of Eating Disorders*, Current Topics in Behavioral Neurosciences 6, DOI 10.1007/7854_2010_92

Abstract We provide a general framework to guide the development of interventions that aim to address persistent features in eating disorders that may preclude effective treatment. Using perfectionism as an exemplar, we draw from research in cognitive neuroscience regarding attention and reinforcement learning, from learning theory and social psychology regarding vicarious learning and implications for the role modeling of significant others, and from clinical psychology on the importance of verbal narratives as barriers that may influence expectations and shape reinforcement schedules.

Keywords Anorexia nervosa · Attention · Eating disorders · Perfectionism · Reinforcment · Temperament · Treatment

I took the road less traveled by, and that has made all the difference.
Robert Frost
Don't consider your reputation and you may do anything you like.
Chinese Proverb

1 Introduction

Clinical perfectionism in those with anorexia nervosa (AN) is proposed to motivate extremes of performance at the expense of health even prior to illness onset. Indeed, the rigid structure (e.g., repetitive behavioral routines, constrained behavioral options) self-imposed by those with AN is juxtaposed upon a relentless drive that, even prior to initial manifestations of extreme dietary restriction, serves to neglect biological needs if pitted against a perceived failure to meet stated objectives (Zucker et al. 2007). In fact, characterizations of the childhoods of those with AN, developmental periods prior to onset of severe malnutrition, are uncanny in their consistency (Wonderlich et al. 2005). Neglecting sleep to complete homework, turning in twice the length of papers requested by teachers, and practicing beyond the recommendations of coaches, pose a dilemma for parents who eventually are forced to "punish" these conscientious children by imposing unwanted limits on individuals who are just "too good," (i.e., follow rules without limit). In fact, so strikingly consistent are these narratives, that a wise clinician can leverage knowledge of this relentless perfectionistic drive and rigid behavioral repertoire to help convince reluctant parents, that is, parents having difficulty recognizing their child has a diagnosis of AN, that their child "fits the mold"; and thus the abrupt changes in nutritive habits are, indeed, but one sign of a larger constellation of features that may be cause for concern. Perfectionism is thus an important feature not only because of its impact on functioning, but also because it may provide an opportunity: an acceptable entry point to speak to parents and children about the biological and behavioral processes that may reinforce the ill state of AN.

In this chapter, we use perfectionism as an exemplar of a trait feature that would benefit from targeted intervention, and as a springboard to illustrate a general framework for addressing trait features in AN. There are several content areas and interpersonal processes we consider in providing this general framework. For example, we address the foci and integrity of attention neural networks, explore interpersonal reinforcement contingencies that may inadvertently exacerbate systematic modes of responding, and consider how reputation narratives may constrain experience and limit knowledge acquisition. We then describe a treatment model that attempts to address these domains. To operationalize the term "trait feature," we adopt as our definition that stated by Tellegen et al. (1988), "a psychological (therefore) organismic structure underlying a relatively enduring behavioral disposition, i.e., a tendency to respond in certain ways under certain circumstances." An extensive developmental literature supports the relationship between temperament and the emergence of behavior problems (Graham et al. 1973; Eisenberg et al. 2000, 2005). A comparison of the predictive validity of personality, socioeconomic status, and cognitive variables on such important life outcomes as mortality, divorce, and occupational attainment revealed that personality demonstrated equal predictive power (Roberts et al. 2007). Against a clinical background, not only do trait features, by definition, persist beyond acute psychiatric symptom expression, but trait features and frank personality disorders have been reported to negatively impact treatment prognosis and illness course (Bizeul et al. 2001; Nilsson et al. 2008; Skodol et al. 2005; Barber et al. 1997). Given the seeming constancy and potential deleterious effects of certain trait modes of responding, the need to incorporate trait features into intervention strategies is apparent. The manner in which to do so is far less clear.

2 Developmental Considerations in Addressing Trait Features

Trait features show different degrees of continuity across the lifespan, developmental variation that can inform intervention approaches (Caspi et al. 2005). For the clinician, temperament and personality research is informative in delineating the structure of personality to focus intervention targets, the stability of personality to inform reasonable intervention goals, and the relative influence of environmental versus biological influences on personality change to inform the strength of expected treatment effects. Before exploring these issues, we begin with some definitions. Temperament, defined as constitutionally based individual differences in reactivity and self-regulation, influenced over time by heredity and experience (Derryberry and Rothbart 1997), had long been considered the domain of researchers investigating biologically influenced modes of responding in early childhood, while personality had been considered the adult expression of stable patterns. However, these differentiations are gradually morphing (McAdams and Olson 2010). Longitudinal studies are increasingly examining the stability of both the structure of personality in childhood and adolescence relative to adulthood as well as the intraindividual stability of features across the lifespan. Such studies are crucial if we are to take adolescent-defined vulnerabilities of pathology and relate them to their childhood equivalents for early intervention and prevention efforts, as well as capitalize on peak periods of plasticity to maximize intervention effectiveness.

Not only may personality evidence greater variability during certain developmental periods, but also average levels of trait features may differ over time. Meta-analyses summarizing longitudinal investigations of personality stability demonstrate developmental changes in stability coefficients within individuals across time as well as developmental shifts in mean levels of trait features across development (Roberts and DelVecchio 2000). For example, in their widely-cited meta-analytic review of 152 longitudinal studies of personality stability, Roberts and Del Vecchio (2000) report rank-order correlation coefficients that demonstrated variability throughout the lifespan with stability increasing from a low of 0.31 in childhood, to 0.54 in college, and stabilizing to 0.74 in late adulthood. We illustrate such developmental variation to highlight an important balance that must be bridged between the expectations for change and the acknowledgement of latent tendencies. For example, research examining the teaching practices of educators demonstrates that children will rise to expectations: they will perform better academically if more is expected of them academically and they are treated as if they can meet those expectations (Page and Rosenthal 1990). There is a clinical corollary to this narrative in the study of anxious children. Study of parenting practices with an anxious child indicates that when parents treat their anxious child tentatively, the child is less likely to approach a subsequent novel task than when parents convey their confidence in their child's ability (Gruner et al. 1999). Such examples highlight how expectations of the abilities of others can influence both the opportunities provided to build behavioral repertoires and shape the behavioral patterns that are reinforced: when we expect error, we look for error, and when we doubt capacities, we limit opportunities. Such examples are not intended to implicate influential members of an individual's social environment as *causing* trait modes of responding, but rather to highlight the importance of incorporating these social contingencies in affecting change.

Individuals form a self-narrative ("I am lazy"), and may imagine (or verify) the narratives that others (e.g., parents, teachers, friends) hold of them ("everyone expects me to be perfect"). Further, whatever the congruence between the expectations of others and an individual's conceptualization of these expectations, in fact, others do form narratives ("Mary never makes mistakes"). Such storied expectations of others and the self have implications for change. Individuals will work to maintain a verbal and felt sense of cohesion of self-definition (Swann and Read 1981). Others may exhibit biased processing of information that confirms the narrative they hold of others [see well-established constructs in social psychology such as the "halo effect" (Nisbett and Wilson 1977)]. Thus, the degree to which an individual's social support network inadvertantly challenges change efforts by reinforcing an individual's tendency to respond in certain ways or alternatively, encourages change by altering reinforcement contingencies, has been of limited study in the management of trait features in eating disorders intervention research.

Yet, it is also true, that when individuals have actively begun to engage in change efforts, they often repeatedly re-experience the strong urges, intrusive cognitions, or other remnants of trait modes of responding (e.g., a sudden urge to escape from an ambiguous situation as in those with elevated levels of harm avoidance) (Marlatt and Gordon 1985). Such urges to perform a repetitive behavior pattern have been robustly reported (Sobik et al. 2005) and strategies to address such urges have been incorporated into relapse prevention methods for binge eating (Sobik et al. 2005; Allen and Craighead 1999) in the context of eating disorder treatment [also see literature on relapse prevention in addiction, (Marlatt and Gordon 1985)]. If not educated about these patterns, individuals attempting to address trait modes of responding may fear that they are relapsing or that change efforts are hopeless, when, in fact, such reemergence of well-established, maladpative patterns is a well-known phenomenon in the emotional learning and memory literature (Kelley et al. 2005; Archbold et al. 2010). Thus, preparing individuals for such psychological triggers as well as giving them tools to respond to such cues is also an essential aspect of treatment of trait modes of responding. Balancing the optimal balance between acceptance and change efforts in the management of trait features is thus a much needed area of inquiry.

3 The Role of Attention in Addressing Trait Features

The harmonic integration of sensation and attention is the bedrock of experience. Thus, characterizing sensory and related attention capacities can help inform the phenomenology of psychiatric disturbance, in general, and AN, in particular. Further, targeting such basic capacities may be a particularly critical domain to address in interventions designed to target trait features. Such a perspective is not new. Posner and Rothbart (2007) in their highly cited review of the neural networks that subserve attention, characterize attention as the critical foundation for self-regulation, processes that modulate reactivity to sensory stimuli. Posner and Rothbart describe three broad attention networks: the *alerting and orienting* network involves maintaining a sense of alertness or awareness to incoming sensory input (from the internal and external environment) combined with capacities to align attention to the sources of this sensory input, for example, guiding visual attention towards a novel movement you notice in a field, towards a thin individual who enters a room, or towards movement in the gut. Thus, these first two networks are conceptualized as reactivity to stimuli (Posner and Rothbart 2007). The third network, the *executive attention* network, is described as resolving conflict between other neural networks, a task which entails both the upregulation and downregulation of various neural activities. Of importance, the efficiency of this latter network demonstrates change with experience, and has been linked with the childhood temperament factor of effortful control, defined as the ability to inhibit a prepotent response in order to activate a weaker response, as well as capacities to plan behavior and detect error (Posner and Rothbart 2009). The efficient function of this network has been linked to a broad range of adaptive outcomes including conscience, theory of mind, emotion regulation, and socialization (Simonds et al. 2007; Kieras et al. 2005; Rothbart et al. 1994). Given the aforementioned definition of trait features as typical modes of responding to stimuli, targeting the efficiency of specific attention networks would seem to be critical in interventions designed to address these stable characteristics.

It is perhaps surprising then that adult models of personality do not incorporate attention. Caspi et al. (2005) in their comprehensive review of developmental variation in personality, note the confusing absence of attention in models of adult personality given the pivotal role of attention in child models of temperament. In contrast, the role of attention and more specifically, attention biases, have been very influential in models of psychiatric disorder maintenance. For example, in major depressive disorder, information-processing models highlight how systematic biases in orienting to certain types of content (e.g., personal failures, past errors) decrease the breadth of information entering awareness and negatively influence subsequent adaptive responding, thus potentially contributing to the maintenance of psychiatric disturbance (Beck 1979). Similarly, early information-processing models of AN and other eating disorders focused on biased orienting and memory for illness-related content (Williamson et al. 1999). Neuropsychological investigations have provided fairly consistent evidence demonstrating that performance on neuropsychological measures of set-shifting are impaired in those with AN (Roberts et al. 2007; Tchanturia et al. 2004). While seemingly implicating impairment in the distributed neural circuitry that supports the attention executive network, neuroimaging data are limited, but provide preliminary support for aberrant functioning in behavioral set-shifting (Zastrow et al. 2009) supporting deficits in inhibiting a prepotent behavioral response in AN. Combined, the available data support the potency of incorporating improvement in attention efficiency into interventions aiming to influence trait modes of responding, a strategy particularly efficacious for AN given prior evidence of impaired performance in tasks that require shifting of attention sets. Further elaboration of attention capacities in those with AN combined with knowledge of how existing treatment models address modes of attention may best facilitate our ability to match patients to treatment and develop novel strategies.

4 Personality and Social Reinforcement Contingencies

4.1 Reputation Narratives

The majority of individuals will work to be understood by others, even when such efforts are detrimental. Self-verification theory is a model proposed by Swann and Read (1981) which posits that individuals will work to maintain a cohesive narrative that corresponds with one's self-definition: they will ignore or negate information that conflicts with their own self-definition and will work to elicit reactions from others that confirm their self-definition. In fact, so powerful is this motive that individuals will pursue consistency over valence: those with a negative self-concept will seek out confirming evidence even when this evidence validates a negative self-concept (North and Swann 2009). In other words, consistency can be more reinforcing than positively valenced information. Of interest, a related finding has been reported in studies examination mechanisms of change in psychotherapy research (Koerner and Linehan 2000; Lynch et al. 2006). Study of process issues, those factors related to the delivery of therapy including the interpersonal transactions of therapist-client, indicates that the experience of validation from the therapist is a potent hypothesized mechanism of change in therapeutic interactions and may contribute to better retention in therapeutic interventions (Linehan et al. 2002). A review by North and Swann (2009) proposes several benefits to self-verification including anxiety reduction, improved health, and psychological coherence. We argue, consistent with the theoretical writings of Linehan (1997) and Greenberg and Paivio (2003) regarding emotional validation and the vast literature on responsive parenting (Propper and Moore 2006), that such self-verification is critical to adaptive self-regulation, as verification of one's experiences is necessary to acquire the ability to discriminate the motivational salience of different internal states. If an actor experiences gut motility, a beating heart, sweaty palms, and other obvious signs of arousal, but is told by significant others that "nothing is wrong," over time, the actor may lose the ability to recognize and discriminate the importance of different arousal states and would fail to link specific arousal states to adaptive modes of responding (i.e., a pounding heart becomes a source of confusion rather than an adaptive response to potential danger). The need for validation thus seems to be essential for achieving adaptive self-regulation, but poses a seemingly complex dilemma for addressing persisting maladaptive modes of responding, as in pathological trait features. A proposed solution, incorporated explicitly into models of cognitive-behavioral therapies (Linehan 1993), is to validate emotional experience and gently shape more adaptive modes of responding via positive and negative reinforcement (e.g., acknowledge the actor is upset, but inform that it is difficult to listen to what they have to say when he or she speaks in an abusive tone). Achieving the balance between validation and change may thus be a critical process in interventions that seek to target trait modes of responding.

4.2 Niche-Building Processes

People may seek out, influence, or create environments that are associated with and may reinforce their trait dispositional tendencies (Caspi et al. 2005). Explanatory models derived from study of environmental and genetic influence on the emergence and maintenance of trait modes of responding share common themes with theoretical models of self-definition from fields of social and clinical psychology such as self-verification theory (Swann and Read 1981). Concepts such as "niche-building processes" (Caspi et al. 2005) incorporate a developmental framework to explain how trait dispositions may be maintained as they elicit certain characteristic responses from others and may further lead to the selection or creation of environments consistent with these typical response patterns. For example, an anxious child may naturally elicit a protective posture from a responsive parent, a sensitive pattern of behavior that is reinforced as the mother witnesses the decreased arousal in the child when novel stimuli are avoided. If such avoidance is taken to extremes, the benevolent desire and parental instinct to comfort and protect may unintendingly backfire: resulting in an increasingly limited range of exposure to novel experiences for the child with associated narrowed opportunities for practice for the child to cope with change and develop an adaptive response repertoire. Anxious parents may further potentiate this avoidance due to a parallel desire to minimize their own arousal by constraining access to unknown environments as well as via role-modeling of fear responses to uncertain situations. Thus, the parent's personality and parent responsivity to the child's temperament are influential features among a complex array of biological and situational variables that shape the form of adaptive parent-child interactions and may contribute to patterns of responding that help shape a persistent self-narrative (Propper and Moore 2006).

Irrespective of cause, over time a narrative may evolve about the child's capacities that is reinforced by others and eventually endorsed by the child. "I am shy" thus becomes a verbal shorthand that encapsulates a broad, yet consistent, behavioral repertoire characterizing the child's interactions with others. This narrative may further facilitate the recall of information as the child is able to differentially retrieve information consistent with this self-narrative, potentially via increased elaboration of self-referential information at encoding. In the present, conceptualizing oneself as shy and fearful, the child may come to doubt her capacities to cope with novelty, influencing subsequent choices when faced with interactions with novel environments. Thus, such transactions exemplify *niche building processes*, a term used to explain influences that constrain variability in human personality (Caspi et al. 2005). Trait features elicit responses from others and influence the environment; individuals with particular trait features may seek out certain environments, and individuals may build situations in which their latent tendencies become reinforced. Such influential transactions between parent and child or between individuals with important social influences illustrate the potential potency of incorporating relevant others into treatments that attempt to manage detrimental trait dispositions.

4.3 Role Modeling

Vicarious or observational learning is a powerful mechanism whereby individuals acquire novel information yet circumvent the necessity of experiencing events directly (Bandura et al. 1963). Perhaps not surprisingly, the influence of vicarious learning has been widely researched in the arena of fear-acquisition as there is an inherent adaptive advantage to learning about potential harm via verbal reports or watching others, thus obviating the need for direct experience (Guzman et al. 2009; Askew and Field 2008). Thus, it is not hard to imagine that if one were attempting to influence a typical mode of responding with the substitution of an alternate pattern, as in treatments that address trait features, role modeling target alternative behaviors by influential others would be a critical domain of inclusion. Otherwise, if individuals in the environment model divergent information from that which is instructed in the therapeutic setting, then behavioral intervention may be rendered obsolete. An excellent example of interventions in which the role-modeling of significant others is incorporated is in the realm of pediatric obesity (Wrotniak et al. 2005). In family-based interventions for pediatric obesity, the inclusion of parents has been found to be essential to reinforce behavior changes in the child (Epstein et al. 2007). However, not only is the child's eating a target of intervention, the parents' eating is also a treatment target as substantial research documents the importance of social influence on food selection and food quantity (Shutts et al. 2009; Redd and de Castro 1992). Further, direct manipulations of the home environment that further reinforce the behavioral changes targeted in treatment have also been essential in achieving and sustaining weight loss maintenance in children (Epstein et al. 2007). Thus, in interventions that aim to target typical modes of responding, incorporation of significant others is essential not only in establishing adaptive reinforcement contingencies of the patient's behavior (e.g., applauding demonstrations of bravery in the harm avoidant child), but also in role modeling such changes to emphasize the critical importance of targeted adaptations in behavior.

4.4 Interim Summary

In the prior sections, we discussed the importance of addressing or acknowledging trait modes of responding in the context of therapeutic interventions. We described a general framework for addressing trait features and highlighted certain domains that would be important for inclusion on the basis of existing evidence. In the next section, we introduce the trait feature of clinical perfectionism and describe how it has been defined in the literature.

5 The Example of Perfectionism

Perfectionism has been associated with increased suicidal ideation (Hamilton and Schweitzer 2000; Hewitt et al. 1997), suicidality (Jacobs et al. 2009), self-injury (Hoff and Muehlenkamp 2009), and completed suicides, often as individuals seemingly reach the pinnacle of achievement in their domains of importance (Blatt 1995). In AN, perfectionism predicts poor treatment prognosis (Bizeul et al. 2001) and persists with illness remission (Nilsson et al. 2008), though exceptions have been noted (Bardone-Cone et al. 2010). Further, those individuals with AN whose mothers also endorse elevated levels of perfectionism exhibit the highest levels of eating disorder symptomatology (trios study) and may thus index a subset of those with AN who evidence a particularly biologically influenced form of the disorder that may be further maintained by social reinforcement contingencies (e.g., modeling of perfectionistic striving in parents). Yet, despite the clinical significance of perfectionism, there has been limited work examining the degree to which the impact of this feature can be minimized.

To be sure, the very nature of the construct of perfectionism continues to be debated (Hewitt et al. 2003; Shafran et al. 2002). Even such basic considerations as whether perfectionism is uni-or multi-dimensional (Shafran et al. 2002; Hewitt and Flett 1991a), whether perfectionism is best conceptualized as a trait disposition or a behavioral pattern (Slade and Owens 1998), or even if there can be both adaptive and maladaptive forms of perfectionism remains an issue (Slade and Owens 1998; Terryshort et al. 1995). Recent critiques of the nature of this construct as it pertains to individuals with eating disorders reject the notion of the multi-dimensional nature of perfectionism and posit that the core aspect of perfectionism is a unidimensional feature that relates to decrements in self-esteem resulting from perceived failure; that is, failure to meet ideals of extreme performance is relevant only so far as such failures influence the individuals' self-evaluation (Shafran et al. 2002). We concur with this emphasis and build on the definition of Shafran et al. (2002) by further operationalizing perfectionism in relation to attention, learning, and reinforcement contingencies, incorporating, in part, elements of the model of perfectionism posited by Slade and Owens (1998) which conceptualizes perfectionism as a behavioral pattern that is maintained by both positive and negative reinforcement. The model of Slade and Owens (1998) adopts a developmental framework to understand the shift of perfectionism from a positively to a negatively reinforced behavioral pattern. In other words, striving for outcomes that were positively reinforced can morph into strivings maintained by negative reinforcement contingencies (e.g., someone who initially experiences pleasure at an accomplishment can come to experience only a temporary decrease in guilt or shame when receiving results because they are striving to avoid disappointing themselves or others rather than working to achieve mastery). Thus, both the definitions of Slade and Owens (1998) and Shafran et al. (2002) address attention by characterizing perfectionism as biased orienting to outcomes of performance. Building on these definitions, we reiterate our definition of perfectionism as

> Perfectionism is the unrelentless striving to avoid the guilt or shame of not working towards the inevitable and infinite next step of a rigid policy designed to avoid error

– a definition that we feel incorporates the definition of Shafran et al. (2002) and Slade and Owens (1998), but attempts to elaborate these definitions by positing specific relationships to the self-conscious emotions and incorporates predictions about modes of approaching learning tasks.

This definition would lead to distinct predictions about the relationship of perfectionism to attention networks and learning strategy. We would expect elevations on measures of perfectionism to be related to biased attention orienting to outcomes of performance, reinforcement of this perseverative focus by transient reductions in negative affect, and subsequent failure to alter strategy on the basis of trial and error learning. Hopelessness, a frequent associated feature of perfectionism, is a logical, but unfortunate, result when biased orienting towards a specific outcome limits the number of options deemed acceptable. Such fixed orienting may increase vulnerability to experience failure and/or may limit sustained feelings of success as the focus of attention immediately moves to the next opportunity for potential failure.

However, perfectionism is also rather unique among trait features in that the role of interpersonal factors is incorporated into many widely-accepted multi-factorial definitions (Hewitt and Flett 1991a). Hewitt and Flett's multi-dimensional perfectionism scale contains three factors: other-oriented perfectionism, self-oriented perfectionism, and socially prescribed perfectionism (Hewitt and Flett 1991b; Cox et al. 2002). While other-oriented perfectionism captures an individual's perfectionist expectations for the behavior of others, socially prescribed perfectionism is the belief that others expect perfectionism from the actor. Relatedly, Frost et al. (1990) developed a multi-dimensional scale of perfectionism with dimensions capturing parental criticism and parental concerns over mistakes. The integration of these social and individually-focused dimensions has been examined in recent factor analyses and adds to the current debate about whether perfectionism can have both adaptive and maladaptive dimensions (O'Connor et al. 2009). O'Conner and Dixon describe a 2-factor solution, one factor associated with striving and thus putative positive reinforcement contingencies and one factor associated with perfectionistic concerns or putative negative reinforcement contingencies. More specifically, *self-oriented perfectionism-striving*, is a factor associated with personal standards, organization, and self-oriented perfectionism and is associated with trait measures of consciousness; *self-oriented perfectionism-concerns* is a factor incorporating concern over mistakes, parental expectations, parental criticism, doubts about actions, and socially-prescribed perfectionism (beliefs that others expect one to be perfect). Other authors (McCreary et al. 2004) have reported a 3-factor solution that has the social factors that maintain perfectionist behavioral patterns load on a separate factor. It is important to reiterate that while the construct of perfectionism is unique in incorporating interpersonal factors into several definitions, as we have previously mentioned, interpersonal factors are influential in the maintenance of many trait patterns of responding. Combined, this body of data suggests that interventions that solely target the individuals may be insufficient, rather, incorporating relevant members of an individual's social network so that reinforcement contingencies can be altered, may be in a particularly efficacious manner to manage deleterious forms of perfectionism, and in strategies to address trait features, in general.

Interim Summary. In the previous section, we discussed prior definitions of perfectionism and attempted to synthesize these definitions with learning theory. We described the role of interpersonal factors in the maintenance of trait features, including perfectionism, and argued that treatment include the social network as a target in addition to the individual. In the following sections, we use clinical perfectionism as a specific example of trait modes of responding and illustrate how it is addressed in an intervention that involves both parents and their adolescents with AN.

6 Perfectionism Addressed Via Group Parent Training

We have developed a group parent-training program (GPT) designed to treat adolescent AN, a treatment influenced by extant models of cognitive behavioral and emotion-focused therapies (Greenberg and Paivio 2003; Linehan 1993). This treatment is designed to enhance adaptive self-regulation in adolescents diagnosed with AN by using parents as role models of responsivity to their own basic biological needs and attentive to the motivations conveyed by emotional experience. Development of this intervention considered the mode of treatment delivery, the content of treatment, and the critical role of therapist as role model. GPT incorporates the elements we have designated as being relevant for addressing trait features (attention, reputation narratives, social modeling, social reinforcement contingencies). In the sections that follow, we describe the theoretical model that guided the development of this intervention, delineate how each domain is addressed, and describe preliminary data on the treatment outcomes.

6.1 Theoretical Model Guiding Intervention Content

Responsive parenting has been an influential construct across psychological disciplines. Developmental psychologists emphasize the importance of a responsive style of parenting in the development of adaptive self-regulation in children, social psychologists emphasize the importance of responsive parenting on theory of mind and other constructs necessary for empathic attunement to others, and clinical psychologists emphasize the link between responsive parenting and mental health (Propper and Moore 2006; Smith et al. 2006; TamisLeMonda et al. 1996). Responsive parenting has numerous definitions, but we adopt the definition of Propper and Moore (2006) as they define the parenting behaviors that may influence infant emotionality: “Caregivers facilitate the establishment of physiological homeostasis as they assist in attaining a balance between endogenous needs and exogenous stimuli” (p. 435). We consider this construct as critically important to incorporate into interventions designed to address AN. Whatever be the complex array and combination of influences that impact the emergence of AN, at its essence, individuals with AN have failed to incorporate the task of responsively "parenting" themselves, they neglect all needs: be they basic sustenance, fatigue, or the motivations expressed by emotional experience. Rather, they are at war with their bodies: trying to manipulate and control what should be a seamless dance between bodily arousal and adaptive response.

We consider *self-knowledge* and *self-trust* (lay concepts we introduce in the intervention) as products of the trial and error learning that necessarily occurs as an individual attempts to decipher and respond to internal experience and to define the meaning of that experience in relation to environmental events. We define self-knowledge as the emerging and evolving product of the ability to decipher motivated states such as hunger, fatique, or emotional experience. We define self-trust as the establishment of a secure base with oneself, that is, the product of adaptively responding to one's basic needs, a combination of interoceptive sensitivity, interoceptive accuracy, visceral conditioning, and abstract cognition. Over time, such dynamic transactions may result in a felt sense of self-knowledge as the individual comes to recognize patterns of arousal, the situations that provoke them, and the actions that dampen them (e.g., "I usually feel this when I see that and doing this makes me feel better"). Thus, repeated and prolonged failure to respond to such internal signals would result in an utter dearth of self-knowledge and failure to develop self-trust, resulting ultimately in a vulnerability to operationalize one's self-defintion in only the most concrete and external of terms: achievement. In other words, perfectionism and related negatively-reinforced achievement striving may be reinforcing, in part, because it obviates the need to decipher internal experience: the goal is what is important irrespective of the cost or limit imposed by somatic needs. Further, operationalizing an abstract notion of the "self" in terms of concrete benchmarks of achievement naturally coincides with either an inability (or unwillingness) to respond to the complex, fluctuating, and constantly unpredictable internal milieu of somatic experience.

Of interest, data from our lab demonstrate that certain aspects of perfectionism are associated with deficits in abstract reasoning. In a study of social cognition in adult women with AN, we studied a sample of 21 women currently diagnosed with AN; 21 women who had a history of AN but were weight-restored for a minimum of 12 months, and 23 adult women who served as the healthy control reference group. An interesting interaction emerged between facets of perfectionism that have been reported to be distinctly elevated in those with eating disorders relative to psychiatric control groups: the concern over mistakes subscale. This subscale was negatively associated with nonverbal abstract reasoning as indexed by subtests on the Wechsler Adult Intelligence Scale (Matrix Subtest) in both groups with AN, with small to medium effects ($r^2 = 0.24$, $p = 0.02$ for current, $r^2 = 0.33$, $p = 0.01$ for weight-restored). However, this relationship was non-siginificant (and positive) in the healthy control group ($r^2 = 0.01$, $p = 0.59$). On measures of verbal abstract reasoning, concerns over mistakes was negatively related to this construct in individuals weight-restored with a history of AN ($r^2 = -0.31$, $p = 0.006$), while healthy controls demonstrated a significant positive relationship ($r^2 = 0.23$, $p = 0.02$). Such preliminary findings highlight that certain aspects of perfectionism such as flaw detection may function very differently against the clinical background of perfectionism. Whereas for healthy control subjects, concern over mistakes may be indexing flaw detection, for those with AN, such flaw detection may be an adaptation to deficits in abilities to derive higher-order abstract concepts from exemplars of both verbal and nonverbal sets. Further study is needed to replicate these findings in a larger sample. This pattern of results has potentially fascinating implications for the role of perfectionism as a compensatory strategy to ameliorate deficits in the ability to understand the abstract, complex, and nuanced notion of the self.

Combined, such data support the importance of enhancing response flexibility as a target of treatment when addressing clinical perfectionism. Further, utilizing strategies that build on concrete examples rather than abstract concepts may be a necessary therapeutic adaptation, particularly in younger aged groups. The strategy that we employed in the development of this intervention was to use parents as role models of adaptive self-regulation. In addition to teaching parents skills to manage their child's disorder, our focus was on capitalizing on parents as influential role models of behavior change. Our reasoning was that by teaching or improving the parents' capacity to be responsive to their own needs, they would serve as a concrete example of "self-responsive parenting," that is, adaptive self-regulation. In addition to potentially benefiting the parent's own mental health and adaptive function, such a strategy would have the benefit of circumventing putative deficits in abstract reasoning in their children by direct role modeling of the development of the self. Further, advances in the study of empathic responding are increasingly supporting that individuals understand the experiences of others via a virtual embodiment of the felt sense of others (i.e., your feeling of sadness provokes similar somatic and cognitive experiences and thus I apprehend your experience) (Ochsner 2008). Thus, improving parents' ability to be responsive to their own needs may enhance their ability to sense the needs of their children.

6.2 Attention Focus in Group Parent Training

By definition, individuals high in clinical perfectionism focus on the outcomes of their efforts. Such a perseverative focus, potentially supported by superior sustained attention (or intriguingly, deficits in set-shifting), may appear advantageous in that the individual seems impervious to goals other than the self-declared objective. Notwithstanding how rewarding (or "in control") this perseverative focus may be experienced by those with AN, in particular, or in those with elevations in clinical perfectionism, in general, in fact, such a strategy does not result in optimal learning (Sutton and Barto 1998). An optimal learner does what works most of the time (Montague et al. 2006). However, every now and then, an optimal learner must sample from novel domains (Sutton and Barto 1998). For example, suppose you have a favorite restaurant and a favorite item that you order at your favorite restaurant. Well, most of the time when you dine at this restaurant, you order your favorite item because you know that you enjoy it – it is a certain quantity. However, there may be items on the menu that are even more delicious than your favorite item. If you don't branch out and try new dishes every now and then, you may not be making the optimal choice. Such a dilemma constantly interferes with adaptive problem-solving in those with elevations in clinical perfectionism. Not only is there a fixation on the outcome of efforts, but there is fixation on a *particular* outcome and a *particular* strategy to achieve that outcome. Once a perfectionist declares the stated objective and the manner in which that objective will be achieved, the proverbial gauntlet has been thrown. No other outcome is acceptable and failure to achieve this stated goal results in elevations in the self-conscious emotions [guilt, shame; i.e., a negatively reinforced contingency as predicted by Owens and Slade (Owens and Slade 2008) and decrements in self-esteem as predicted by Shafran et al. (2002)]. In terms of attention, this may be operationalized as biased orienting to outcomes, advanced abilities to ignore or disregard conflicting information as reflected in advanced capacities to maintain a particular set (or failure to shift sets), and failure to benefit from feedback.

We adopted a "process versus outcome" approach, grounded in mindfulness techniques, to address biased orienting and inflexible responding in AN. Mindfulness skills, and the corresponding acceptance-based philosophies that support these approaches, have been increasingly integrated into recent evolutions of interventions based on behavioral learning theories (Linehan 1993; Hayes et al. 2006). Mindfulness is born of Eastern spiritual practices and emphasizes the importance of increasing contact with the present moment unfettered by beliefs or judgments that would otherwise shape the direction and sustain fixation of attention (Kabat-Zinn 1990). Motivated states naturally and adaptively direct visually guided attention to objects most salient to satiate the needs of the individual at that moment. While adaptive, such foci may restrict the field of vision and limit behavioral options. Thus, one technique to facilitate development of focused, purposeful, and flexible attention is mindfulness practice. In a typical mindfulness exercise, an individual is asked to shift the perspective from which private events are viewed, from a participant, in whom there is no separation between self and experience, to an observer, who observes one's thoughts dispassionately – as one would observe any other object in the environment. Individuals learn to describe internal and external events as a means to disconnect themselves from the literal interpretation of their experience. That is, thoughts and feelings are labeled and experienced as thoughts and feelings. Thus, the thought "I am a failure" becomes "I am having the thought that I am a failure."

The philosophy and related strategies of mindfulness-based approaches may be particularly suited for individuals whose biased focus towards outcomes may compromise effectiveness. However, we were concerned that "marketing" the proposed strategies as mindfulness-based approaches may be unpalatable to parents whose own extreme achievement-striving makes them biasedly oriented to the most concrete of outcomes. Thus, in accordance with the strategies of motivational interviewing (Miller and Rollnick 2002), we join with parents in validating the utility and ease of an outcome-focused approach, and gradually highlight the differential benefit of a more balanced, or "process"-focused approach. As explained in the context of this intervention, while an initial focus on the outcome of efforts may dictate the selection of responses and the dictated goal, a *process approach* values each step along the path (i.e., attempts towards the goal). The value of every behavioral attempt towards a goal is framed as providing new information that advances knowledge, thereby constantly updating the utility of the selected path. Thus, while a clinical perfectionist may view any failure to adhere to a course of action as a signal of weakening "will-power" or a decrement in self-discipline, the process philosophy allows parents to alter their mode of learning while preserving their personal dignity: by switching to a process approach, they may maximize outcomes by becoming more flexible responders to momentary fluctuations in experience. This is merely a way of framing the tenets of reinforcement learning in a way that is accessible in the context of a clinical intervention. We are teaching parents (and their children) to be optimal learners (Sutton and Barto 1998). In fact, prior research demonstrates decrements in problem-solving in individuals with elevated clinical perfectionism (Stoeber and Eysenck 2008). Further, given evidence of prior decrements in abstract reasoning, we attempt to frame the process ideology very concretely. Parents are given behavioral examples that operationalize the adoption of an attentional stance that moves from biased orienting towards outcomes to more flexibly orienting to the vast variety of stimuli in the environment, that is, the momentary fluctuations in experience that inform optimal decision-making. Table 1 provides some examples that parents are given to differentiate these different attention foci. Parents are assigned homework assignments in which they must practice this shift of focus (Zucker et al. 2005). Thus, they are not directly trying to change the adolescent by manipulating her or his attention; rather, they are gently shaping this shift in focus via their own role modeling and via shifts in reinforcement contingencies as described below.

6.3 Shaping as Attention Retraining to Goal Approximations

Parents sometimes (or often) inadvertently reinforce an outcome focus. Questions about their child's academic grades, their race time during track events, or other strictly outcome-focused parameters of performance have the effect of narrowing both the breadth and depth of conversations. We propose that such shaping via the focus of conversations gradually orients the child's attention to these concrete indexes of achievement. Yet, parents may direct the orienting of attention in ways other than the topic of conversation, but also by the manner in which they reinforce the behavior of their child. For example, we have operationalized a behavioral pattern we refer to as the "yes, but" phenomenon among parents with elevations in clinical perfectionism. When enacting this behavioral pattern, parents, in a seemingly benevolent gesture to advance their child's performance, only partially reinforce efforts by praising, but then immediately couple that praise with a suggestion for superior efforts on the next learning occasion: "Great job on

Table 1 Parent Handout of a Process vs. Outcome Approach

Baby Steps Towards a Process Approach

1. Choose one event every day to focus on with your PROCESS outlook. Be in the moment, do only one thing, be an observer and describer – not a judger and multi-tasker.
2. Set aside a time each day where the family is together for at least a 15 min period to touch base and connect. If possible, make it a mealtime.
3. Do one thing at a time
4. Learn from an experience, don't judge
5. Trust that if you focus on being effective in the moment, the rest will follow.

	Process	Outcome
Athletics	• Enjoyment of the event • Using performance to become a better athlete	• Score
Academics	• Learning • Using grades as data to improve study strategies • Using grades as data to enhance learning	• Grades, class rank
College choice	• Consideration of your child's personality, academic interests, and extracurricular interests • Finding a school that matches your child	• Highest ranked • Finding a school that matches the neighbor's expectations
Conversation	• Listening to the other person • Honoring your opinions • Being genuine in your feedback	• Focusing on how the other person will feel about you after the conversation • Thinking about what you should be doing rather than sitting and talking • Planning the next thing you will say
Mealtimes	• You are eating and enjoying the food • You are enjoying the company of your family • You are focusing conversation on experiences of the day	• You are doing a million other things while you eat • You are watching television while you eat, so you don't miss anything • You are watching television while you eat so you can avoid talking
Physical appearance	• You are appreciating what you look like in the moment • You take pride in your appearance • You focus on what your body can do • You accept those aspects of your appearance that you do not control • You invest a reasonable amount of time in your appearance	• You compare yourself to an ideal • You berate yourself for the discrepancy between your ideal and you • You refuse to accept your current appearance if it is not the ideal

this paper, *but* next time you should really try to work on your transitions." There are several processes that may be interacting to negatively impact improved performance via this behavioral pattern. First, emotional tone as embodied in factors such as voice prosody is important. When such feedback is delivered against a negative emotional background, such seemingly benevolent feedback can be interpreted as shaming and belittling. Second, such change efforts inadvertently ignore the vast amount of research supporting the utility of shaping change efforts (Skinner 1953; Skinner 1975; Morris et al. 2005). In other words, when individuals are rewarded for approximating a desired outcome, receipt of the reward is reinforcing enough to persist with change efforts. In contrast, parents with elevations in clinical perfectionism, in a benevolent desire to improve their child's outcomes, may persistently correct their child, a behavioral pattern that may inadvertently stifle subsequent strivings.

6.4 Role Modeling

Parent role modeling of a desired change in behavior is a potentially potent influence on adolescent behavior change, in general. In the case of AN, in which adolescents have at best mixed motivations to attend treatment, role modeling of parents may be one of the few vehicles available to clinicians to influence behavior change in the initial stages of treatment. The debate between the existence of negative and positive forms of perfectionism becomes particularly informative in determining what aspects of parent behavior should be modeled, and what aspects should be addressed in treatment. In accordance with prior reviews of perfectionism, we concur that it is not striving for extreme achievement that is detrimental. Rather, there are specific aspects of such striving that may impede psychological and physical health, not only in those with AN, but in clinical perfectionism more generally.

Shafran et al. (2002) propose that it is reactions to failure that may influence psychological and physical health in those with elevated clinical perfectionism. We further contend that not all striving is adaptive and provide parents with several parameters to examine their chosen path of perfectionistic strivings: health and authenticity. First, we give parents the guideline that strivings for achievement should not interfere with health. While parents can readily recognize the dangerous consequences of such relentless striving on the health of their ill child, they may be less aware of the chronic impact of sleep deprivation, skipped meals, lack of balance between leisure and work, et cetera, on their own mental health and ultimately, productivity.

Of importance, perfectionism is notoriously difficult to target in treatment for many reasons, not the least of which is the presence of intermittent reinforcement schedules in which the desired perfectionistic outcome is achieved, albeit fleetingly. Targeting perfectionism in those parents who likely have been successful (indeed, sometimes incredibly successful) in obtaining outcomes may seem a daunting task, indeed. Yet, multiple factors support why addressing trait features in parents in the context of eating disorder treatment may afford a particularly efficacious milieu for such efforts. First, parents will do for their children what they will not do for themselves. Thus, parents may be more willing to attempt to alter a typical mode of responding if such change may potentially benefit their child than they would be to address this behavior pattern without this social contingency in place. Second, we do not focus on the extent of striving but rather on reactions to failure. Fairburn and Shafran and others have posited that perfectionism was toxic to those individuals who interpreted perceived failures in relationship to evaluations of the self (Shafran et al. 2002). We concur with this definition as failures to achieve a self-proclaimed outcome may result in increases in the self-conscious emotions, a potent form of aversive conditioning. As those with elevations in clinical perfectionism may be particularly vulnerable to define self-worth in concrete terms, such cognitive features may eventually provide an additional target of intervention. To alter this influence on self-definition, we considered several elements as critical. We had to change the context of a failure to achieve desired outcomes from a perceived threat to self-worth to an opportunity for self-growth. Fortunately, a vast literature on reinforcement learning supported our case. Research on problem-solving in those with elevations in clinical perfectionism supported a perseverative problem-solving style, inefficiency on cognitive tasks, and difficulty re-engaging on a task with which they had experienced prior failure (Stoeber and Eysenck 2008; Egan et al. 2007; Saddler and Sacks 1993). Study of the neurobiology of reinforcement learning, in general, highlighted the changing rate of dopaminergic firing, a marker of increased potentiation of synaptic communication, in response to unexpected outcomes: that is, we learn more when things do not go as planned (Schultz et al. 2008). Parents are taught that a focus on experiential learning may have an unintended effect: it may improve outcomes. Armed with data and the putative benefit of their child's health, introduction of content to address perfectionism in parents of those with AN had a surprising effect – shifting focus from the intermittent reinforcement of rare outcomes to the constant reinforcement afforded by new learning.

Finally, we desired to have parents examine whether such strivings were authentic reflections of personal values. As mentioned, for many, perfectionistic strivings are intended to influence public opinion and are less about what is personally meaningful for the individual in question. As forging an identity distinct from their eating disorder may be an essential component of effective treatments for AN (Serpell et al. 1999), parental role modeling of the search for self-motivated (rather than reputation-mediated) goals may be another critical social influence to shape their child's own self-selection of goal-directed behavior. Role modeling of such complex constructs by the parents further provided a concrete example that would hopefully bypass any deficits in abstract reasoning in the adolescent, so that he or she would fully comprehend what the parent was modeling. Combined, we intended for such treatment foci to not only alter parent role modeling of behavior critical to influence the biased outcome strivings of the child, but also that the parent's own mental health may be improved.

6.5 *Social Reinforcement Contingencies*

People value success and attune to authenticity. As one may imagine, the desire to change the context and nature of perfectionistic strivings has multiple social barriers – not just those presented by the role modeling of parents. Rather, an outcome-focused perspective provides a simple vehicle to rank and judge others and is a natural part of our evolutionary history (Zink et al. 2008). Wisely, we do not take up this argument. Rather, we focus on the realities of intimate human relationships and the essential role of vulnerability in the formation of trusted bonds. In economic games of trust, the individual who is voracious in his or her strivings is rarely trusted, while players who sacrifice part of their hand or are willing to risk evidence of a vulnerability are often described as more trustworthy and suitable teammates (King-Casas et al. 2005). Such findings from the field of neuroeconomics contradict several well-replicated behavioral patterns in those with elevations in clinical perfectionism: the belief that failure will distance them from others (Hewitt et al. 1998). Rather than attempt to shift such contingencies, the focus of treatment is on operationalizing the limits of control: those aspects that parents can control (their own behavior) and those they cannot (everything else- including the opinions of others). Thus, parents are encouraged to risk the authentic display of mistakes and to examine the outcomes.

6.6 Niche Environments

As individuals seek out and create environment that reinforce trait modes of responding, it would not be surprising to see individuals with elevations on measures of perfectionism seek out competitive environments. Surprisingly, the data are rather mixed and this may speak to the debate regarding whether there are adaptive and maladaptive forms of perfectionism (Slade and Owens 1998; Flett and Hewitt 2006). For example, Stornelli et al. (2009) examined differences in perfectionism and negative affect between students in academically gifted programs, arts programs, and regular academic programming finding no differential distributions of levels of perfectionism, but a consistent relationship of perfectionism with negative affect. However, research has also supported elevated levels of perfectionism in increasingly competitive environments such as dance companies (Thomas et al. 2005). Though the direction of causality is necessarily complex, such a relationship may be an excellent exemplar of the operation of niche environments: individuals' trait features becoming intensified via the type of environments they choose combined with the tendency to seek out those environments that reinforce these tendencies. Certainly, it is not hard to imagine how challenging it would be for individuals with elevated levels of clinical perfectionism when surrounded by individuals who perpetually feel guilty about their failures to achieve desired outcomes. In intensive treatments for substance abuse, two complementary social models are recommended in relation to the niche environments of individuals with a former addiction. Alcoholics Anonymous, a widely disseminated and influential program for the treatment of alcohol abuse and dependence, has individuals with shared experiences participate in group meetings and engage in peer mentoring to support behavior change. Accumulating evidence supports the efficacy of this model (Laffaye et al. 2008; Blonigen et al. 2009; McKellar et al. 2003). In addition, individuals with prior addiction are often encouraged to change their prior social context to the extent that former partners and/or friendships help to facilitate abuse (McKellar et al. 2008). Our goal was to model the creation of adaptive social networks while not inadvertently reinforcing maladaptive behavior in the context of GPT. The group context itself provides a potent environment for the formation of adaptive social contingencies. Each week, social dynamics within the group demand that parents report mistakes, group leaders are encouraged to share their own social mishaps, and particularly emotional, volatile, and seemingly ineffective interactions are celebrated with humor (Zucker et al. 2005). Despite the reluctance or disbelief with which parents may regard changes in social status resulting from confessions of failure, the group context provides a virtual experiment of these beliefs: not only is status not injured, but relationships are enhanced (Kawamura and Frost 2004). Further, via repeated exposures not only to their own, but to the mistakes of other parents, parents come to appreciate what they have been instructed about the educational value of mistakes: they learn from the mistakes of other parents: when things go well, they pay less attention. Our adolescent group establishes similar expectations for failed performance.

6.7 *Changes in Perfectionism in Adolescent Anorexia Nervosa*

We examined changes in self-report measures of perfectionism in the context of a treatment trial of adolescent AN. In this trial, families were randomized to receive either the Maudsley model of family therapy as manualized by Lock et al. (2001), or Group Parent Training and an Adolescent Interoceptive-Skills Group (adolescents in a group with other adolescents with AN). Our interest was examining change in the frequency of perfectionistic cognitions relative to changes in the experienced distress regarding perfectionistic cognitions. In other words, with treatment we expected individuals to become less distressed by the presence of perfectionistic cognitions (e.g., thought content related to fear of failure). As our interest is in the capacity for change in trait features in adolescents for the purposes of this chapter, we combined our treatment groups to maximize our power to detect change. We employed the Perfectionistic Cognitions Inventory (PCI), a self-report measure of the frequency of perfectionistic cognitions (e.g., "I need to be perfect. I should never make the same mistake twice. I have to be the best."), designed by Flett et al. (1998). We adapted this measure by adding an item that assessed the level of distress about having this thought for each item that assessed frequency. We totaled these distress items separately from the frequency items. The change in PCI distress from baseline to 6 months was significant ($n = 17$, $p < 0.05$), and there were trends from 3 months to 6 months ($n = 16$, $p = 0.08$) and from baseline to 12 months ($n = 10$, $p = 0.07$). However, the frequency of perfectionistic cognitions also changed, from baseline to 3 months, baseline to 6 months, and baseline to 12 months. The PCI total change from 3 months to 6 months was not significant. Controlling for level of baseline symptoms, the PCI sum change score from baseline to 6 months was correlated with EDE shape concern at 6 months (controlling for baseline EDE shape concern, $r = 0.52$, $p < 0.05$; $n = 13$). The PCI distress change score from 6 to 12 months was correlated with the change in fasting from 6 to 12 months ($p = -0.68$, $p < 0.05$; $n = 7$). To be sure, our sample size and subsequent power limit our ability to fully address the capacity for change. We provide these data merely to illustrate that when designing interventions that aim to address or incorporate trait modes of responding, our choice of outcome measures may need to reflect alternatives to symptom reduction, but rather changes in the impact of symptoms on functioning.

7 Summary

We attempted to provide a general framework to guide the development of interventions that aim to address persistent features in eating disorders that may preclude effective treatment. Using perfectionism as an exemplar, we drew from research in cognitive neuroscience regarding attention and reinforcement learning, from learning theory and social psychology regarding vicarious learning and implications for the role modeling of significant others, and from clinical psychology on the importance of verbal narratives as barriers that may influence expectations and shape reinforcement schedules. Exciting advances in the neuroscience are rapidly propelling our understanding of brain function in AN, findings that will greatly augment our capacity to develop novel intervention strategies. Yet, all interventions are embedded in a social context and thus, the thoughtful incorporation and leveraging of this social context may greatly potentiate our capacity to significantly change the course of these severe disorders.

References

Allen HN, Craighead LW (1999) Appetite monitoring in the treatment of binge eating disorder. Behav Ther 30:253–272

Archbold GEB, Bouton ME, Nader K (2010) Evidence for the persistence of contextual fear memories following immediate extinction. Eur J Neurosci 31:1303–1311

Askew C, Field AP (2008) The vicarious learning pathway to fear 40 years on. Clin Psychol Rev 28:1249–1265

Bandura A, Ross D, Ross SA (1963) Vicarious reinforcement and imitative learning. J Abnorm Soc Psychol 67:601–607

Barber JP, Morse JQ, Krakauer ID, Chittams J, CritsChristoph K (1997) Change in obsessive-compulsive and avoidant personality disorders following time-limited supportive-expressive therapy. Psychotherapy 34:133–143

Bardone-Cone AM, Sturm K, Lawson MA, Robinson DP, Smith R (2010) Perfectionism across stages of recovery from eating disorders. Int J Eat Disord 43:139–148

Beck AT (1979) Cognitive therapy for depression. Guilford, New York

Bizeul C, Sadowsky N, Rigaud D (2001) The prognostic value of initial EDI scores in anorexia nervosa patients: a prospective follow-up study of 5–10 years. Eur Psychiatry 16:232–238
Blatt SJ (1995) The destructiveness of perfectionism – implications for the treatment of depression. Am Psychol 50:1003–1020
Blonigen DM, Timko C, Moos BS, Moos RH (2009) Treatment, alcoholics anonymous, and 16-year changes in impulsivity and legal problems among men and women with alcohol use disorders. J Stud Alcohol Drugs 70:714–725
Caspi A, Roberts BW, Shiner RL (2005) Personality development: stability and change. Annu Rev Psychol 56:453–484
Cox BJ, Enns MW, Clara IP (2002) The multidimensional structure of perfectionism in clinically distressed and college student samples. Psychol Assess 14:365–373
Derryberry D, Rothbart MK (1997) Reactive and effortful processes in the organization of temperament. Dev Psychopathol 9:633–652
Egan SJ, Piek JP, Dyck MJ, Rees CS (2007) The role of dichotomous thinking and rigidity in perfectionism. Behav Res Ther 45:1813–1822
Eisenberg N, Fabes RA, Guthrie IK, Reiser M (2000) Dispositional emotionality and regulation: their role in predicting quality of social functioning. J Pers Soc Psychol 78:136–157
Eisenberg N, Zhou Q, Spinrad TL et al (2005) Relations among positive parenting, children's effortful control, and externalizing problems: a three-wave longitudinal study. Child Dev 76:1055–1071
Epstein LH, Paluch RA, Roemmich JN, Beecher MD (2007) Family-based obesity treatment, then and now: twenty-five years of pediatric obesity treatment. Health Psychol 26:381–391
Flett GL, Hewitt PL (2006) Positive versus negative perfectionism in psychopathology – a comment on Slade and Owens's dual process model. Behav Modif 30:472–495
Flett GL, Hewitt PL, Blankstein KR, Gray L (1998) Psychological distress and the frequency of perfectionistic thinking. J Pers Soc Psychol 75:1363–1381
Frost R, Marten P, Lahart C, Rosenblate R (1990) The dimensions of perfectionism. Cognit Ther Res 14:449–468
Graham P, Rutter M, George S (1973) Temperamental characteristics as predictors of behavior disorders in children. Am J Orthopsychiatry 43:328–339
Greenberg LS, Paivio SC (2003) Working with emotions in psychotherapy. Guilford Press, New York
Gruner K, Muris P, Merckelbach H (1999) The relationship between anxious rearing behaviours and anxiety disorders symptomatology in normal children. J Behav Ther Exp Psychiatry 30:27–35
Guzman YF, Tronson NC, Guedea A et al (2009) Social modeling of conditioned fear in mice by non-fearful conspecifics. Behav Brain Res 201:173–178
Hamilton TK, Schweitzer RD (2000) The cost of being perfect: perfectionism and suicide ideation in university students. Aust N Z J Psychiatry 34:829–835
Hayes SC, Luoma JB, Bond FW, Masuda A, Lillis J (2006) Acceptance and commitment therapy: model, processes and outcomes. Behav Res Ther 44:1–25
Hewitt PL, Flett GL (1991a) Perfectionism in the self and social contexts: conceptualization, assessment, and association with psychopathology. J Pers Soc Psychol 60:456–470
Hewitt PL, Flett GL (1991b) Dimensions of perfectionism in unipolar depression. J Abnorm Psychol 100:98–101
Hewitt PL, Newton J, Flett GL, Callander L (1997) Perfectionism and suicide ideation in adolescent psychiatric patients. J Abnorm Child Psychol 25:95–101
Hewitt PL, Norton GR, Flett GL, Callander L, Cowan T (1998) Dimensions of perfectionism, hopelessness, and attempted suicide in a sample of alcoholics. Suicide Life Threat Behav 28:395–406
Hewitt PL, Flett GL, Besser A, Sherry SB, McGee B (2003) Perfectionism is multidimensional: a reply to Shafran, Cooper and Fairburn (2002). Behav Res Ther 41:1221–1236

Hoff ER, Muehlenkamp JJ (2009) Nonsuicidal self-injury in college students: the role of perfectionism and rumination. Suicide Life Threat Behav 39:576–587
Jacobs RH, Silva SG, Reinecke MA et al (2009) Dysfunctional attitudes scale perfectionism: a predictor and partial mediator of acute treatment outcome among clinically depressed adolescents. J Clin Child Adolesc Psychol 38:803–813
Kabat-Zinn J (1990) Full castastrophe living: using the wisdom of your body and mind to face stress, pain, and illness. Bantam Dell, New York
Kawamura KY, Frost RO (2004) Self-concealment as a mediator in the relationship between perfectionism and psychological distress. Cognit Ther Res 28:183–191
Kelley AE, Schiltz CA, Landry CF (2005) Neural systems recruited by drug- and food-related cues: studies of gene activation in corticolimbic regions. Physiol Behav 86:11–14
Kieras JE, Tobin RM, Graziano WG, Rothbart MK (2005) You can't always get what you want – effortful control and children's responses to undesirable gifts. Psychol Sci 16:391–396
King-Casas B, Tomlin D, Anen C et al (2005) Getting to know you: reputation and trust in a two-person economic exchange. Science 308:78–83
Koerner K, Linehan MM (2000) Research on dialectical behavior therapy for patients with borderline personality disorder. Psychiatr Clin North Am 23:151–167
Laffaye C, McKellar JD, Ilgen MA, Moos RH (2008) Predictors of 4-year outcome of community residential treatment for patients with substance use disorders. Addiction 103:671–680
Linehan M (1993) Cognitive-behavioral treatment of borderline personality disorder. Guilford Press, New York
Linehan MM (1997) Self-verification and drug abusers: implications for treatment. Psychol Sci 8:181–183
Linehan MM, Dimeff LA, Reynolds SK et al (2002) Dialectical behavior therapy versus comprehensive validation therapy plus 12-step for the treatment of opioid dependent women meeting criteria for borderline personality disorder. Drug Alcohol Depend 67:13–26
Lock J, Le Grange D, Agras WS, Dare C (2001) Treatment manual for anorexia nervosa: a family-based approach. Guilford Press, New York
Lynch TR, Chapman AL, Rosenthal MZ, Kuo JR, Linehan MM (2006) Mechanisms of change in dialectical behavior therapy: theoretical and empirical observations. J Clin Psychol 62:459–480
Marlatt GA, Gordon JR (1985) Relapse prevention: maintenance strategies in the treatment of addictive behaviors. Guilford Press, New York
McAdams DP, Olson BD (2010) Personality development: continuity and change over the life course. Annu Rev Psychol 61:517–542
McCreary BT, Joiner TE, Schmidt NB, Ialongo NS (2004) The structure and correlates of perfectionism in African American children. J Clin Child Adolesc Psychol 33:313–324
McKellar J, Stewart E, Humphreys K (2003) Alcoholics anonymous involvement and positive alcohol-related outcomes: cause, consequence, or just a correlate? A prospective 2-year study of 2, 319 alcohol-dependent men. J Consult Clin Psychol 71:302–308
McKellar J, Ilgen M, Moos BS, Moos R (2008) Predictors of changes in alcohol-related self-efficacy over 16 years. J Subst Abuse Treat 35:148–155
Miller WR, Rollnick S (2002) Motivational interviewing, 2nd edn. Guilford Press, New York
Montague PR, King-Casas B, Cohen JD (2006) Imaging valuation models in human choice. Annu Rev Neurosci 29:417–448
Morris EK, Smith NG, Altus DE (2005) B. F. Skinner's contributions to applied behavior analysis. Behav Anal 28:99–131
Nilsson K, Sundbom E, Hagglof B (2008) A longitudinal study of perfectionism in adolescent onset anorexia nervosa-restricting type. Eur Eat Disord Rev 16:386–394
Nisbett RE, Wilson TD (1977) Halo effect – evidence for unconscious alteration of judgments. J Pers Soc Psychol 35:250–256
North RJ, Swann WB (2009) Self-verification 360 degrees: illuminating the light and dark sides. Self Identity 8:131–146

Ochsner KN (2008) The social-emotional processing stream: five core constructs and their translational potential for schizophrenia and beyond. Biol Psychiatry 64:48–61
O'Connor RC, Dixon D, Rasmussen S (2009) The structure and temporal stability of the child and adolescent perfectionism scale. Psychol Assess 21:437–443
Owens RG, Slade PD (2008) So perfect it's positively harmful? Reflections on the adaptiveness and maladaptiveness of positive and negative perfectionism. Behav Modif 32:928–937
Page S, Rosenthal R (1990) Sex and expectations of teachers and sex and race of students as determinants of teaching behavior and student performance. J Sch Psychol 28:119–131
Posner MI, Rothbart MK (2007) Research on attention networks as a model for the integration of psychological science. Annu Rev Psychol 58:1–23
Posner MI, Rothbart MK (2009) Toward a physical basis of attention and self-regulation. Phys Life Rev 6:103–120
Propper C, Moore GA (2006) The influence of parenting on infant emotionality: a multi-level psychobiological perspective. Dev Rev 26:427–460
Redd M, de Castro J (1992) Social facilitation of eating: effects of social instruction on food intake. Physiol Behav 52:749–754
Roberts BW, DelVecchio WF (2000) The rank-order consistency of personality traits from childhood to old age: a quantitative review of longitudinal studies. Psychol Bull 126:3–25
Roberts ME, Tchanturia K, Stahl D, Southgate L, Treasure J (2007) A systematic review and meta-analysis of set-shifting ability in eating disorders. Psychol Med 37:1075–1084
Rothbart MK, Ahadi SA, Hershey KL (1994) Temperament and social-behavior in childhood. Merrill Palmer Q – J Dev Psychol 40:21–39
Saddler CD, Sacks LA (1993) Multidimensional perfectionism and academic procrastination: relationships with depression in university students. Psychol Rep 73:863–871
Schultz W, Preuschoff K, Camerer C et al (2008) Explicit neural signals reflecting reward uncertainty. Philos Trans R Soc B Biol Sci 363:3801–3811
Serpell L, Treasure J, Teasdale J, Sullivan V (1999) Anorexia nervosa: friend or foe? Int J Eat Disord 25:177–186
Shafran R, Cooper Z, Fairburn CG (2002) Clinical perfectionism: a cognitive-behavioural analysis. Behav Res Ther 40:773–791
Shutts K, Kinzler KD, McKee CB, Spelke ES (2009) Social information guides infants' selection of foods. J Cogn Dev 10:1–17
Simonds J, Kieras JE, Rueda MR, Rothbart MK (2007) Effortful control, executive attention, and emotional regulation in 7–10-year-old children. Cogn Dev 22:474–488
Skinner B (1953) Science and human behavior. Free Press, New York
Skinner BF (1975) Shaping of phylogenic behavior. J Exp Anal Behav 24:117–120
Skodol AE, Pagano ME, Bender DS et al (2005) Stability of functional impairment in patients with schizotypal, borderline, avoidant, or obsessive-compulsive personality disorder over two years. Psychol Med 35:443–451
Slade PD, Owens RG (1998) A dual process model of perfectionism based on reinforcement theory. Behav Modif 22:372–390
Smith KE, Landry SH, Swank PR (2006) The role of early maternal responsiveness in supporting school-aged cognitive development for children who vary in birth status. Pediatrics 117:1608–1617
Sobik L, Hutchison K, Craighead L (2005) Cue-elicited craving for food: a fresh approach to the study of binge eating. Appetite 44:253–261
Stoeber J, Eysenck MW (2008) Perfectionism and efficiency: accuracy, response bias, and invested time in proof-reading performance. J Res Pers 42:1673–1678
Stornelli D, Flett GL, Hewitt PL (2009) Perfectionism, achievement, and affect in children: a comparison of students from gifted, arts, and regular programs. Can J Sch Psychol 24:267–283
Sutton RS, Barto AG (1998) Reinforcement learning: an introduction. MIT Press, Boston
Swann WB, Read SJ (1981) Self-verification processes – how we sustain our self-conceptions. J Exp Soc Psychol 17:351–372

TamisLeMonda CS, Bornstein MH, Damast AM (1996) Responsive parenting in the second year: specific influences on children's language and play. Early Dev Parent 5:173–183
Tchanturia K, Morris R, Anderluh M et al (2004) Set shifting in anorexia nervosa: an examination before and after weight gain, in full recovery and relationship to childhood and adult OCPD traits. J Psychiatr Res 38(5):545–552
Tellegen A, Lykken D, Bouchard T Jr et al (1988) Personality similarity in twins reared apart and together. J Pers Soc Psychol 54:1031–1039
Terryshort LA, Owens RG, Slade PD, Dewey ME (1995) Positive and negative perfectionism. Pers Individ Dif 18:663–668
Thomas JJ, Keel PK, Heatherton TF (2005) Disordered eating attitudes and behaviors in ballet students: examination of environmental and individual risk factors. Int J Eat Disord 38:263–268
Williamson DA, Muller SL, Reas DL, Thaw JM (1999) Cognitive bias in eating disorders: implications for theory and treatment. Behav Modif 23:556–577
Wonderlich SA, Lilenfeld LR, Riso LP, Engel S, Mitchell JE (2005) Personality and anorexia nervosa. Int J Eat Disord 37(Suppl):S68–S71, discussion S87–S89
Wrotniak BH, Epstein LH, Paluch RA, Roemmich JN (2005) The relationship between parent and child self-reported adherence and weight loss. Obes Res 13:1089–1096
Zastrow A, Kaiser S, Stippich C et al (2009) Neural correlates of impaired cognitive-behavioral flexibility in anorexia nerovsa. Am J Psychiatry 166:608–616
Zink CF, Tong YX, Chen Q et al (2008) Know your place: neural processing of social hierarchy in humans. Neuron 85:273–283
Zucker N, Ferriter C, Best S, Brantley A (2005) Group parent training: a novel approach for the treatment of eating disorders. Eat Disord 13:391–405
Zucker N, Losh M, Bulik C et al (2007) Anorexia nervosa and autism spectrum disorders: guided investigation of social cognitive endophenotypes. Psychol Bull 133:976–1006

Index

R.A.H. Adan and W.H. Kaye (eds.), *Behavioral Neurobiology of Eating Disorders*,
Current Topics in Behavioral Neurosciences 6, DOI 10.1007/7854_2010

MIX
Papier aus verantwortungsvollen Quellen
Paper from responsible sources
FSC® C105338

If you have any concerns about our products,
you can contact us on
ProductSafety@springernature.com

In case Publisher is established outside the EU,
the EU authorized representative is:
Springer Nature Customer Service Center GmbH
Europaplatz 3, 69115 Heidelberg, Germany

Printed by Libri Plureos GmbH
in Hamburg, Germany